*ARTHUR EDWIN KENNELLY*



# THE APPLICATION OF HYPERBOLIC FUNCTIONS TO ELECTRICAL ENGINEERING PROBLEMS

Elibron Classics
www.elibron.com

Elibron Classics series.



ISBN 0-543-96404-3 (paperback)
ISBN 0-543-96403-5 (hardcover)

This Elibron Classics Replica Edition is an unabridged facsimile of the edition published in 1916 by University of London Press, Ltd., London.

# THE APPLICATION OF HYPERBOLIC FUNCTIONS TO ELECTRICAL ENGINEERING PROBLEMS

# THE APPLICATION OF HYPERBOLIC FUNCTIONS TO ELECTRICAL ENGINEERING PROBLEMS

BY

A. E. KENNELLY, M.A., D.Sc.

PROFESSOR OF ELECTRICAL ENGINEERING AT HARVARD UNIVERSITY

*SECOND EDITION*

London: University of London Press, Limited

AT ST. PAUL'S HOUSE, WARWICK SQUARE

1916

HODDER AND STOUGHTON

PUBLISHERS TO

THE UNIVERSITY OF LONDON PRESS

# PREFACE TO FIRST EDITION

HYPERBOLIC functions have numerous, well recognized uses in applied science, particularly in the theory of charts (Mercator's projection), and in mechanics (strains). But it is only within recent years that their applications to electrical engineering have become evident. Wherever a line, or series of lines, of uniform linear constants is met with, an immediate field of usefulness for hyperbolic functions presents itself, particularly in high-frequency alternating-current lines.

The following pages are intended to cover the scope and purport of five lectures given for the University of London, at The Institution of Electrical Engineers, Victoria Embankment, by kind permission of the Council, May 29 to June 2, 1911, bearing the same title as this book.

The central ideas around which those lectures, and this presentation, have been framed are—

(1) That the engineering quantitative theories of continuous-currents and of alternating-currents are essentially one and the same; all continuous-current formulas for voltage, current, resistance, power and energy being applicable to alternating-current circuits, when complex numbers are substituted for real numbers. Thus there appears to be only one continuous-current formula in this book (277) which is uninterpretable vectorially in alternating-current terms; namely, as shown in Appendix J, that which deals with the mechanical forces developed in a telegraph receiving instrument, such forces being essentially "real" and not complex quantities.

(2) That there is a proper analogy between circular and hyperbolic trigonometry, which permits of the extension of the notion of an "angle" from the circular to the hyperbolic sector. The conception of the "hyperbolic angle" of a continuous-

current line is useful and illuminating, leading immediately, in two-dimensional arithmetic, to an easy comprehension of alternating-current lines.

The subject, which is very large, very useful, and very beautiful, is only outlined in the following pages. There are many directions in which accurate and painstaking research is needed, in the laboratory, the factory, and the field. Fortunately there are already a number of workers in this field, and good progress is, therefore, to be looked for. It is earnestly hoped that this book may serve as an additional incentive to such research.

The author desires to acknowledge his indebtedness to the writings of Heaviside, Kelvin, J. A. Fleming, C. P. Steinmetz, and many others. A necessarily imperfect bibliography of the subject, in order of date, is offered in an Appendix. He is also indebted to the Engineering Departments of the British Post Office, the National Telephone Company and Mr. B. S. Cohen, the Eastern Telegraph Company and Mr. Walter Judd, also the American Telegraph and Telephone Company and Dr. F. B. Jewett, for data and information; likewise to Mr. Robert Herne, Superintendent of the Commercial Cable Company, in Rockport, Massachusetts, for kind assistance in obtaining measured cable signals. He also has to thank Professor John Perry, Professor Silvanus P. Thompson, and Mr. W. Duddell for valued suggestions. In particular, he is indebted to the great help and courtesy of Dr. R. Mullineux Walmsley, in the presentation of the lectures, and in the publication of this volume.

Although care has been taken to secure accuracy in the mathematics, yet errors, by oversight, may have crept in. If any should be detected by the reader, the author will be grateful for criticisms or suggestions.

A. E. K.

*Cambridge, Mass. (U.S.A.),*
*December 1911.*

# PREFACE TO SECOND EDITION

Now that fairly extensive Tables, and curve-sheet charts for their rapid interpolation, have been sent to press,* it may be said that hyperbolic functions applied to alternating-current circuits have risen from the stage of theory outlined in the first edition of this book, to a stage of practical utility; because problems which would take hours of labor to solve by other methods, may be solved in a few minutes by the use of the hyperbolic Tables and curve sheets. In fact, with the atlas open at the proper chart, any complex hyperbolic function can be read off within a few seconds of time, ordinarily, to at least such a degree of precision as is offered by a good 25-centimeter slide rule. Consequently, hyperbolic trigonometry becomes a practical engineering tool of great swiftness and power, in dealing with alternating-current circuits having both series impedance and shunt admittance.

Since the publication of the first edition, a considerable number of tests, made in the laboratory, on alternating-current artificial lines, at various frequencies up to 1000 cycles per second, have demonstrated the practical serviceability of the hyperbolic analytical methods presented to the reader. No intimation has been received by the author as to inaccuracies in the original text, which had to be proof-read from across the Atlantic Ocean. A few typographical errors have, however, been eliminated from the text in this edition, a few additional formulas offered, and two new appendices added. The most important addition is the proposition that on any and every uniform section of line AB, in the steady single-frequency state, there exists a hyperbolic angle $\theta$ subtended by the section, and also definite hyperbolic

* Bibliography, 92 and 93.

position-angles $\delta_A$ and $\delta_B$ at the terminals, which depend upon the power delivery, and which differ by $\theta$. The potential at any point P of the line is always directly proportional to the sine, and the current to the cosine, of the position angle $\delta_P$ of the point, which position-angle is in direct intermediate relation to the distances of P from the terminals. Consequently, as soon as the power distribution over an alternating-current line-system has become steady, each and every point of the system virtually acquires a hyperbolic position-angle, such that along any uniform line-section in the system, the potential and current are respectively simple sine and cosine properties of that position-angle.

A. E. K.

*Harvard University,*
*Cambridge, Mass. (U.S.A.).*

# TABLE OF CONTENTS

## APPENDIX K



## APPENDIX L



## APPENDIX M



## APPENDIX N

# THE APPLICATION OF HYPERBOLIC FUNCTIONS TO ELECTRICAL ENGINEERING PROBLEMS

## CHAPTER I

### ANGLES IN CIRCULAR AND HYPERBOLIC TRIGONOMETRY

*Generation of Circular Angles.*—If we plot to Cartesian co-ordinates the locus of $y$ ordinates for varying values of $x$ abscissas in the equation—

$$y^2 + x^2 = 1 \quad . \quad . \quad \text{(cm., or units of length)}^2 \qquad (1)$$

we obtain the familiar graph of a circle, as indicated in Fig. 1; where O is both the origin of co-ordinates $x$, $y$, and the centre of the portion of a circle $f'Ag$. The radius OA, on the axis of abscissas, is taken as of unit length. As $x$ diminishes from $+1$ to 0, $y$ increases from 0 to $+1$, and the radius-vector OE moves its terminal E over the circular arc AE$g$. At any position such as OE, the tangent E$f$ to the path of the moving terminal is perpendicular to the radius-vector. As the radius-vector rotates * about the centre O, it describes a circular sector AOE and a circular angle, $\beta =$ AOE. The magnitude of this circular angle may be defined in either of two ways, namely—

(1) By the ratio of the circular arc length $s$ described, during the motion, by the terminal E, to the length $\rho$ of the radius-vector;

* Only the positive root of equation (1) is here considered, with the corresponding positive or counter-clockwise rotation of the radius-vector. In what follows, a hyperbola may be understood to be in all cases a rectangular hyperbola.

(2) By the area of the circular sector AOE swept out by the radius-vector during the motion.

*Generation of Hyperbolic Angles.*—If we plot to Cartesian co-ordinates the locus of $y$ ordinates for varying values of $x$ abscissas in the equation—

$$y^2 - x^2 = 1 \quad . \quad . \quad (\text{cm., or units of length})^2 \qquad (2)$$

we obtain the familiar graph of a rectangular hyperbola, as indicated in Fig. 2; where O is both the origin of co-ordinates

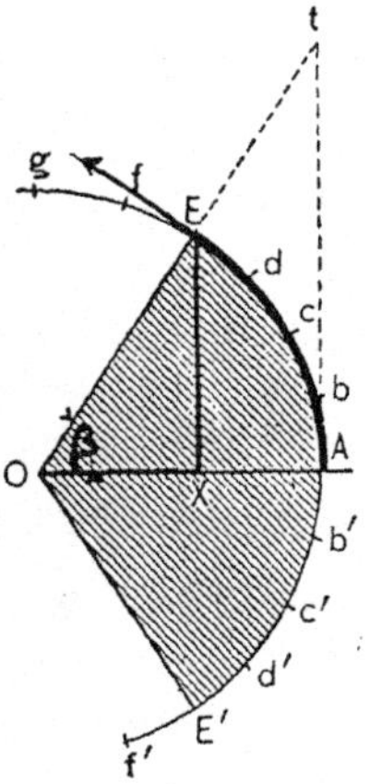


Fig. 1.—Circular Sector and Circular Functions.

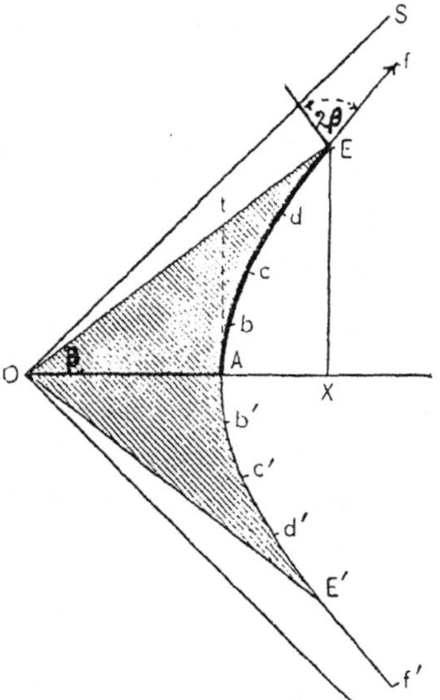


Fig. 2.—Hyperbolic Sector and Hyperbolic Functions.

$x$, $y$, and the centre of the hyperbola branch $f'\mathrm{A}f$. The radius or semi-axis OA, on the axis of abscissas, is taken as of unit length. As $x$ increases from 1 to $\infty$, $y$ increases from 0 to $\pm\infty$, and the radius-vector OE moves its terminal E over the hyperbolic arc A E$f$. At any position, such as OE, at which the radius-vector makes a circular angle $\beta$ with the X axis, the tangent E$f$ to the path of the moving terminal makes a circular angle $\beta$ with the Y axis; or a circular angle of $2\beta$ with a perpendicular to the radius-vector. As the radius-vector rotates about the centre O, it describes * a hyperbolic sector AOE, a circular angle $\beta$=AOE, and also a *hyperbolic angle* AOE. The magnitude of this hyperbolic angle may be defined in either of two ways, namely—

(1) By the ratio of the hyperbolic arc distance $s$ described, during the motion, by the terminal E, to the length $\rho$ of the radius-vector;

(2) By the area of the hyperbolic sector AOE swept out by the radius-vector during the motion.

*Algebraic Definition of any Angle, Circular or Hyperbolic.*— In the circular locus AE$g$ of Fig. 1, or in the hyperbolic locus AE$f$ of Fig. 2, let the rotating radius-vector OE generate at any time an element of arc of length—

$$ds = \sqrt{(dy)^2 + (dx)^2} \quad . \quad . \quad . \quad . \quad . \quad \text{cm.} \quad (3)$$

and let

$$\rho = \sqrt{y^2 + x^2} \quad . \quad . \quad . \quad . \quad . \quad . \quad . \quad . \quad \text{"} \quad (3a)$$

be the corresponding instantaneous value of the radius-vector length. Then the element of angle described during the motion will be

$$d\beta = d\theta = \frac{ds}{\rho} \quad \text{circular or hyperbolic radian (numeric)} \quad (4)$$

That is, the element of angle described in the circular locus of Fig. 1 will be a circular angle element $d\beta$, and will be expressible in units of *circular radians;* while the element of angle described in the hyperbolic locus of Fig. 2 will be a hyperbolic angle element $d\theta$, and will be expressible in units of *hyperbolic radians.*

As the motion proceeds in Figs. 1 and 2 from an initial to a final position of the radius-vector, the total angle described during the motion will be—

$$\beta = \theta = \int\frac{ds}{\rho} \quad \text{circular or hyperbolic radians} \quad (5)$$

In the case of the circular locus of Fig. 1, the radius-

* Only the positive root of equation (2) is here considered, with the corresponding positive or counter-clockwise rotation of the radius-vector.

vector $\rho$ is a constant, and equal, by assumption, to unity; consequently, equation (5) becomes for the circular case

$$\beta = \int_{s1}^{s2} \frac{ds}{1} = s \quad \text{circular radians (numeric)*} \quad (6)$$

where $s$ is the length of the circular arc described between the limits $s_1$ and $s_2$, while $\beta$ is the corresponding angle in circular radians.

In the case of the hyperbolic locus of Fig. 2, the radius-vector $\rho$ varies. Consequently, equation (5) becomes for the hyperbolic case

$$\theta = \int_{s_1}^{s_2} \frac{ds}{\rho} = \frac{s}{\rho'}, \quad \text{hyperbolic radians (numeric)} \dagger \ (7)$$

where $s$ is the length of the hyperbolic arc described between the limits $s_1$ and $s_2$; while $\rho'$ is the integrated mean value of $\rho$ during the motion, as defined by (7), and $\theta$ is the corresponding angle in hyperbolic radians.

*Angles in Terms of Sector Area.*—In the circular sector of Fig. 1, or the hyperbolic sector of Fig. 2, the magnitude of the angle described by the radius-vector OE, between an initial and a final position, is numerically twice the area of the sector swept out by the radius-vector during the motion. Thus in Fig. 1, with the radius OA = 1 cm., if the radius-vector describes the heavy arc A, *b*, *c*, *d*, E, then the circular angle $\beta$ will be, in circular radians, double the sector area AOE in sq. cm.; or will be equal to the shaded double-sector area EOE′ in sq. cm. Similarly, in Fig. 2, with the radius OA = 1 cm. if the radius-vector describes the heavy arc A, *b*, *c*, *d*, E, then $\theta$ will be, in hyperbolic radians, double the sector area AOE in sq. cm.; or will be equal to the shaded double-sector area EOE′, in sq. cm. In Fig. 1, the circular angle AOE = $\beta$ is

* The dimensions of all angles, whether circular or hyperbolic, are assumed in this discussion to be zero; so that an angle is accepted as a numerical quantity, notwithstanding the fact that the arc $ds$ has a different direction in the Cartesian plane from that of the radius-vector $\rho$.

† *See* Appendix L.

represented as 1 circular radian, and the double-shaded sector area is 1 sq. cm. if OA = 1 cm. Similarly, in Fig. 2, the hyperbolic angle AOE = $\theta$ is represented as 1 hyperbolic

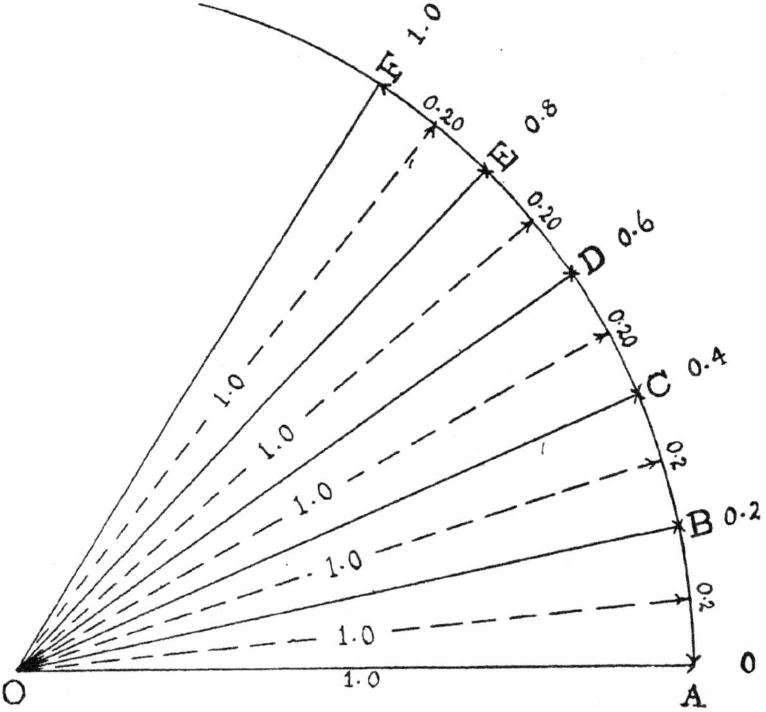


FIG. 3.—A Circular Angle of 1 circular radian, in five sections of 0·2 radian each, expressed as $\beta = \int \frac{ds}{\rho}$.

radian, and the double-shaded sector area is 1 sq. cm. if OA = 1 cm.

It is evident that the hyperbolic angle $\theta$ of the sector AOE in Fig. 2 must be carefully distinguished from the circular angle $\beta$ of the same sector. In the case represented $\theta = 1$ hyperbolic radian; whereas $\beta = 0{\cdot}65087$ circular radian (37° 17′ 33″).

The preceding algebraic relations between arc and radius-vector ratios of circular and hyperbolic angles are illustrated in greater detail by Figs. 3 and 4. In Fig. 3 each of the circular arcs AB, BC, CD, DE, EF possesses a length of 0·2, if the radius OA be taken as of unit length. Consequently, each of the circular angles in the sectors AOB, BOC, COD,

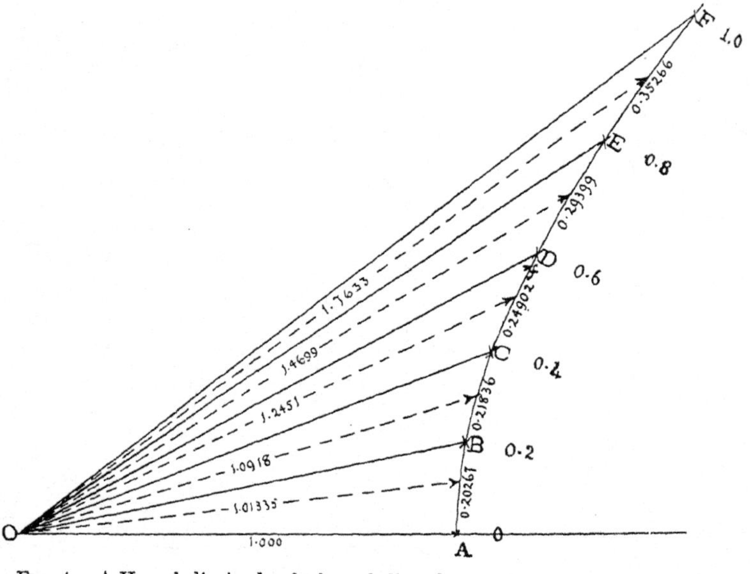


FIG. 4.—A Hyperbolic Angle of 1 hyperbolic radian, in five sections of 0·2 radian each, expressed as $\theta = \int \frac{ds}{\rho}$.

DOE, and EOF is 0·2 circular radian. The total circular angle AOF of the sector AOF is thus 1 circular radian.

In Fig. 4 each of the hyperbolic segments AOB, BOC, COD, DOE, and EOF contains a hyperbolic angle of 0·2 hyperbolic radian, the length of the arcs AB, BC, CD, DE and EF, increasing as the hyperbolic angle increases, and also the lengths of the integrated mean radii-vectores $\rho'$ which are indicated in Fig. 4 for each sector. Consequently, the total hyperbolic angle of the sector AOF is 1 hyperbolic radian, the arc ABCDEF having a total length of 1·3167 units, if the radius OA be taken as of unit length. The integrated mean radius-vector $\rho'$ for the total hyperbolic angle of the sector AOF intersects the curve at *f*.

For brevity, we may use the term "hyp." as an abbreviation for the unit hyperbolic radian; so that in Fig. 4 we may say that each of the sectors contains, and each of the arcs subtends, a hyperbolic angle of 0·2 hyp.; while the total sector AOF contains, and the arc ABCDEF subtends, a hyperbolic angle of 1 hyp.

Hyperbolic angles and hyperbolic trigonometry are of great importance in the theory of electric conductors as used in electric engineering.

## TRIGONOMETRIC FUNCTIONS OF CIRCULAR AND HYPERBOLIC ANGLES.

Trigonometry recognizes certain functions or ratios of lengths in connection with circular and hyperbolic angles. If we retain the initial radius as of unit length, the ratios become simplified into the numerical lengths of certain straight lines. In Figs. 1 and 2, XE is the sine, OX is the cosine, and A*t* the tangent of the angle of the sector, circular or hyperbolic.*

It is evident that when the angle is very small, both the hyperbolic and circular sines are likewise very small; the hyperbolic and circular tangents are likewise very small, while the hyperbolic and circular cosines are very nearly unity. As the angle increases through many radians, the circular sine periodically fluctuates between the limits $+ 1$ and $- 1$, while the hyperbolic sine increases steadily from 0 to $\infty$. The circular cosine periodically fluctuates between $+ 1$ and $- 1$, while the hyperbolic cosine increases steadily from 1 to $\infty$. The circular tangent periodically fluctuates discontinuously between $+ \infty$ and $- \infty$, while the hyperbolic tangent steadily

* Reference is here made only to the *numerical* lengths of these functions; and their proper direction in the plane, real or imaginary, is ignored.

increases from 0 to 1. Fig. 5 shows the graphs of the hyperbolic functions for the first few hyps., the hyperbolic angle

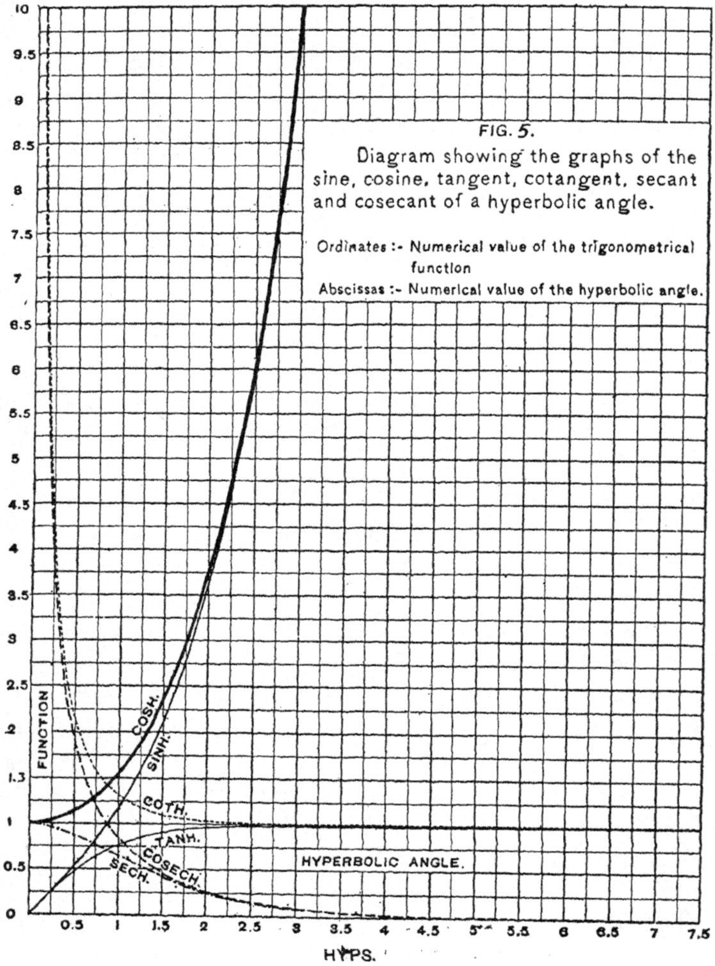


being marked along the axis of abscissas, and the numerical value of the function along the axis of ordinates.

In order to distinguish between hyperbolic and circular

functions, the letter $h$ is affixed to the function when the hyperbolic function is denoted; thus, the sine, cosine, versine, tangent, secant, cosecant, and cotangent of a hyperbolic angle $\theta$ are respectively indicated by the customary notation—

$$\sinh \theta, \cosh \theta, \text{versh } \theta, \tanh \theta, \text{sech } \theta, \text{cosech } \theta, \coth \theta.$$

By the process described in Appendix A, the standard formulas of circular trigonometry may be readily transformed into corresponding formulas of hyperbolic trigonometry. It will be found that circular function formulas involving only first powers, in general transform into corresponding hyperbolic function formulas without change. Thus the formula—

$$\sin 2\beta = 2 \sin \beta \cos \beta \quad . \quad . \quad . \quad \text{numeric} \quad (8)$$

transforms directly into—

$$\sinh 2\theta = 2 \sinh \theta \cosh \theta \quad . \quad . \quad \text{,,} \quad (9)$$

But circular function formulas involving squares, or second powers of functions, usually involve one or more changes of sign in hyperbolic transformation. Thus—

$$\cos^2\beta + \sin^2\beta = 1 \quad . \quad . \quad . \quad . \quad \text{numeric} \quad (10)$$

becomes—

$$\cosh^2\theta - \sinh^2\theta = 1 \quad . \quad . \quad . \quad . \quad \text{,,} \quad (11)$$

With this reservation in mind, it is not worth while preparing a special list of hyperbolic trigonometric formulas. They may be obtained from the corresponding circular trigonometric formulas by transformation on inspection. Consequently, no appreciable additional mental labour is needed for memorizing formulas when learning to apply hyperbolic trigonometry, after the student has learned to apply circular trigonometry. The formulas already learned with the latter suffice for both. A short list of comparative formulas in circular and hyperbolic trigonometry is given in Appendix B.

# CHAPTER II

## APPLICATIONS OF HYPERBOLIC FUNCTIONS TO CONTINUOUS-CURRENT LINES OF UNIFORM RESISTANCE AND LEAKANCE IN THE STEADY STATE

*Perfectly Insulated Lines. Rectilinear Graphs.*—Let us first consider a uniform conducting line such as a telegraph line L kilometers long, but perfectly insulated from the ground and from all other conductors. Such a line will have a uniform linear conductor-resistance of $r$ ohms per km., and its total conductor-resistance will be $Lr$ ohms; but it will, by assumption, be devoid of leakance.

If we free the distant end B of this line AB, Fig. 6, and apply an e.m.f. of $E_A$ volts to the home end A, as by means of the battery shown, it is evident that all parts of the line conductor will take the same electric potential, and the graph of this potential as ordinates, to distance along the line as abscissas, will be the straight line AB parallel to the axis of abscissas.

Again, if we ground the distant end B of the line, as in Figs. 7 and 8, the graph of electric potential will be the inclined straight line AB, Fig. 7, falling from $E_A$ volts at A to zero at B; while at any distance $L_1$ km. from A the potential will be

$$e = E_A - IL_1 r \quad . \quad . \quad . \quad . \quad . \quad \text{volts} \quad (12)$$

Moreover, since there is no current leakage along the line, the current strength will be the same at all points, and the current graph will be the straight line AB, Fig. 8, parallel to the axis of abscissas.

Similarly, if the line instead of being either freed or grounded at B, is grounded there through some constant resistance, it is evident that the graph of potential along the line under an impressed e.m.f. $E_A$ at A, would still be a straight line, but a straight line of lesser inclination than AB in Fig. 7; while the graph of current would still be a straight line parallel to the axis of abscissas, but a straight line lower than AB in Fig. 8.

Similar reasoning applies when an e.m.f. is applied at B, either alone, or in conjunction with an e.m.f. at A.

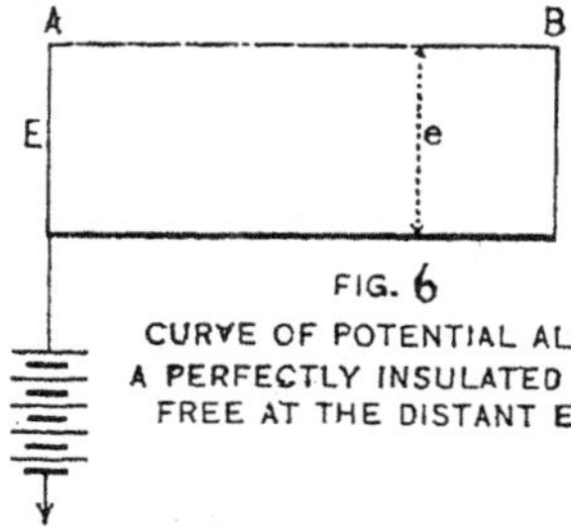


FIG. 6
CURVE OF POTENTIAL ALONG A PERFECTLY INSULATED LINE, FREE AT THE DISTANT END

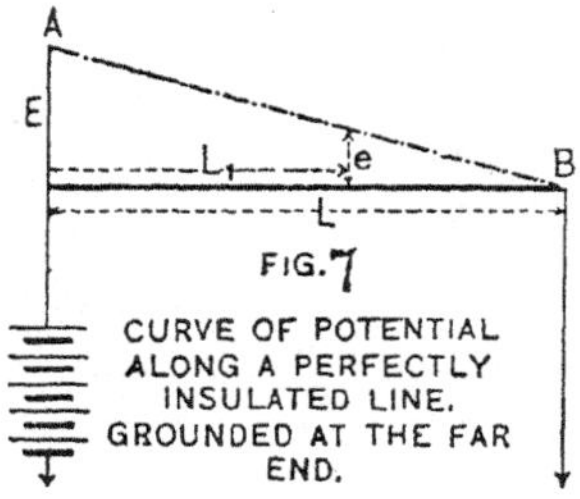


FIG. 7
CURVE OF POTENTIAL ALONG A PERFECTLY INSULATED LINE, GROUNDED AT THE FAR END.

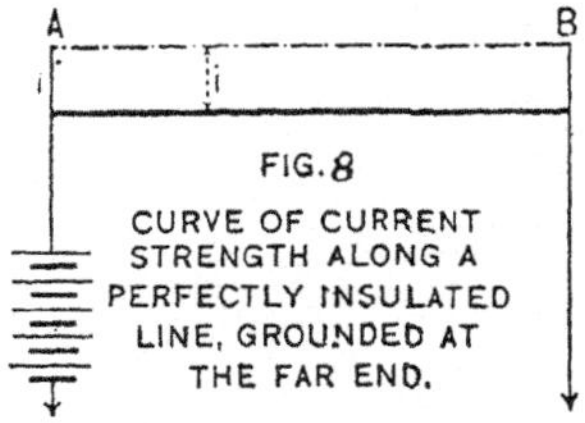


FIG. 8
CURVE OF CURRENT STRENGTH ALONG A PERFECTLY INSULATED LINE, GROUNDED AT THE FAR END.

Consequently we may include all possible conditions under the statement that the graphs of potential and current over any uniform perfectly insulated conductor, in the steady state, are straight lines.*

*Lines of Uniform Resistance and Leakance.*—If now the line, instead of being perfectly insulated, has a uniform linear leakance of $g$ mhos per km.; then if we free the distant end B, and apply an e.m.f. $E_A$ to the home end A, as in Fig. 9, the graph of electric potential along the line will become

* The steady state of current flow will be assumed to have been established in all cases considered.

a catenary, or curve of hyperbolic cosines, such as AB, and the graph of current along the line will be a catenary-slope curve, or curve of hyperbolic sines, such as the dotted line *Ab*, Fig. 9. In the case there represented $L = 500$ km., $r = 10$ ohms per km., and $g = 0{\cdot}5 \times 10^{-6}$ mho per km., or half a micromho per km. corresponding to a linear insulation-resistance of 2 megohm-kilometers.

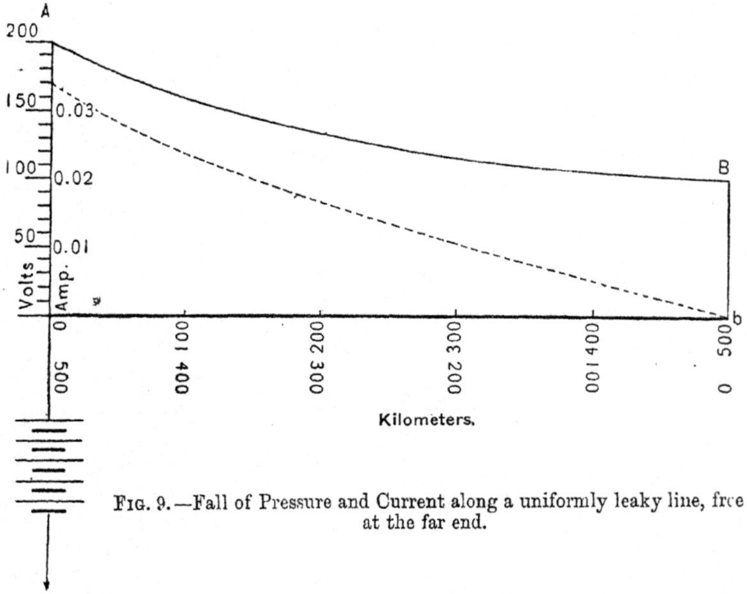


FIG. 9.—Fall of Pressure and Current along a uniformly leaky line, free at the far end.

Under these conditions, as shown in Appendix C, we obtain the following fundamental equations for potential and current. With no leakage—

$$E_A - I_A L_1 r = e = E_B + I_B L_2 r \quad . \quad . \quad \text{volts} \quad (13)$$

$$I_A = i = I_B \quad . \quad . \quad . \quad . \quad \text{amperes} \quad (14)$$

With uniform leakage—

$$E_A \cosh L_1 a - I_A r_\circ \sinh L_1 a = e = E_B \cosh L_2 a + I_B r_\circ \sinh L_2 a \quad \text{volts} \quad (15)$$

$$I_A \cosh L_1 a - \frac{E_A}{r_\circ} \sinh L_1 a = i = I_B \cosh L_2 a + \frac{E_B}{r_\circ} \sinh L_2 a \quad \text{amperes} \quad (16)$$

where $E_A$ and $I_A$ are the e.m.f. and current at A
" $E_B$ " $I_B$ " " " B
" $e$ " $i$ " " " some intermediate point, distant $L_1$ km. from A, and $L_2$ km. from B (Fig. 10)

also—

$$a = \sqrt{rg} \quad . \quad . \quad . \quad . \quad \text{hyp. per km.} \qquad (17)$$

$$r_0 = \sqrt{\frac{r}{g}} \quad . \quad . \quad . \quad . \quad . \quad . \quad . \quad \text{ohms} \qquad (18)$$

The constant $a$ is to be considered as a hyperbolic angle subtended by unit length of line. It is called the *attenuation-constant* of the line. Its dimensions are $\left(\frac{1}{\text{length}}\right)$, or a numeric divided by a length.

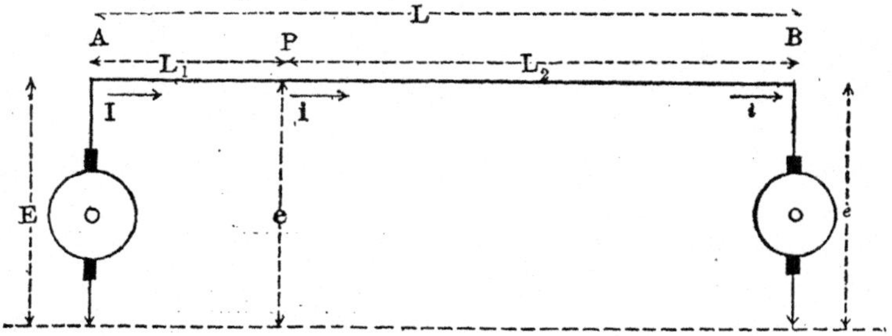


FIG. 10.—Diagram of Simple Ground-Return Circuit: A, Generator End; B, Motor or Receiving End; P, Intermediate Point.

The constant $r_0$ is to be considered as a characteristic resistance pertaining to the line. It is the resistance which an indefinitely long line, of the given linear constants $r$ and $g$, would offer at either end say A, as measured to ground, whether the other end were freed, grounded, or left in any intermediate condition of ground through resistance. It is called the *surge-resistance* of the line. In the case of the line considered with Fig. 9, the surge-resistance is—

$$r_0 = \sqrt{10 \times 2{,}000{,}000} = 4472 \text{ ohms.}$$

The attenuation-constant for the case indicated in Fig. 9 is $a = \sqrt{10 \times 0{\cdot}5 \times 10^{-6}} = 0{\cdot}002236$ hyp. per km. The physical meaning of the constant is that when an impressed e.m.f. E volts is applied to one end of a line, which is either indefinitely long, or is grounded at the distant end through a resistance equal to the surge-resistance of the line, the potential at a distance of one km. from the home end will have fallen from E to $E\varepsilon^{-\alpha}$ volts, where $\varepsilon$ is the base of Naperian logarithms, or 2·71828. In each and every unit length of line the potential will fall by the factor $\varepsilon^{-\alpha}$. Consequently, after $L_1$ km. the potential will have fallen to $E\varepsilon^{-L_1\alpha}$ volts. The factor $\varepsilon^{-L_1\alpha}$ is called the normal attenuation-factor for the length $L_1$. Thus, in the case of the line represented by Fig. 9, with an attenuation-constant of 0·002236 hyp. per km., if an e.m.f. of say $E_A = 200$ volts be impressed at A, and the line is grounded at B through 4472 ohms, the potential at 1 km. from A will have fallen to $200\ \varepsilon^{-0\cdot 002236} = 199\cdot 552$ volts. The potential in each and every km. will fall by 0·2236 per cent., and, after running 500 km., the potential will fall to $200\ \varepsilon^{-1\cdot 118} = 200 \times 0\cdot 3269 = 65\cdot 38$ volts. The normal attenuation-factor for 500 km. of this line is therefore 0·3269. If the line were grounded at B through more or less than $r_0$ ohms, the attenuation-factor would be greater or less than the normal.

*Angle subtended by a Uniform Line.*—A uniform line possesses, or may be said to subtend, a hyperbolic angle—

$$\theta = L\alpha = L\sqrt{rg} = \sqrt{RG} \quad . \quad . \quad . \quad \text{hyps.} \quad (19)$$

where $R = Lr$ is the total conductor-resistance of the line in ohms, and $G = Lg$ is the total dielectric conductance of the line in mhos. That is, the angle of the line in hyps. is the geometric mean of the conductor-resistance and dielectric conductance. The angle of a uniform line increases directly with its length. The attenuation-constant $\alpha$, in overland telegraph lines, varies between the approximate limits of $10^{-5}$ and $10^{-2}$ hyp. per km., according to the condition of insulation. If we take 1000 km. as the greatest length of telegraph line likely to be operated in a single section, without repeaters, the angle of such a line may vary between the limits of 0·01 and 10 hyps. In practice, however, the line would probably cease to be workable telegraphically when the leakance became sufficiently great to bring the line angle to 4 hyps., for which the normal attenuation-factor is $\varepsilon^{-4}$ or 0·018; so that the received current would be only 1·8 per cent. of the current at the sending end, if the receiving end were grounded through a resistance equal to the surge-resistance of the line. A uniform line is a more efficient transmitter of current from the generating to the receiving end as its line angle is reduced; although not in simple proportion.

*Trigonometrical Properties of a Simple Uniform Line in Relation to its Angle. Distant End freed.*—It is easily shown from equations (15) and (16), substituting the proper terminal values of potential and current, that when a line of angle $\theta$ is freed at the distant end, with a steady e.m.f. $E_A$ applied at the home end, the resistance offered by the line at the home end (see Appendix C) is—

$$R_f = r_o \coth\theta \quad . \quad . \quad . \quad . \quad . \quad \text{ohms} \qquad (20)$$

the current entering the line is—

$$I_f = \frac{E_A}{r_o}\tanh\theta = \frac{E_A}{r_o\coth\theta} \quad . \quad . \quad \text{amperes} \qquad (21)$$

and the potential at the distant free end is—

$$E_B = E_A \operatorname{sech}\theta = \frac{E_A}{\cosh\theta} \quad . \quad . \quad . \quad \text{volts} \qquad (22)$$

Table I gives the potential at the distant free end of a line as a decimal fraction of the e.m.f. impressed at the home end, for different lengths of line and various attenuation-constants. Thus, with a line of attenuation-constant 0·0025 hyp. per mile or km., and a length of 500 miles or km., the line angle would be 1·25 hyps., and the Table shows that the voltage at the distant free end would be 0·53, or 53 per cent. of the impressed voltage.

*Distant End grounded.*—Similarly operating upon the fundamental formulas (15) (16) we have, with the distant end of the line grounded, the line resistance offered at the home end—

$$R_g = r_o \tanh\theta \quad . \quad . \quad . \quad . \quad . \quad \text{ohms} \qquad (23)$$

TABLE I

POTENTIAL AT DISTANT FREE END, WITH UNIT E.M.F. IMPRESSED ON LINE AT HOME END. CONTINUOUS-CURRENT CASE.

| Length $a=$ | 0·0001 | 0·0005 | 0·001 | 0·0025 | 0·005 | 0·0075 | 0·01 |
|---|---|---|---|---|---|---|---|
| 50 | 0·9999 | 0·9997 | 0·9988 | 0·992 | 0·970 | 0·930 | 0·8868 |
| 100 | 0·9999 | 0·9987 | 0·9950 | 0·970 | 0·887 | 0·772 | 0·6481 |
| 200 | 0·9998 | 0·9950 | 0·9803 | 0·887 | 0·648 | 0·425 | 0·2658 |
| 300 | 0·9995 | 0·9888 | 0·9566 | 0·772 | 0·425 | 0·209 | 0·094 |
| 400 | 0·9992 | 0·9803 | 0·9250 | 0·648 | 0·266 | 0·094 | 0·037 |
| 500 | 0·9988 | 0·9695 | 0·8868 | 0·530 | 0·163 | 0·048 | 0·014 |
| 600 | 0·9982 | 0·9566 | 0·8435 | 0·425 | 0·094 | 0·022 | 0·005 |
| 700 | 0·9975 | 0·9416 | 0·7967 | 0·337 | 0·060 | 0·011 | 0·002 |
| 800 | 0·9968 | 0·9250 | 0·7477 | 0·266 | 0·037 | 0·005 | 0·001 |
| 900 | 0·9960 | 0·9066 | 0·6978 | 0·209 | 0·022 | 0·002 | 0·0003 |
| 1000 | 0·9950 | 0·8868 | 0·6480 | 0·163 | 0·014 | 0·001 | 0·0001 |

The current entering the line at the home end is—

$$I_g = \frac{E_A}{r_\circ} \coth \theta = \frac{E_A}{r_\circ \tanh \theta} \quad . \quad . \quad \text{amperes} \qquad (24)$$

The current escaping to ground at the distant end is—

$$I_B = \frac{E_A}{r_\circ \sinh \theta} \quad . \quad . \quad . \quad . \quad \text{amperes} \qquad (25)$$

So that the line behaves at the distant end as though it had a line resistance $r_o \sinh \theta$ ohms without leakance. This apparent resistance of the line, as judged at the distant grounded end, is called the *receiving-end resistance* grounded. It is

$$R_l = r_\circ \sinh \theta \quad . \quad . \quad . \quad . \quad . \quad . \quad \text{ohms} \qquad (26)$$

*Apparent Home-End Resistance of a Uniform Line.*—It is evident from formulas (20) and (23) that—

$$r_\circ = \sqrt{R_f . R_g} = \sqrt{r . \frac{1}{g}} = \sqrt{\frac{R}{G}} \quad . \quad . \quad \text{ohms} \qquad (27)$$

or, that the surge-resistance of a uniformly leaky line is the geometrical mean of its apparent resistances when freed and grounded, respectively, at the distant end. When the line is electrically long, *i.e.* when $\theta$ is over 2·5 hyps., tanh $\theta$ ascendingly approaches unity within less than 0·5 per cent., and coth $\theta$ descendingly approaches unity within less than 0·5 per cent.; so that the apparent resistance offered by an electrically long line converges to the surge-resistance $r_0$, whatever the condition of the distant end. Moreover, dividing (23) by (20) obtain—

$$\tanh \theta = \sqrt{\frac{R_g}{R_f}} \quad . \quad . \quad . \quad . \quad . \quad . \quad \text{numeric} \quad (28)$$

so that—

$$\theta = \tanh^{-1}\sqrt{\frac{R_g}{R_f}} \quad . \quad . \quad . \quad . \quad \text{hyps.} \quad (29)$$

and—

$$a = \frac{\theta}{L} \quad . \quad . \quad . \quad . \quad . \quad \text{hyp. per km.} \quad (30)$$

Consequently, if the apparent resistances $R_f$, $R_g$, of the line are correctly measured at either end, the values of $r_0$ and $a$ are determined with the aid of tables of hyperbolic functions,* and from these the corrected linear constants of the line are found by the relations—

$$r = ar_0 \quad . \quad . \quad . \quad . \quad \text{ohms per km.} \quad (31)$$

and—

$$g = \frac{a}{r_0} \quad . \quad . \quad . \quad . \quad . \quad \text{mhos per km.} \quad (32)$$

Thus, if a telegraph line, when freed at the distant end, was observed to offer a resistance $R_f = 5912$ ohms, and when grounded at the distant end, a resistance $R_g = 4434$ ohms, the surge-resistance of the line would be $r_0 = \sqrt{5912 \times 4434} =$ 5120 ohms, and the angle of the line $\theta = \tanh^{-1}\sqrt{\frac{4434}{5912}} =$ $\tanh^{-1} 0{\cdot}86603 = 1{\cdot}317$ hyps. If the line is known to have a length L of 800 km., the attenuation-constant, or linear angle, is $\frac{1{\cdot}317}{800} = 0{\cdot}001646$ hyp. per km. Consequently, the inferred

* The best tables of hyperbolic functions of real hyperbolic angles are probably *Hyperbolic Functions*, by G. F. Becker and C. E. van Orstrand, Smithsonian Institution, Washington (D.C.), 1909; and *Tafeln der Hyperbelfunctionen und der Kreisfunctionen*, by Dr. W. Ligowski (Berlin: Ernst & Korn, 1890). *See also* Appendix to Dr. J. A. Fleming's *The Propagation of Electric Currents* (London: Constable & Co., 1911).

true linear resistance of the line is $r = 5120 \times 0{\cdot}001646 = 8{\cdot}428$ ohms per km., and the inferred true linear leakance is $g = \frac{0{\cdot}001646}{5120} = 0{\cdot}3215 \times 10^{-6}\ \frac{\text{mho}}{\text{km.}}$ (3,110,600 ohm-km. linear insulation resistance). The apparent or uncorrected values would have been $r^1 = \frac{4434}{800} = 5{\cdot}543$ ohms per km., and $g^1 = \frac{1}{5912 \times 800} = 0{\cdot}2114 \times 10^{-6}\ \frac{\text{mho}}{\text{km.}}$.

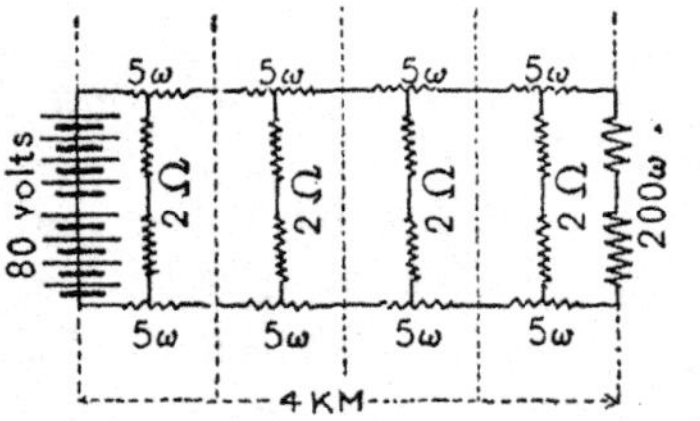


FIG. 11.—Diagram of four kilometers of line with 10ω per loop kilometer and 2 megohm-kilometers.

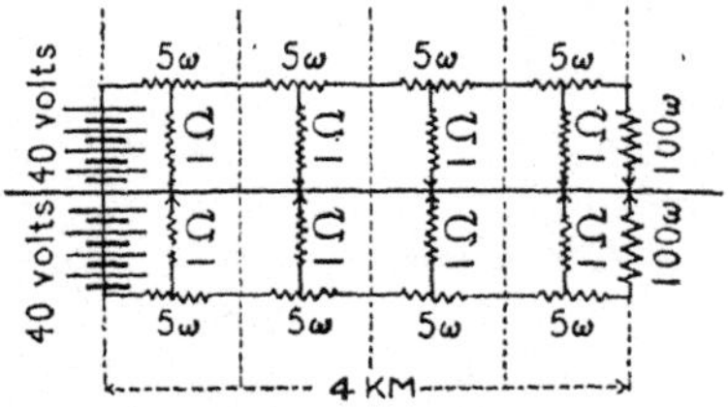


FIG. 12.—Diagram of four kilometers composed of two separate circuits with ground return, each having 5ω per kilometer and 1 megohm-kilometer.

The constants $\alpha$ and $r_0$ are more important in the theory of long uniform electric conductors than the constants $r$ and $g$. The former may therefore be called the *characteristic constants* or *characteristics* of a line; while the latter are the *secondary constants.*

*Characteristics of Loop-Lines and of Wire-Lines.*—We have hitherto discussed the characteristic constants $\alpha$ and $r_0$ of single-wire lines with ground-return circuit, such as are used in wire telegraphy. We now proceed to discuss the characteristics of loop-lines such as are used in wire telephony.

In Fig. 11 a loop-line or metallic circuit is indicated, 4 km. in length, with an e.m.f. of $E_{//} = 80$ volts impressed at the sending end, and a load of $\sigma_{//} = 200$ ohms resistance at the receiving end. The linear conductor-resistance is $r_{//} = 10^{\omega}$ per loop-km., and the linear insulation-resistance is 2 megohm-kilometers representing a linear dielectric leakance $g_{,,} = 0{\cdot}5 \times 10^{-6}$ mho per loop-km. The same system is represented in Fig. 12 with respect to a symmetrical dividing line of zero electric potential, or neutral plane of ground potential. No change in the distribution of potential, current, resistances, or power would be made by connecting the dividing line to ground, or by separating the two halves of the system, and completing each portion by a perfectly conducting ground return. In each half of Fig. 12, we have then an impressed e.m.f. of $E_{,} = E_{,,}/2 = 40$ volts, a load resistance $\sigma_{,} = \sigma_{,,}/2 = 100$ ohms, a linear resistance of $r_{,} = r_{,,}/2 = 5$ ohms per wire-km., and a linear leakance of $g_{,} = 2g_{,,} = 10^{-6}$ mho per wire-km.

The characteristics of each half of the system in Fig. 12 are—

$$a_{,} = \sqrt{r_{,}g_{,}} = \sqrt{5 \times 10_{-6}} = 2{\cdot}236 \times 10^{-3} \text{ hyp. per wire-km.} \quad (33)$$

$$r_{o,} = \sqrt{r_{,}/g_{,}} = \sqrt{5 \times 10^{6}} = 2236 \quad . \quad . \quad . \quad \text{ohms per wire} \quad (34)$$

The characteristics of the double-wire system in Fig. 11 are—

$$a_{,,} = \sqrt{r_{,,}g_{,,}} = \sqrt{10 \times 0{\cdot}5 \times 10^{-6}} =$$

$$2{\cdot}236 \times 10^{-3} \quad . \quad . \quad \text{hyp. per loop-km.} \quad (35)$$

$$= \sqrt{r_{,}g_{,}} = a_{,}$$

$$r_{o,,} = \sqrt{r_{,,}/g_{,,}} = \sqrt{10/0{\cdot}5 \times 10^{-6}} = 4472 \quad \text{ohms per loop} \quad (36)$$

$$= \sqrt{\frac{2r_{,}}{\left(\frac{g_{,}}{2}\right)}} = 2\sqrt{\frac{r_{,}}{g_{,}}} = 2r_{o,}.$$

Consequently, the attenuation-constant has the same value whether computed from the linear resistance and conductance per loop- or per wire-kilometer. The surge-resistance of a loop circuit is twice the surge-resistance of each half considered as a wire circuit to neutral plane. The impressed e.m.f. is, however, also twice as great in the loop circuit as in each wire circuit; so that the current in the loop is the same as the current in each wire.

It is, therefore, a matter of indifference, in computing an attenuation-constant $a$ or a line angle $\theta = La$, whether we use secondary constants $r$ and $g$ per loop-mile or per wire-mile. In computing the surge-resistance, there is an obvious ratio of 2 : 1 in the two cases. We shall, therefore, continue to discuss single-wire circuits, for the greater simplicity of conception and representation, with the understanding that looped circuits are thereby included.

*Characteristics with respect to Unit of Length.*—It is evident, from an inspection of (17) and (18), that if we change the unit of line length, say by using the mile instead of the km., the attenuation-constant will be increased in direct proportion to the size of the unit, whereas the surge-resistance will not be affected. Thus, a line of linear resistance 1·0 ohm per wire-km., and linear leakance of 1 micromho ($10^{-6}$ mho)

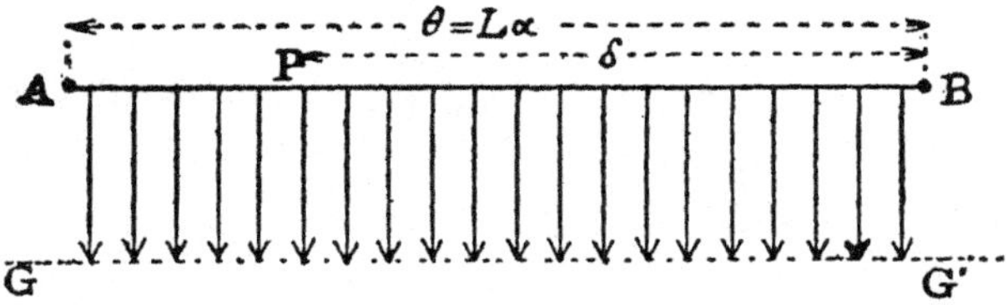


FIG. 13.—Uniform Line with distributed resistance and leakance.

per wire-km., would have an attenuation-constant of 0·001 hyp. per km., or 1 milli-hyp. per km., and a surge-resistance of 1000 ohms. The angle subtended by 1000 km. of such a line would be 1 hyp. The same line would necessarily have a linear resistance of 1·609 ohms per wire statute English mile, and a linear leakance of $1{\cdot}609 \times 10^{-6}$ mhos per wire statute English mile; while its length would be 621·1 English statute miles. On this basis, the attenuation-constant would be $\alpha = 1{\cdot}609 \times 10^{-3}$ hyp. per mile, the angle subtended by the whole line would be 1 hyp., and the surge-resistance would be 1000 ohms as before. That is, attenuation-constants taken with reference to the English statute mile are 1·6093 times those taken with reference to the international kilometer, but neither line-angles nor surge-resistances are affected by such references. The hyperbolic angle and the surge resistance of a line are its primary constants, and are invariants with respect to units of line length.

*Trigonometrical Properties of an Angular Point on a Line.*—If we consider the uniform line AB, Fig. 13, in any steady state with the distant end B either free or to ground, and an electric potential of $U_A$ volts applied at the home end A, we may define the point P on the line as an *angular point*, if its distance from B is measured by the hyperbolic angle $\delta$ hyps. The home end A has the angle $\theta = L\alpha$ hyps. We then find from (15) and (16)—

With B free, $$U_P = U_A \frac{\cosh \delta}{\cosh \theta} \quad \text{volts} \quad (37)$$

$$I_P = I_A \frac{\sinh \delta}{\sinh \theta} \quad \text{amperes} \quad (38)$$

$$R_P = R_A \frac{\coth \delta}{\coth \theta} \quad \text{ohms} \quad (39)$$

With B grounded, $$U_P = U_A \frac{\sinh \delta}{\sinh \theta} \quad \text{volts} \quad (40)$$

$$I_P = I_A \frac{\cosh \delta}{\cosh \theta} \quad \text{amperes} \quad (41)$$

$$R_P = R_A \frac{\tanh \delta}{\tanh \theta} \quad \text{ohms} \quad (42)$$

*Particular Case of Very Short Lines. Approximate Formulas.*—When a uniform conducting line is electrically very short, *i. e.* when its angle $\theta$ is very small, say not exceeding 0·1 hyp., we may without much error substitute—

$\theta$ for $\sinh \theta$
1 ,, $\cosh \theta$
$\theta$ ,, $\tanh \theta$
$\frac{1}{\theta}$ ,, $\operatorname{cosech} \theta$
1 ,, $\operatorname{sech} \theta$
$\frac{1}{\theta}$ ,, $\coth \theta$

We then have with the distant end free, by (20), (21) and (22)—

$$R_f = \frac{r_0}{\theta} = \frac{1}{G} \quad \text{ohms} \quad (43)$$

$$I_f = E_A G \quad \text{amperes} \quad (44)$$

$$E_B = E_A \quad \text{volts} \quad (44a)$$

and, with the distant end grounded, by (23), (24) and (26)—

$$R_g = r_\circ \theta = R \quad . \quad . \quad . \quad . \quad . \quad \text{ohms} \quad (45)$$

$$I_g = E_A/R \quad . \quad . \quad . \quad . \quad . \quad \text{amperes} \quad (46)$$

$$R_l = r_\circ \theta = R \quad . \quad . \quad . \quad . \quad \text{ohms} \quad (47)$$

Or the conditions become equivalent to those of a very short line of resistance R ohms, and of very small leakage conductance G mhos.

*Particular Case of Short Lines. Approximate Formulas.*—We may regard a line as a short line, although not a very short line, when its angle lies between 0·1 and 0·5 hyp. In the case of short lines, we may use two terms in the expansions by series of the trigonometric functions (Appendix B) and substitute—

$$\theta + \frac{\theta^3}{3!} \text{ for } \sinh \theta$$

$$1 + \frac{\theta^2}{2!} \quad \text{,,} \quad \cosh \theta$$

$$\theta - \frac{\theta^3}{3} \quad \text{,,} \quad \tanh \theta$$

$$\frac{1}{\theta} - \frac{\theta}{3!} \quad \text{,,} \quad \text{cosech } \theta$$

$$1 - \frac{\theta^2}{2!} \quad \text{,,} \quad \text{sech } \theta$$

$$\frac{1}{\theta} + \frac{\theta}{3} \quad \text{,,} \quad \coth \theta$$

We then have with the distant end free, by (20), (21) and (22)—

$$R_f = \frac{1}{G}\left(1 + \frac{\theta^2}{3}\right) = \frac{R}{3} + \frac{1}{G} \quad . \quad . \quad . \quad . \quad \text{ohms} \quad (48)$$

$$I_f = E_A G\left(1 - \frac{\theta^2}{3}\right) = \frac{E_A}{\frac{R}{3} + \frac{1}{G}} \quad . \quad . \quad \text{amperes} \quad (49)$$

$$E_B = E_A\left(1 - \frac{\theta^2}{2}\right) = E_A\left(1 - \frac{RG}{2}\right) \quad . \quad \text{volts} \quad (50)$$

That is, a short line offers a resistance at the home end when freed at the distant end, as though all its leakance were applied as a single leak one-third of the line length away from the home end. The potential at the far end also behaves as though the leakance were lumped and applied as a single leak half-way along the line, the drop of pressure in the line being then $E_A G \cdot \frac{R}{2}$ volts.

Similarly, we have with the distant end grounded, by (23), (24), and (26)—

$$R_g = R\left(1 - \frac{\theta^2}{3}\right) = R\left(1 - \frac{RG}{3}\right) \quad . \quad . \quad \text{ohms} \quad (51)$$

$$I_g = \frac{E_A}{R}\left(1 + \frac{\theta^2}{3}\right) = E_A\left(\frac{1}{R} + \frac{G}{3}\right) \quad . \quad \text{amperes} \quad (52)$$

$$R_l = R\left(1 + \frac{\theta^2}{6}\right) = R\left(1 + \frac{GR}{6}\right) \quad . \quad . \quad \text{ohms} \quad (53)$$

That is, a short line offers a resistance at the home end, when grounded at the distant end, as though the leakance were withdrawn and one-third of it were applied as a single leak at the home end. The current escaping to ground at the far end behaves as though two-thirds of the lumped leakance were applied as a single leak at the middle of the line.

*Angle subtended by a Terminal Load.*—If instead of grounding a line at the distant end directly, we ground it through a resistance $\sigma$ ohms, the effect is the same as though a certain angle were added to the line at the distant end. Thus, Fig. 14 represents a uniform line AB of angle $\theta$, grounded at B through a terminal load resistance $CD = \sigma$ ohms. The angle $\theta'$ subtended by this load is such that

$$\tanh \theta' = \frac{\sigma}{r_o} \quad . \quad . \quad . \quad . \quad . \quad . \quad . \quad . \quad \text{numeric} \quad (54)$$

or—

$$\theta' = \tanh^{-1}\left(\frac{\sigma}{r_o}\right) \quad . \quad . \quad . \quad . \quad . \quad \text{hyp.} \quad (55)$$

This load angle therefore depends not upon the absolute value of the load-resistance, but upon the ratio which the load-resistance bears to the surge-resistance of the line to which it is applied. Three cases have to be distinguished with continuous-current lines, according as $\frac{\sigma}{r_o}$ is less than, greater than, or equal to, unity. With alternating-current lines, the distinction becomes unnecessary.

*First Case. Terminal Load-Resistance less than Surge-Resistance.*—If $\sigma$ is less than $r_o$, the equivalent terminal load angle $\theta'$ is easily found from a table of hyp. tangents. Thus, if a

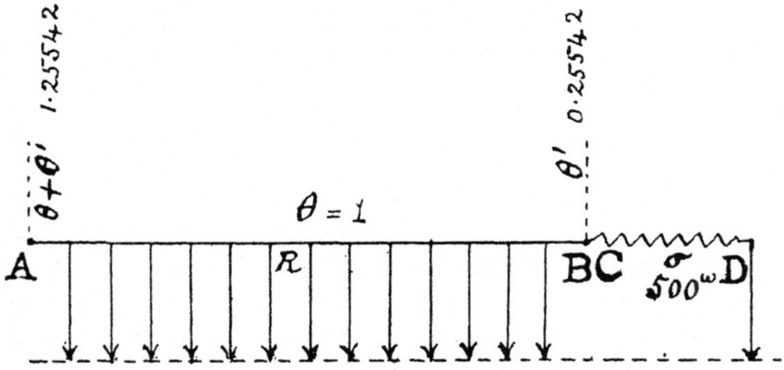


FIG. 14.—Uniform Line grounded through a terminal load with an increase in line-angle.

line AB, 500 km. long, has $r = 4$ ohms per km., and $g = 10^{-6}$ mho per km., its conductor-resistance R will be 2000 ohms and its dielectric leakance G will be $5 \times 10^{-4}$ mho. Its angle will be $\theta = \sqrt{2000 \times 5 \times 10^{-4}} = 1$ hyp., and its surge-resistance $r_o = \sqrt{\frac{2000}{5 \times 10^{-4}}} = 2000$ ohms. If this line is grounded through a terminal load-resistance (CD, Fig. 14) of, say, $\sigma = 500$ ohms, the angle of this load will have as its tangent $\sigma/r_o = 500/2000 = 0{\cdot}25$. The angle is thus found by tables to be 0·25542 hyp. The angle at A subtended by the terminally loaded line is $\delta_A = \theta + \theta' = 1{\cdot}25542$ hyps.

The resistance offered by the terminally loaded line at A is therefore by (23) through extension of the line angle—

$$R_g = r_o \tanh \delta_A = r_o \tanh (\theta + \theta') \quad . \quad \text{ohms} \quad (56)$$

The current escaping to ground at distant end is by (25) (41)—

$$I_B = \frac{E_A \cosh \theta'}{r_o \sinh \delta_A} = \frac{E_A \cosh \theta'}{r_o \sinh (\theta + \theta')} \quad . \quad . \quad \text{amperes} \quad (57)$$

so that by inserting the resistance $\sigma$ in the ground path the receiving-end resistance of the line is increased from $R_l = r_o \sinh \theta$ to—

$$R_l = \frac{r_o \sinh (\theta + \theta')}{\cosh \theta'} \quad . \quad . \quad . \quad . \quad \text{ohms} \quad (58)$$

$$= r_o \sinh \theta + \sigma \cosh \theta \, . \quad . \quad . \quad \text{ohms} \quad (59)$$

Formulas (40), (41), and (42) will be found to apply for any point P of the line AB; that is, for any angle $\delta$ between $\theta'$ and $(\theta + \theta')$. The terminal load $\sigma$ has no angle of its own, but gives to the end B of the line a virtual angle of $\theta'$. Formulas (40) to (42) will therefore not hold for values of $\delta$ less than $\theta'$.

The potential at B is by (40)—

$$U_B = E_A \frac{\sinh \theta'}{\sinh (\theta + \theta')} \quad . \quad . \quad . \quad . \quad \text{volts} \quad (60)$$

In the case considered, the terminally loaded line would offer at A a resistance of $2000 \times 0{\cdot}84979 = 1699{\cdot}58$ ohms. The current which would escape to ground would be the current escaping through the junction BC, and by (41) or (57), would be, with 100 volts applied at A: $\dfrac{100 \times 1{\cdot}03279}{2000 \times 1{\cdot}61216} = 0{\cdot}032$ ampere, and the potential at B, $100 \times \dfrac{0{\cdot}25817}{1{\cdot}61216} = 16{\cdot}0$ volts.

*Second Case. Terminal Load-Resistance greater than Surge-Resistance.*—If $\sigma$ is greater than $r_o$, we apply formulas (37), (38), and (39), instead of (40), (41), and (42). That is, the line condition is considered as though modified from the freed state.

The angle of the terminal load at B is now—

$$\theta' = \coth^{-1}\left(\frac{\sigma}{r_o}\right) = \tanh^{-1}\left(\frac{r_o}{\sigma}\right) \quad . \quad . \quad . \quad \text{hyp.} \quad (61)$$

which is found by a table of tangents. The angle at the home end A is then $\delta_A = \theta + \theta'$ as before.

The resistance offered by the line as measured at A is by (20)—

$$R_f = r_0 \coth \delta_A = r_0 \coth (\theta + \theta') \quad . \quad . \quad \text{ohms} \quad (62)$$

The potential at B is by (37)—

$$U_B = E_A \frac{\cosh \theta'}{\cosh (\theta + \theta')} \quad . \quad . \quad . \quad . \quad . \quad . \quad . \quad \text{volts} \quad (63)$$

The current escaping at B through the load to ground is by (38)—

$$I_B = \frac{E_A \sinh \theta'}{r_0 \cosh (\theta + \theta')} = \frac{E_A}{r_0 \sinh \theta + \sigma \cosh \theta} \quad . \quad \text{amperes} \quad (64)$$

That is, the formulas of (37), (38) and (39) apply between the limits $\delta = \theta'$ and $\delta = \theta + \theta'$; but fail for lower values, or in attempting to apply an angle within the load-resistance CD.

*Third Case. Terminal Load equal to Surge-Resistance.*—In this case the angle given to the end B by the terminal load is infinite, either by (55) or (61). This means that the voltage and current fall off exponentially. The resistance offered by the line at A, or at any point on the line beyond A, is $r_0$ ohms, as already noticed on page 14. The potential at any point P, whose angular distance from A is $\xi$ hyps., is—

$$U_P = U_A \varepsilon^{-\xi} \quad . \quad . \quad . \quad . \quad . \quad . \quad \text{volts} \quad (65)$$

the current at the same point is also—

$$I_P = I_A \varepsilon^{-\xi} \quad . \quad . \quad . \quad . \quad \text{amperes} \quad (66)$$

the current at the sending end is—

$$I_A = \frac{E_A}{r_0} \quad . \quad . \quad . \quad . \quad . \quad \text{amperes} \quad (67)$$

and at the receiving end—

$$I_B = I_A \varepsilon^{-\theta} = I_A \varepsilon^{-La} \quad . \quad . \quad \text{amperes} \quad (68)$$

This is the case of normal attenuation referred to on

page 14. In the case of a very long line, when $\theta$ is over 4 hyps., sinh $\theta$ becomes very nearly $\frac{\varepsilon^\theta}{2}$, and the receiving-end impedance direct to ground $\frac{r_0 \varepsilon^\theta}{2}$. The current escaping to ground over a very long line is thus—

$$I_B = 2I_A \varepsilon^{-\theta} \quad . \quad . \quad . \quad . \quad \text{amperes} \quad (69)$$

or double the current through a terminal load equal to the surge-resistance.

## CHAPTER III

### EQUIVALENT CIRCUITS OF CONDUCTING LINES IN THE STEADY STATE.

It is evident, and has been quantitatively demonstrated in the last chapter, that any given uniform conducting line in the steady state offers a certain resistance at the sending end, and also offers a certain receiving-end resistance at the receiving end. That is, the line system, with its distributed resistance and leakance, may be replaced by a certain equivalent resistance, or group of resistances, at the sending end, and the same proposition applies also to the conditions at the receiving end. Pursuing this inquiry, it may be proved that there exist an infinite number of groups of resistances which, in the steady state, may replace the actual uniform line, with its distributed constants, both at the sending end and at the receiving end. So that, if we select any one of this infinite number of resistance groups and substitute it for the line, there will be no change made, by the substitution, in the distribution of potentials, currents, or powers external to the group or equivalent line conductor. Such an equivalent group of resistances, capable of being substituted for the line without disturbing the electrical conditions outside the line, is called an *equivalent circuit* of the line.

Although an infinite number of equivalent circuits, made up of four resistances, exist for any given line, and *a fortiori* of more than four resistances, yet there are only two equivalent circuits that can be made up of three resistances, and none of less than three. In the case of a single uniform line, either unloaded or symmetrically loaded, one of these triple groups of resistances is a star group of three branches, or a $T$ of resistances, two *arms* of the $T$, or two branches, being equal, and disposed in series to represent the line resistance, and the third branch, or *staff* of the $T$, a resistance in derivation, acting as a leak, or leakage conductance. The other triple group of resistances is a delta group, or triangle group, of three branches, or a $\Pi$, one branch, the *architrave*, being disposed in series, to represent the line resistance, and the two other resistances, which are equal, are in derivation, and form the *pillars* of the $\Pi$, acting as equal leaks or leakage conductances. In Fig. 15 we have at AB a diagram of a single uniform line of total actual conductor-resistance R ohms and total actual leakance G mhos. The angle of the line is $\theta$ hyps. The surge-resistance of the line is $z_0$ ohms.* The surge admittance of the line is $y_0 = 1/z_0$ mhos.

At A′B′ (Fig. 15) is shown the equivalent $T$ of the actual line AB, consisting of the two arms A′O and OB′, each equal to a certain resistance of $\rho'$ ohms, and the staff OG′ of a certain resistance R′, with corresponding conductance $g'$.

At A″B″ (Fig. 15) is shown the equivalent $\Pi$ of the actual line AB, consisting of the architrave resistance $\rho''$ ohms in the line or series circuit, and of two pillar resistances each equal to R″ ohms, one connected in derivation or as a leak at A″, and the other in derivation or as a leak at B″, each having a conductance of $g''$ mhos.

It is shown in Appendix D that the $T$ group A′OB′G′ is the equivalent of the uniform line AB, when the equal resistances of the arms and the conductance of the staff are each adjusted to a certain definite function of the line characteristics. What is true of the equivalent $T$ in this respect is necessarily true of the equivalent $\Pi$, because, by a known theorem, a certain resistance delta is always the external equivalent of a given resistance triple star, as shown in Appendix E.

Although, as has been said, an infinite number of four or more branched equivalent circuits exist for any given line, yet the term "equivalent circuit" may be understood in what follows to denote one or other of the two sole equivalent

* The symbol $z_0$ is here used for surge-resistance, instead of $r_0$, in anticipation of the alternating-current case to be discussed later on.

circuits with three branches, viz. the equivalent $T$ and the equivalent $\Pi$. Sometimes it is convenient to replace a line by its equivalent $T$; at other times it is preferable to replace it

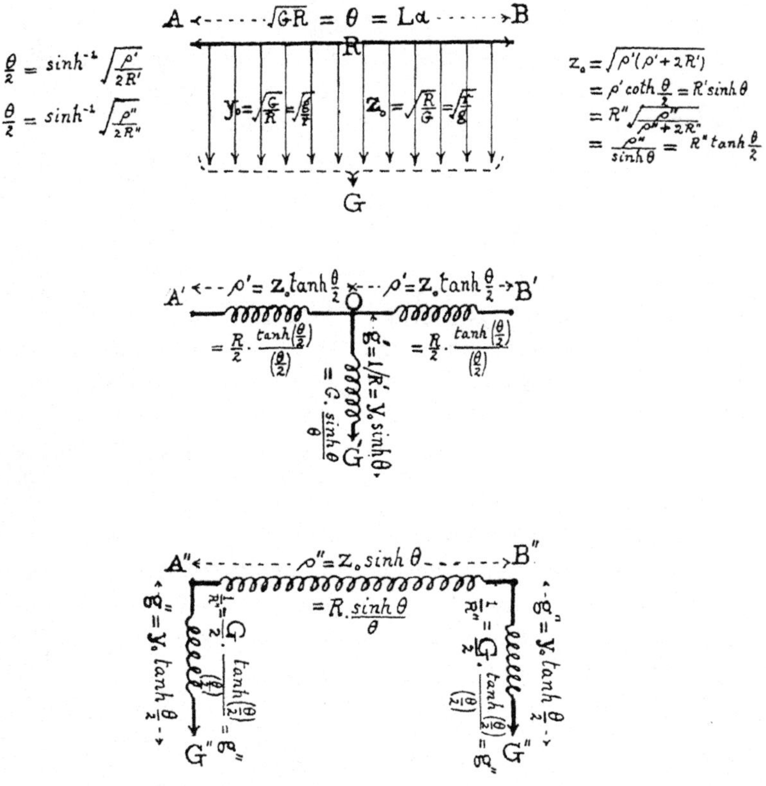

FIG. 15.—A Single Uniform Line and its equivalent circuits with the mutual quantitative relations.

by its equivalent $\Pi$; consequently, both equivalent circuits will be discussed at length, especially because, although in the continuous-current case, both the $T$ and the $\Pi$ equivalents of a line have physical significance and can be actually constructed of resistance coils; yet in the alternating-current case, either the equivalent $T$ or the equivalent $\Pi$ may have arithmetical meaning only, and may be devoid of simple physical significance; so that it cannot be constructed out of a simple association of resistances and reactances.

*Equivalent T.*—As indicated in Fig. 15, the following values must be assigned to the parts of the $T$ in order that it may be externally equivalent to the uniform line AB. Calling $\rho'$ the value of each arm resistance, and $g'$ the value of the staff conductance—

$$\rho' = z_0 \tanh \frac{\theta}{2} = z_0 \coth \theta - R' \quad . \quad . \quad . \quad . \quad \text{ohms} \quad (70)$$

$$g' = y_0 \sinh \theta \quad . \quad . \quad . \quad . \quad . \quad . \quad . \quad \text{mhos} \quad (71)$$

or, since the total conductor-resistance R of the uniform line is by (19) and (27)—

$$R = z_0\, \theta \quad . \quad . \quad . \quad . \quad . \quad . \quad . \quad . \quad \text{ohms} \quad (72)$$

and the total leakance of the uniform line, by (32), is—

$$G = y_0\, \theta \quad . \quad . \quad . \quad . \quad . \quad . \quad . \quad . \quad \text{mhos} \quad (73)$$

we may construct the *nominal* $T$ of the line, by placing in each arm half the actual resistance of the uniform line, that is—

$$\frac{R}{2} = z_0 \frac{\theta}{2} \quad . \quad . \quad . \quad . \quad . \quad . \quad . \quad \text{ohms} \quad (73a)$$

and in the staff, the actual total leakance G of the uniform line. This nominal $T$ will, however, fail to be equivalent to the actual uniform line; because, although the total conductor resistance and the total leakance of the nominal $T$ are respectively the same as the total conductor-resistance and leakance of the line; yet in the nominal $T$ the leakance is collected into one lump and placed at the middle of the line; whereas in the line the leakance is distributed. The nominal $T$ has, therefore, an error due to the lumpiness of the leakance; whereas in the equivalent $T$ this error is eliminated. The correcting factor which must be applied to each arm of the nominal $T$, in order to convert it into the corrected value, or arm of the equivalent $T$, is—

$$k_{\rho,} = \frac{\tanh\left(\frac{\theta}{2}\right)}{\left(\frac{\theta}{2}\right)} \quad . \quad . \quad . \quad . \quad \text{numeric} \quad (74)$$

This factor approaches the value 1 as $\theta$ becomes very small. It tends to diminish with increasing real values of $\theta$. That is, the arms of an equivalent $T$ are always smaller than those of the corresponding nominal $T$ of a continuous-current line, and tend to become smaller as the angle of the line increases.

The numerical values of $k_{,}$ are, however, subject to great fluctuations in the alternating-current case.

Similarly, the correcting factor which must be applied to the conductance in the leak of the nominal $T$ is—

$$k_{g,} = \frac{\sinh\theta}{\theta} \quad . \quad . \quad . \quad . \quad . \quad . \quad \text{numeric} \quad (75)$$

a quantity which commences at 1 when $\theta = 0$, and increases indefinitely with $\theta$. Consequently, the correcting factor of both leak and arms is unity for a very short line; or the equivalent $T$ degrades into the nominal $T$ for $\theta = 0$; but as the line angle increases, the equivalent $T$ diverges from the nominal $T$, the equivalent $T$ arms diminishing relatively in resistance, and the equivalent $T$ leak increasing relatively in conductance, in the case of a continuous-current system.

*Equivalent Π.*—As indicated in Fig. 15, the following values must be assigned to the parts of the $\Pi$ in order that it may be externally equivalent to the uniform line AB. Calling $\rho''$ the value of the architrave-resistance, and $g''$ the value of the conductance in each pillar—

$$\rho'' = z_{\circ} \sinh\theta \quad . \quad . \quad . \quad . \quad . \quad . \quad \text{ohms} \quad (75a)$$

$$g'' = y_{\circ} \tanh\frac{\theta}{2} \quad . \quad . \quad . \quad . \quad . \quad . \quad \text{mhos} \quad (76)$$

or, since the total conductor-resistance R of the uniform line is $z_{\circ}\theta$, and the total leakance $y_{\circ}\theta$, we may construct the *nominal Π* of the line by placing in the architrave the total actual line-resistance $z_0\theta$, and half the total leakance in each pillar; that is, we assign a conductance of $y_0\frac{\theta}{2}$ to each pillar of the nominal $\Pi$. As before, the nominal $\Pi$ will have a lumpiness error, and this error is avoided in the equivalent $\Pi$. The correcting factor which must be applied to the nominal architrave-resistance in order to convert it into the equivalent $\Pi$ architrave is—

$$k_{p_{\prime\prime}} = \frac{\sinh \theta}{\theta} \quad . \quad . \quad . \quad . \quad . \quad . \quad \text{numeric} \quad (77)$$

Similarly, the correcting factor which must be applied to the conductance in each pillar of the nominal $\Pi$ in order to convert it into the equivalent $\Pi$ pillar is—

$$k_{g_{\prime\prime}} = \frac{\tanh\left(\frac{\theta}{2}\right)}{\frac{\theta}{2}} \quad . \quad . \quad . \quad . \quad . \quad \text{numeric} \quad (78)$$

It is evident that the correcting factors of the nominal $T$ and $\Pi$ are the same, but mutually inverted, the correction tending in the continuous-current case ($\theta$ a real numeric) to increase the line-resistance of the $\Pi$ and to diminish the leaks in the pillars.

As a numerical instance, we may take a uniform line of $L = 100$ km., $r = 20$ ohms per km., $g = 20 \times 10^{-6}$ mho per km.; so that $\theta = 2$ hyps., and $r_0 = 1000$ ohms. Fig. 16 shows the nominal $T$ and $\Pi$ of this line, as well as the equivalent $T$ and $\Pi$. It will be seen that the nominal circuits each place 2000 ohms in the line, and 2 millimhos of leakance in total derivation. The correcting factor $\frac{\sinh 2}{2}$ is 1·81343, and the correcting factor $\frac{\tanh 1}{1}$ is 0·761594. Applying these correcting factors, we obtain the equivalent circuits. Either of these equivalent circuits may be substituted for the uniform line AB, without disturbing in any way the external conditions, whether e.m.f. is applied at either or both ends of the line.

*Assigning a Uniform Line Equivalent to a Symmetrical T or Π. Reversion of Equivalent T.*—Just as we have seen that to any given uniform line there corresponds one, and only one, symmetrical $T$ or $\Pi$ which is its perfect external equivalent in the

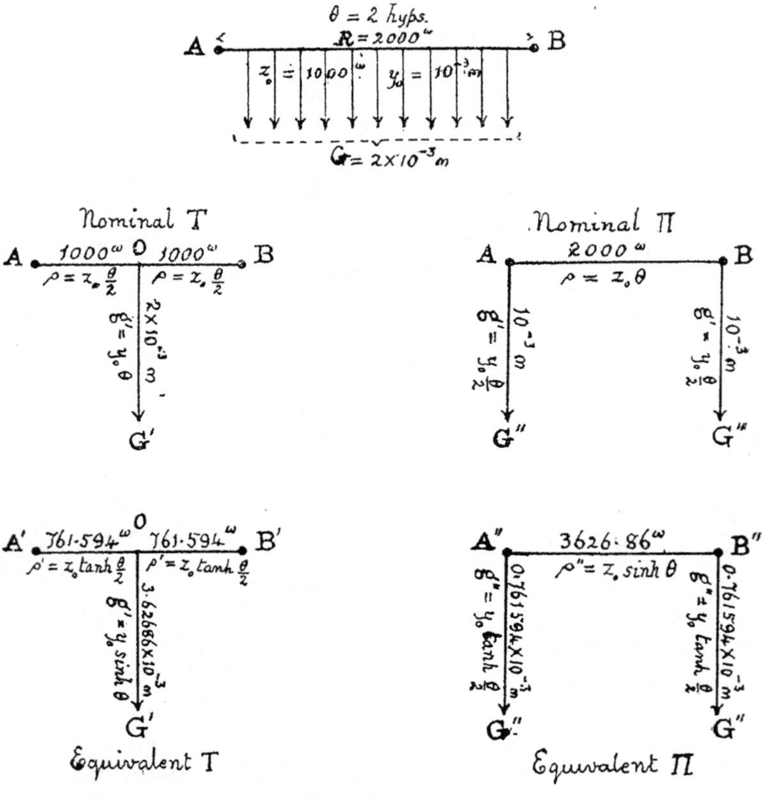


FIG. 16.—Nominal and Equivalent T's and Π's of a Uniform Line.

steady state; so, conversely, to any given symmetrical $T$ or $\Pi$, there corresponds one, and only one, uniform line with distributed constants.

Suppose a symmetrical $T$ given, as in Fig. 15, A′OB′G′, with two equal arms of $\rho'$ ohms, and a leak of $g'$ mhos. Then the apparent angle of this line, treating it as a nominal $T$ of total line resistance $R = 2\rho'$ ohms, and total line leakance $G = g'$ mhos, would be by (19)—

$$\theta = \sqrt{2\rho' . g'} \quad . \quad . \quad . \quad \text{apparent hyps.} \quad (79)$$

and the apparent semi-angle or apparent angle of half the line would be—

$$\frac{\theta}{2} = \sqrt{\frac{\rho' . g'}{2}} \quad . \quad . \quad . \quad \text{apparent hyps.} \quad (80)$$

But this semi-angle cannot be correct, because it is based on a lumped leakance instead of a distributed leakance as assumed in (19). The correction-factor by which the left-hand member of (80) must be multiplied, in order to eliminate this error of assumption, is—

$$k_{\theta'} = \frac{\sinh\left(\frac{\theta}{2}\right)}{\left(\frac{\theta}{2}\right)} \quad . \quad . \quad . \quad . \quad . \quad \text{numeric} \quad (81)$$

That is, the formula (80) becomes on correction—

$$\sinh\frac{\theta}{2} = \sqrt{\frac{\rho' g'}{2}} \quad . \quad . \quad . \quad . \quad . \quad \text{numeric} \quad (82)$$

or the angle of the required equivalent uniform line is—

$$\theta = 2\sinh^{-1}\sqrt{\frac{\rho' g'}{2}} \quad . \quad . \quad . \quad . \quad . \quad \text{hyps.} \quad (83)$$

$$= 2\sinh^{-1}\sqrt{\frac{\rho'}{2R'}} \quad . \quad . \quad . \quad . \quad . \quad \text{"} \quad (84)$$

if $R' = \frac{1}{g'}$ is the resistance of the staff-leak in ohms.

Again, if we proceed to form the surge-resistance of the line corresponding to a nominal *T*, we have, by (27)—

$$r_{\circ} = \sqrt{\frac{R}{G}} \quad . \quad . \quad . \quad . \quad . \quad . \quad \text{ohms} \quad (85)$$

But applying this process to the equivalent *T*, Fig. 15, with $2\rho'$ instead of R and $g'$ instead of G, we obtain an apparent surge-resistance $r_{\circ}'$; namely—

$$r_{\circ}' = \sqrt{\frac{2\rho'}{g'}} = \sqrt{2\rho' R'} \quad . \quad . \quad \text{apparent ohms} \quad (86)$$

which contains a lumpiness error. The correcting factor which must be applied to the left-hand member of (86), in order to arrive at the true surge-resistance $r_o$ of the equivalent uniform line, is—

$$k_{r'_o} = \cosh\frac{\theta}{2} = \sqrt{1 + \frac{\rho' g'}{2}} \quad . \quad . \quad . \quad \text{numeric} \quad (87)$$

where $\theta$ is obtained from (83) or (84). Consequently—

$$r_o = r'_o \cosh\frac{\theta}{2} = \cosh\frac{\theta}{2}\sqrt{\frac{2\rho'}{g'}} = \text{R}' \sinh\theta \quad \text{ohms} \quad (88)$$

$$= \rho'\sqrt{1 + \frac{2}{g'\rho'}} = \rho'\sqrt{1 + \frac{2\text{R}'}{\rho'}} = \rho' \coth\frac{\theta}{2} \quad \text{"} \quad (89)$$

*Reversion of Equivalent Π.*—If a symmetrical *Π* is given, like that in Fig. 15, we may proceed to determine the uniform line which is its external equivalent. If the *Π* were a nominal *Π*, we should have $\text{R} = \rho''$, and $\text{G} = 2g''$. In that case, the angle of the line would be, by (19)—

$$\theta = \sqrt{\rho'' 2g''} \quad . \quad . \quad . \quad . \quad \text{apparent hyps.} \quad (90)$$

and the semi-angle of the line would be—

$$\frac{\theta}{2} = \sqrt{\frac{\rho'' . g''}{2}} \quad . \quad . \quad . \quad \text{apparent hyps.} \quad (91)$$

But owing to the fact that the *Π* is an equivalent, and not a nominal *Π*, this equation contains a lumpiness error. The correction-factor which must be applied to the left-hand member of (91), in order to eliminate this error, is—

$$k_{\theta''} = \frac{\sinh\left(\frac{\theta}{2}\right)}{\left(\frac{\theta}{2}\right)} \quad . \quad . \quad . \quad . \quad \text{numeric} \quad (92)$$

That is, the formula (91) becomes on correction—

$$\sinh\frac{\theta}{2} = \sqrt{\frac{\rho'' g''}{2}} \quad . \quad . \quad . \quad . \quad \text{numeric} \quad (93)$$

or the angle of the required equivalent uniform line is—

$$\theta = 2\sinh^{-1}\sqrt{\frac{\rho'' g''}{2}} = 2\sinh^{-1}\sqrt{\frac{\rho''}{2\text{R}''}} \quad . \quad . \quad . \quad \text{hyps.} \quad (94)$$

$$= 2\tanh^{-1}\sqrt{\frac{\rho'' g''}{2 + \rho'' g''}} = 2\tanh^{-1}\sqrt{\frac{\rho''}{2\text{R}'' + \rho''}} \quad \text{"} \quad (95)$$

Again, if we proceed to form the surge-resistance of the line corresponding to a nominal $\Pi$, we have as in (85)—

$$r_\circ'' = \sqrt{\frac{\rho''}{2g''}} = \sqrt{\frac{\rho'' R''}{2}} \quad . \quad \text{apparent ohms} \quad (96)$$

which contains a lumpiness error. The correcting factor which must be applied to the left-hand member of (96), in order to arrive at the true surge-resistance $r_o$ of the equivalent uniform, is—

$$k_{r_\circ''} = \text{sech}\,\frac{\theta}{2} = \sqrt{\frac{2}{2 + g''\rho''}} \quad . \quad . \quad \text{numeric} \quad (97)$$

where $\theta$ is obtained from (94) or (95).

Consequently—

$$r_\circ = r_\circ'' \,\text{sech}\,\frac{\theta}{2} = \text{sech}\,\frac{\theta}{2}\sqrt{\frac{\rho''}{2g''}} \quad . \quad . \quad . \quad . \quad . \quad \text{ohms} \quad (98)$$

$$= R''\sqrt{\frac{\rho''}{2\,R'' + \rho''}} = R''\tanh\frac{\theta}{2} = \frac{\rho''}{\sinh\theta} \quad \text{,,} \quad (99)$$

*General Equivalent Circuit Formulas.*—We may write down the values of the elements in the equivalent circuits of a line in terms of the observed resistances $R_f$ and $R_g$ of the line at the home end when freed and grounded respectively at the distant end. For the equivalent $T$ we have—

$$\rho' = R_f\left(1 - \sqrt{1 - \frac{R_g}{R_f}}\right) \quad . \quad . \quad . \quad \text{ohms} \quad (100)$$

$$R' = R_f\sqrt{1 - \frac{R_g}{R_f}} \quad . \quad . \quad . \quad . \quad . \quad \text{,,} \quad (101)$$

and for the equivalent $\Pi$, we have—

$$\rho'' = R_g \Big/ \sqrt{1 - \frac{R_g}{R_f}} \quad . \quad . \quad . \quad . \quad . \quad \text{,,} \quad (102)$$

$$R'' = R_f\left(1 + \sqrt{1 - \frac{R_g}{R_f}}\right) \quad . \quad . \quad \text{,,} \quad (103)$$

from which—

$$\frac{\rho' + R''}{2} = R_f \quad . \quad . \quad . \quad . \quad . \quad \text{,,} \quad (104)$$

$$\rho'/g'' = \rho''/g' = R''\rho' = R'\rho'' = R_f R_g = r/g = R/G = r_\circ^2 \ \text{ohms}^2 \quad (105)$$

$$\rho' g' = \rho'' g'' = \cosh\theta - 1 = \text{versh}\,\theta = 2\sinh^2\frac{\theta}{2} \ \text{numeric} \quad (106)$$

*Artificial Lines.*—Artificial lines are either single- or double-wire lines, according as they are of the telegraphic or telephonic type. All double-wire artificial lines may be divided symmetrically into a pair of single-wire lines, each operated to neutral or zero-potential plane in the manner indicated in Figs. 11 and 12. Consequently, it is only necessary to discuss the behaviour of single-wire artificial lines. In the continuous-current case, such lines are made up of uniform sections of line resistance in series, and leakance in derivation. A particular case of such an artificial line of five sections is illustrated in Figs. 17, 18 and 19. Each section consists of a symmetrical $T$ having 250 ohms in each arm, and a leak of 250 micromhos in the staff. Consequently, by what has been seen in relation to the equivalence of a uniform line to a given symmetrical $T$, it follows that each such section is externally equivalent to a corresponding section of uniform line with distributed resistance and leakance. Applying formulas (83) and (88) to the case of Fig. 17, we find for the equivalent line section $\theta = 0{\cdot}35172$ hyp. and $r_0 = 1436{\cdot}13$ ohms. The whole line, therefore, subtends an angle of 1·7586 hyps. at A. The secondary constants of the equivalent uniform line are by (72) and (73), R = 2525·5 ohms, and G = 1·22495 millimho. In the artificial line there are actually 2500 ohms in series-resistance and 1·25 millimhos of total leakance.

The artificial line, therefore, behaves externally in the steady state exactly like a uniform distributed-constant line of 1·7586 hyps. and 1436·13 ohms surge-resistance. In Fig. 17 the artificial line is grounded at the distant end B, and 100 volts potential is applied at A. The currents and potentials are indicated in the various elements of the line by Ohm's law computation; but they also conform to the hyperbolic trigonometry of a single uniform line at the ends A and B.

Moreover, at each junction between sections, marked respectively 1, 2, 3 and 4 in Fig. 17, the hyperbolic angle is indicated, and corresponds to like symmetrical points along the equivalent uniform line; so that the potential and current distributions along the artificial line conform precisely to those on the

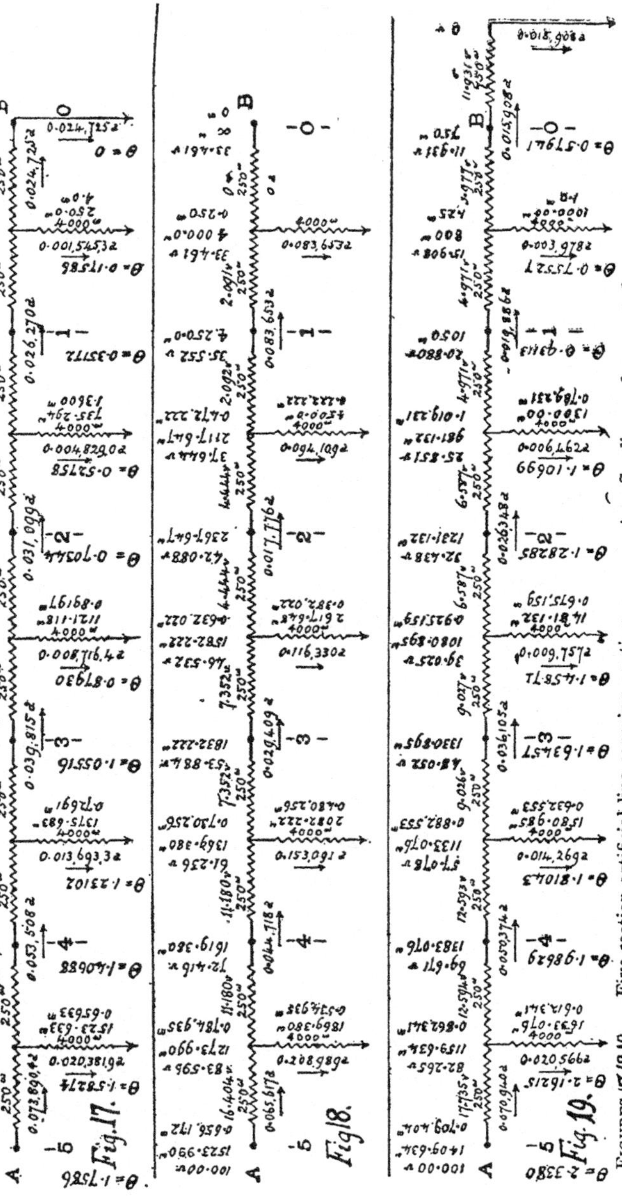


FIGURES 17, 18, 19. — Five-section artificial line carrying continuous currents. Sending end connected to 100 volts e.m.f. Distant end grounded, freed, and to ground through resistance. In Figure 18 the indicated leak currents should be divided by ten.

equivalent uniform line, not merely at the two ends A and B, but also at the four section-junctions; or at six points in all. In order, therefore, to compute trigonometrically the distribution of current or potential at any point on the artificial line, say at the second leak, it is only necessary to compute the corresponding condition at the nearest equiangular point on the uniform equivalent line, by formulas already given, and then to apply Ohm's law over the intervening semi-section of line-resistance or arm of a *T*. This trigonometrical method of computing the distribution is, in general, much less laborious than that of working along the artificial line from either end by Ohm's law—that is, the trigonometrical method is a labour-saving device in the continuous-current case, and would also be so in any alternating-current case, if the proper hyperbolic-function Tables were available.

In Fig. 18 the artificial line is freed at the end B with an impressed potential of 100 volts at A. The distribution of potential and current over the artificial line agrees with that of the above-mentioned equivalent distributed-constant line at the ends, and at the four section-junctions.

If we plot the potential along the line as ordinates to line angle or length as abscissas, we shall have a true catenary falling from 100 to 33·461 volts, for the graph in the case of the real line; whereas we shall have a broken line or series of descending straight lines falling from 100 to 33·461 volts for the graph in the case of the artificial line. The true catenary will cut these straight lines at points corresponding to section-junction points. (See Appendix F.)

The broken straight line is, in fact, the funicular polygon obtained by lumping the masses symmetrically at the centres of the equivalent catenary-sections.

In Fig. 19 the artificial line is grounded at the end B through a resistance of 750 ohms. The effect is just the same as though the equivalent real line were so grounded. The equivalent angle subtended by the terminal load resistance is $\theta' = 0{\cdot}57941$ hyp., and the angles of all equivalent points along the artificial and equivalent real lines are increased by this amount in the manner indicated.

An alternative method of dealing trigonometrically with the distribution of potential and current over an artificial line, by the use of continued fractions, is discussed in Appendix F.

# CHAPTER IV

## REGULARLY LOADED UNIFORM LINES

A LOAD in a line may be defined as an element of resistance inserted in the line, or an element of leakance inserted in derivation on the line; or a combination of both; in such a manner as to introduce a discontinuity into the line, or break up its uniform distribution of line constants. Loads inserted in series into a line may be called *series* loads, or *impedance* loads; while those added in derivation may be called *leak* loads. Loads at the end of a line are called *terminal* loads; while those elsewhere are called *intermediate* loads. When loads are inserted at regular intervals along a line, or according to some assigned law of succession, they are called *regular* loads. When inserted otherwise, they are called either *irregular* loads or *casual* loads.

*Regular Series Loading.*—We may first consider the effect of loading a line at regular intervals with equal resistances. This is the simplest and most important case of regular loading.

On the first row of Fig. 20, we have a line indicated which at uniform angular intervals of $\theta = 0{\cdot}35172$ hyp. has series resistances inserted, each of $\Sigma = 200$ ohms. The surge-resistance of the line before the loading is $r_0 = 1436{\cdot}13$ ohms. Each section of line between adjacent loads might be 100 km. It is required to find the characteristics of the loaded line.

In the process represented by Fig. 20, the first step is to find the equivalent $T$ of a line section. This is done with the aid of formulas (70) and (71). This gives us a resistance $\rho'$ of 249·985 ohms in each arm of the $T$, and a resistance R′ of 4000·22 ohms in the staff of the $T$, corresponding to a leakance $g' = 249{\cdot}92$ micromhos. The angle subtended by this

$T$ will be the same as the angle of the line-section to which it corresponds ($\theta = 0{\cdot}35172$ hyp.). The actual resistance R in

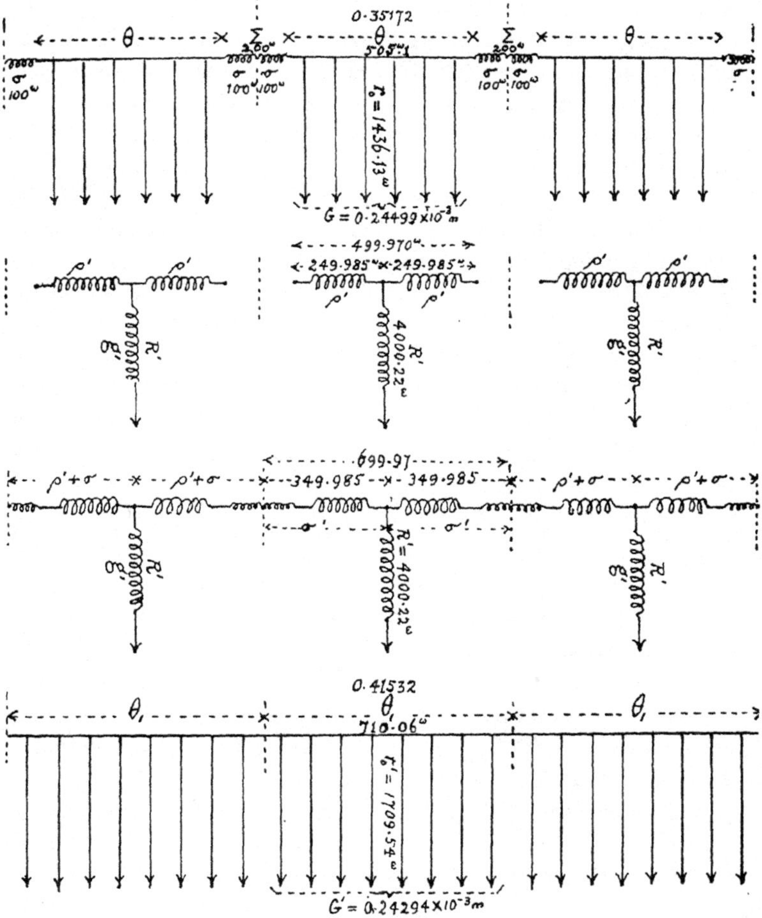


FIG. 20.—Reduction of a Uniform Actual Line with loads in series to an Equivalent Unloaded Uniform Line.

a section of the line would be 505·1 ohms, and the actual leakance G in the section 244·99 micromhos.

The next step is to divide the loads $\Sigma$ into two halves, of $\sigma$ ohms each, and to add $\sigma$ on to each arm of the equivalent $T$, as indicated on the third row of Fig. 20. The extended $T$ which now includes half a load at each end, has a resistance of $\rho' = 349{\cdot}985$ ohms in each arm, and the same leak resistance as before, $R' = 4000{\cdot}22$ ohms. The entire loaded line may now be regarded as made up of sections of these extended $T$'s. We now proceed to find by formulas (84) and (89) the characteristics of the sections of uniform line equivalent to the extended $T$'s. As shown on the lowest row of Fig. 20 this is found to be $\theta_{,} = 0{\cdot}41532$ hyp., and $r_0' = 1709{\cdot}54$ ohms. The conductor-resistance corresponding to such a section is $r_0'\,\theta_{,} = 710{\cdot}06$ ohms, a virtual increase of $204{\cdot}96$ ohms; while the corresponding section leakance is $\theta_{,}/r'_0 = 242{\cdot}94$ micromhos, or a virtual reduction of 2·05 micromhos.

It is self-evident that if the 200 additional ohms of load-resistance per section had been inserted by uniform distribution instead of in lumps, there would have been only 200 ohms increase in the section R, with no change in the section G. Consequently, the effect of inserting the added resistance in lumps at 100 km. intervals is a virtual increase of 4·96 ohms of line-resistance per section, together with a slight virtual reduction in leakance. The steps of the process also indicate that for a given amount of additional resistance to be inserted at regular intervals into a line, the shorter the interval, the smaller is the inserted load, and the smaller the change in the extension of the equivalent $T$'s; whereas with long sections, large loads must be inserted at their junctions. The extensions of the equivalent $T$'s are then correspondingly enlarged, with increased effect on the line angle, and on the virtual extra resistance of the loads due to their lumpiness.

If $\theta$ be the angle of a section before loading (hyps.)
" $\theta_{,}$ " " " " after " "
" $z_0$ , surge-resistance of the line before loading (ohms)
" $z_0'$ " " " " after " "
" $\Sigma = 2\sigma$ be the resistance of each series load (ohms),

then the process above outlined leads to the following results—

$$\sinh\frac{\theta_{\prime}}{2} = \sinh\frac{\theta}{2}\sqrt{1+\frac{\sigma\coth\frac{\theta}{2}}{z_{0}}} \quad . \quad . \quad . \quad . \quad . \quad \text{numeric} \quad (107)$$

$$\cosh\frac{\theta_{\prime}}{2} = \cosh\frac{\theta}{2}\sqrt{1+\frac{\sigma\tanh\frac{\theta}{2}}{z_{0}}} \quad . \quad . \quad . \quad . \quad \text{"} \quad (108)$$

$$\tanh\frac{\theta_{\prime}}{2} = \tanh\frac{\theta}{2}\sqrt{\frac{1+\frac{\sigma\coth\frac{\theta}{2}}{z_{0}}}{1+\frac{\sigma\tanh\frac{\theta}{2}}{z_{0}}}}$$

$$= \sqrt{\tanh\frac{\theta}{2}\cdot\tanh\left(\frac{\theta}{2}+\delta\right)} \quad . \quad . \quad . \quad \text{numeric} \quad (109)$$

if $\frac{\sigma}{z_{0}} = \tanh\delta$.

In the continuous-current case, if $\sigma > z_{0}$, let $\frac{\sigma}{z_{0}} = \coth\delta$.

$$\text{Then}\ \coth\frac{\theta_{\prime}}{2} = \sqrt{\coth\frac{\theta}{2}\cdot\coth\left(\frac{\theta}{2}+\delta\right)} \quad . \quad \text{numeric} \quad (110)$$

Also—

$$\sinh\theta_{\prime} = \sinh\theta\sqrt{1+\frac{2\sigma}{z_{0}}\coth\theta+\left(\frac{\sigma}{z_{0}}\right)^{2}} \quad . \quad \text{"} \quad (111)$$

$$\cosh\theta_{\prime} = \cosh\theta+\frac{\sigma}{z_{0}}\sinh\theta \quad . \quad . \quad . \quad . \quad . \quad \text{"} \quad (112)^{*}$$

$$z_{0}' = z_{0}\sqrt{\left(\tanh\frac{\theta}{2}+\frac{\sigma}{z_{0}}\right)\left(\coth\frac{\theta}{2}+\frac{\sigma}{z_{0}}\right)} \quad . \quad \text{ohms} \quad (113)$$

$$= z_{0}\sqrt{1+\frac{2\sigma}{z_{0}}\coth\theta+\left(\frac{\sigma}{z_{0}}\right)^{2}} \quad . \quad . \quad . \quad . \quad \text{"} \quad (114)$$

$$\text{or—} \quad \frac{z'_{0}}{z_{0}} = \frac{\sinh\theta_{\prime}}{\sinh\theta} \quad . \quad . \quad . \quad . \quad . \quad . \quad . \quad . \quad . \quad \text{numeric} \quad (115)$$

Consequently, the characteristics $\theta_{\prime}$ and $z'_{0}$ of a section after loading can always be computed when the value $\Sigma$ of each

* Formula 112 was first published by Dr. G. A. Campbell. *See* Bibliography, 27.

resistance-load is given, as well as the characteristic constants $\theta$ and $z_0$ of the section before loading. On the other hand, if the resistance of the load were distributed uniformly over the section, the effect would evidently be to increase $\theta$ and $z_0$ in the same ratio; that is—

$$\frac{\theta_{,}}{\theta} = \frac{z_0'}{z_0} = \sqrt{\frac{\Sigma + R}{R}} = \sqrt{1 + \frac{\Sigma}{R}} \quad . \quad \text{numeric} \quad (116)$$

*Regular Leak Loading.*—If as indicated on the first row of Fig. 21, equal leakances of conductance $\Gamma$ mhos are inserted at uniform angular distances $\theta$ hyps. along a line, the surge-resistance of which is $z_0$ ohms; we may proceed to determine the effect of this loading on the line characteristics.

The first step is to break each load into halves of conductance $\gamma$ mhos ($\Gamma = 2\gamma$), as on the second row of Fig. 21.

The next step is to find the equivalent $\Pi$ of each unloaded section of line, as on the third row of Fig. 21, with the aid of formulas (75) and (76).

We now extend each section $\Pi$ by adding the leak $\gamma$ to each pillar conductance, as indicated in Fig. 21, at the fourth row; where—

$$g_{,,} = g'' + \gamma \quad . \quad . \quad . \quad . \quad . \quad \text{mhos} \quad (117)$$

The extended $\Pi$'s will now subtend a new angle $\theta_{,}$, which we proceed to find by reverting them to their corresponding uniform line-sections, using (95) and (99).

In the case represented in Fig. 21, each leak $\Gamma$ has a conductance of 0·096971 millimho, and the line-sections are the same as in Fig. 20. The final equivalent line-section angle is $\theta_{,} = 0{\cdot}41532$ hyp., and the surge-resistance 1206·4 ohms.

With the notation of (107) we find—

$$\tanh\frac{\theta_{,}}{2} = \tanh\frac{\theta}{2}\sqrt{\frac{1 + \gamma z_0 \coth\frac{\theta}{2}}{1 + \gamma z_0 \tanh\frac{\theta}{2}}} \quad . \quad \text{numeric} \quad (118)$$

$$= \sqrt{\tanh\frac{\theta}{2}\tanh\left(\frac{\theta}{2} + \delta\right)} \quad . \quad . \quad \text{,,} \quad (119)$$

if $\gamma z_{\circ} = \tanh \delta$, and—

$$z_{\circ}' = \frac{z_{\circ}}{\sqrt{\left(1 + \gamma z_{\circ} \tanh \frac{\theta}{2}\right)\left(1 + \gamma z_{\circ} \coth \frac{\theta}{2}\right)}} \quad \text{ohms} \quad (120)$$

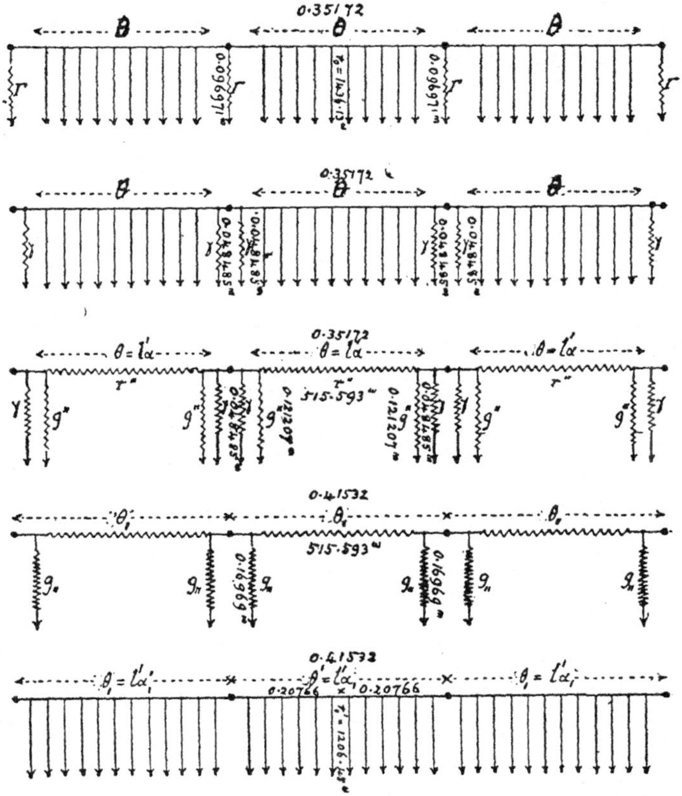


FIG. 21.—Reduction of a Uniform Actual Line with loads in derivation to an Equivalent Unloaded Actual Line.

Regular leak loads $\gamma$ produce the same effect on the line section-angle as regular series loads $\sigma$, when—

$$\frac{\sigma}{\gamma} = z_{\circ}^{2} \quad . \quad . \quad . \quad . \quad . \quad \text{ohms} \quad (121)$$

The final surge-resistances in the two cases will not, however, be the same, but their geometrical mean will be the original or unloaded surge-resistance.

It follows from the preceding formulas that every regularly loaded line can be replaced by an equivalent uniform unloaded line of the same length, which, in turn, can be completely replaced for all external purposes by a certain equivalent $T$, and corresponding equivalent $\Pi$.

The effect of loading a line at regular intervals, either with series resistances, or with leaks, is always to increase the angle $\theta$ of the line, and reduce its normal attenuation-factor $e^{-\theta}$, in the continuous-current case; besides the effect of long intervals between loads, or lumpiness, which exaggerates the influence. In the alternating-current case, it will be seen that loading always increases the numerical hyperbolic angle of the line; but by no means always reduces the normal attenuation-factor.

It may also be pointed out that, as a consequence deducible from (120)—

$$\frac{z_o'}{z_o} = \frac{\sinh \theta}{\sinh \theta_1} \quad . \quad . \quad . \quad . \quad . \quad . \quad . \quad \text{numeric } \angle \quad (120a)$$

or—

$$\frac{y_o'}{y_o} = \frac{\sinh \theta_1}{\sinh \theta} \quad . \quad . \quad . \quad . \quad . \quad . \quad . \quad \text{,,} \quad \angle \quad (120b)$$

a relation inverse to that which exists, by (115), for the case of regular series loading. In language, when a line is subjected to regular series loading, the ratio of the new to the old surge impedance, at any midload point, is the ratio of the sine of the new, to the sine of the old section-angle. Whereas, when a line is subjected to regular leak loading, the corresponding ratio is the ratio of the sine of the old to the sine of the new section-angle. In other words, the ratio of surge admittances with leak loading corresponds to the ratio of surge impedances with series loading.

# CHAPTER V

## COMPLEX QUANTITIES

In the discussion which has preceded, all the quantities employed have been simple "real" numerical quantities, comprised between the limits of $-\infty$ and $+\infty$; so that they are capable of being represented geometrically by their positions on a single graduated straight line; *i.e.* by their assignment in one dimensional space.

It is, however, a seemingly universal and a wonderful law, that all the numerical formulas and rules of quantitative behaviour for continuous-current circuits, or conductors, are exactly the same for single frequency alternating-current circuits or conductors, in respect to potentials and currents as also (with minor reservations) to power and energy, if these formulas and rules are interpreted as relating to complex numbers; or such numbers as are represented by their positions on a single graduated plane; *i.e.* by their assignment in two-dimensional space.*

The importance of this law will be evident, when it is recognized that each and all of the formulas hitherto discussed in relation to continuous-current lines and systems, are immediately applicable, without any change in notation, to alternating-current lines and systems, provided that we extend the meaning of the notation to include two-dimensional numbers instead of one-dimensional numbers. From this standpoint, it will be seen that we have already dealt with alternating-current lines and systems unawares, and that the continuous-current case is merely the particular case, in each formula, when the numerical quantities it employs degrade into ordinary one-dimensional numbers.

* *See* Bibliography, 6, 9, 9*a*, 10, 17, 66.

There is, moreover, a great advantage in dealing first with the hyperbolic trigonometry of continuous-current lines; because one-dimensional arithmetic is easier to grasp and follow than two-dimensional arithmetic. On this account, our ideas are limned more clearly, and our advance proceeds more surely, when we deal first with the problem on the one-dimensional aspect which the continuous-current case supplies. After having mastered the subject in one dimension, we are in a strong position to attack the vastly wider two-dimensional fields, which alternating-current cases offer successively to our mental vision. We need never have doubts or fears as to the safe road

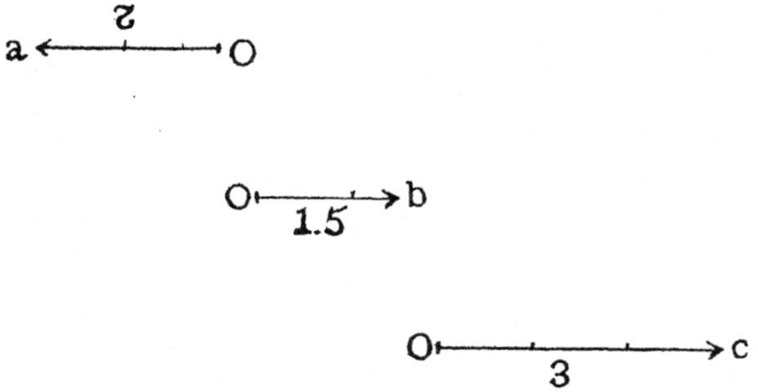


FIG. 22.—Geometrical Representation of Simple Numbers.

to pursue in our advance on a two-dimensional problem, if our formulas and weapons of attack have been forged on the one-dimensional hearth.

Whenever a new problem arises in alternating-current technology, whether it is more conveniently dealt with by hyperbolic trigonometry or not, the safe rule is to find the corresponding problem in continuous-current technology, and solve it there by simple one-dimensional arithmetic. The equations and formulas of the continuous-current solution then apply to the alternating-current solution, by extending their meaning into two-dimensional arithmetic.

*Simple and Complex Numbers.*—In Fig. 22, three simple numbers are represented geometrically, namely $Oa = -2$, $Ob = +1{\cdot}5$, and $Oc = +3$. These numbers all lie in one direction, across the page, in the plane of the paper, and positive numbers are directed towards the right hand. All simple arithmetical operations upon such one-dimensional numbers, such as addition, subtraction, multiplication, division, powers, roots, etc., beget other one-dimensional numbers, so that all simple arithmetic belongs to the one-dimensional system, or to a single straight line in space. Any convenient straight line in space will serve as the line of reference. It is convenient to

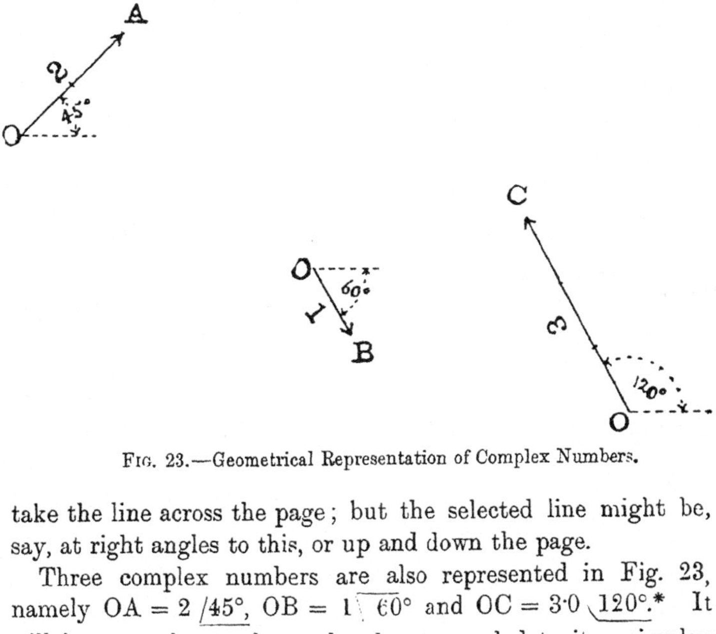


FIG. 23.—Geometrical Representation of Complex Numbers.

take the line across the page; but the selected line might be, say, at right angles to this, or up and down the page.

Three complex numbers are also represented in Fig. 23, namely OA = $2\,\underline{/45^\circ}$, OB = $1\,\overline{\backslash 60^\circ}$ and OC = $3{\cdot}0\,\underline{\backslash 120^\circ}$.* It will be seen that each number has appended to it a circular angle, which defines the direction of the line representing the number in the single plane of reference, with respect to an initial direction of reference therein. We are at liberty to choose any convenient plane, and any convenient direction of

* In what follows, $A\,\underline{/\beta}$ has the same meaning as A cis $\beta$ or $A\epsilon^{j\beta}$ in regular mathematical notations, and may be regarded as a convenient abbreviation for either of these forms of expression. The *modulus* A is the length-factor, or tensor, and the *argument* $\beta$ is the circular-angle-factor, or versor, of the complex number.

reference in the same. It is convenient to select the plane of the page, and the direction parallel to the lines across the page as the direction of reference. From this standpoint, all simple numbers, as in Fig. 22, are numbers having the angle $/\underline{0^\circ}$.

Complex numbers are sometimes called plane-vectors or, for brevity, vectors. They must, however, be carefully distinguished from three-dimensional vectors.

It is also to be noted that negative signs are not indispensable when writing or specifying individual complex numbers. The

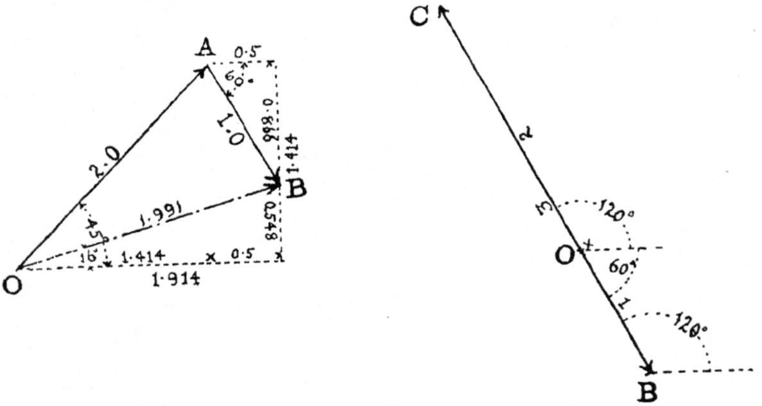


FIG. 24.—Vector Sum of Two Complex Numbers.

angle inseparably attached to each number is sufficient to define a negative direction. Thus $10\backslash\underline{180^\circ}$ is the same as $-10\ /\underline{0^\circ}$. We may, therefore, use negative signs or not, as may be convenient. In Fig. 23, we might express O C as $-3{\cdot}0 \times$ O B, if desired.

*Addition of Complex Numbers.*—When two complex numbers are added together geometrically, one of them is transferred to the end of the other, and the new number, or sum, is that corresponding to a line drawn from the origin of the latter to the free end of the former so transferred. Thus in Fig. 24 $2\ /\underline{45^\circ} + 1\ \overline{\backslash 60^\circ} = \text{OB} = 1{\cdot}991\ /\underline{16^\circ}$. Again $1\ \overline{\backslash 60^\circ} + 3{\cdot}0\backslash\underline{120^\circ} = \overline{\text{OC}} = 2{\cdot}0\backslash\underline{120^\circ}$. This process of adding on one vector to the end of another is aptly described as *geometrical addition.*

By successive additions of complex numbers, any or all parts of the plane of reference may be invaded.

The process of vector addition, or geometrical addition, of complex numbers is easily conducted by a draughtsman at the drawing-board; but in order to be carried out numerically, it is desirable to *analyse the complex numbers into components.* A complex number is most conveniently analysed into two mutually perpendicular components, one in the direction of reference, or across the page, and the other up and down the page. These components are called respectively the *real* and *imaginary* components of the complex number, the terms having been bestowed by algebraists from the standpoint of one-dimensional arithmetic. The rule for the analysis of any number of length or *modulus* A and angle or *argument* $\beta$ is by circular trigonometry—

$$A\,\underline{/\beta} = A\cos\beta + j^{*} A\sin\beta \quad . \quad . \quad . \quad . \quad \text{cm.} \angle \quad (122)$$

it being understood that we are discussing the complex number in geometrical terms. Thus, in Fig. 23, the number OA is 1·414 $+ j\,$1·414; OB is 0·5 $- j\,$0·866; OC is $-$ 1·50 $+ j\,$2·6. The symbol $j$ here indicates that the quantity to which it is prefixed is to be measured upwards along the "imaginary" axis. The sign $-j$ prefixed to a number means that it is to be measured downwards along the imaginary axis. The $j$ "operator" is thus an operator which, when applied to a line, representing a number, rotates it in the positive or counter-clockwise direction through 90°, or $\pi/2$ circular radians, in the plane of reference.

Two successive applications of the $j$ operator thus reverse the direction of a line, or rotate it through 180°; so that $j \times j$ or $j^2$ is equivalent to giving the negative sign to a complex number without changing its angle. Thus we have the well-known relation—

$$j = \sqrt{-1} \quad . \quad . \quad . \quad \text{numeric} \angle \quad (123)$$

and— $\qquad j^2 = -1,\ j^4 = +1,\ j^3 = -j,$ and so on.

To find the sum of a plurality of complex numbers, we add their real components, and also their imaginary components.

* $j$ is used in electrotechnics, instead of $i$ as in mathematics, to avoid confusion with currents.

Thus in Fig. 23, OA + OB + OC, becomes in Fig. 25, OA + AB + BC = (1·414 + 0·5 − 1·50) + $j$(1·414 − 0·866 + 2·60)

$$= 0{\cdot}414 + j3{\cdot}148$$
$$= \text{OC} = 3{\cdot}175 \;\underline{/82{\cdot}5^\circ},$$

where OC is the vector sum of OA, OB, and OC, the three complex numbers in Fig. 23.

In order to resolve the components of a complex quantity into a plane vector, let $\pm x$ be the real component and $\pm jy$

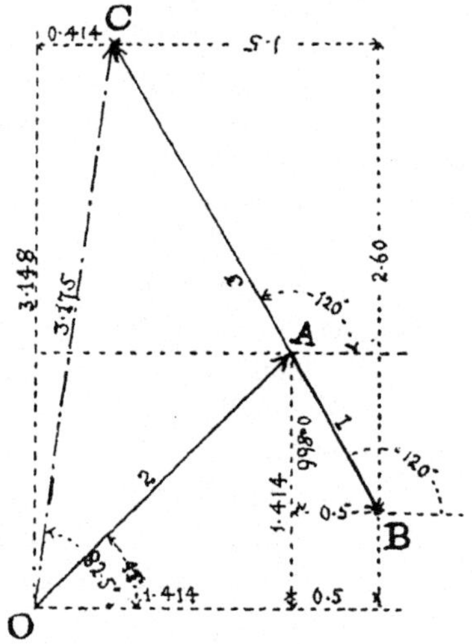


Fig. 25.—Vector Sum of Three Complex Numbers.

the imaginary component, A $\underline{/\beta}$, the resultant complex number. Then by circular trigonometry and (122)—

$$A = \sqrt{x^2 + y^2} \quad . \quad . \quad \text{modulus, cm.} \quad (124)$$

and—
$$\beta = \tan^{-1}\left(\frac{\pm y}{\pm x}\right) \text{ argument, radians} \quad (125)$$

Thus in Fig. 25, with $x = +\,0{\cdot}414$, and $y = +\,j3{\cdot}148$, we have

$$A = \sqrt{0{\cdot}414^2 + 3{\cdot}148^2} = \sqrt{0{\cdot}1714 + 9{\cdot}9099} = \sqrt{10{\cdot}0813}$$
$$= 3{\cdot}1751,$$

and $\beta = \tan^{-1}(+\,3{\cdot}148 / +\,0\;414) = \tan^{-1}(+\,7\dot{\cdot}604)$

$$= 82^\circ\;30' \text{ or } 82{\cdot}5^\circ.$$

*Subtraction of Complex Numbers.*—Subtraction of one complex number $A_2 \underline{/\beta_2}$ from another $A_1 \underline{/\beta_1}$ merely requires that the negative sign be given to the former, or that, without change of sign, its angle $\underline{/\beta_2}$ be changed by 180°, and addition then follows. On the drawing-board, this is performed by first laying off $A_1 \underline{/\beta_1}$ and then adding $-A_2 \underline{/\beta_2}$ to the end of it. In order to perform the operation arithmetically, each vector is analysed into its components, and the subtraction then proceeds along each axis, as in one-dimensional arithmetic. Thus if $A_1 \underline{/\beta_1} = \pm x_1 \pm jy_2$ and $A_2 \underline{/\beta_2} = \pm x_2 \pm jy_2$. Then $A_1 \underline{/\beta_1} - A_2 \underline{/\beta_2} = \pm(x_1 - x_2) \pm j(y_1 - y_2) = \pm X \pm jY$.

*Multiplication of Complex Numbers.*—The multiplication of complex numbers in any order is effected by multiplying together their lengths or moduli, as in one-dimensional arithmetic, and adding their angles.

Thus if $A_1 \underline{/\beta_1}$ and $A_2 \underline{/\beta_2}$ are the two numbers to be multiplied together, and $A \underline{/\beta}$ the product, we have—

$$A \underline{/\beta} = A_1 \underline{/\beta_1} \times A_2 \underline{/\beta_2}$$
$$= A_1 A_2 \underline{/\beta_1 + \beta_2}. \quad \text{numeric } \angle \qquad (126)$$

Thus if $A_1 \underline{/\beta_1} = OA$ Fig. 23 $= 2 \underline{/45°}$,

and $A_2 \underline{/\beta_2} = OB$ „ „ $= 1 \overline{\backslash 60°}$,

then the product $A \underline{/\beta}$ is $2 \overline{\backslash 15°}$.

If the two complex numbers to be multiplied are analysed into components, the multiplication may be effected, although less conveniently, by the rules of algebra. Thus if $(\pm x_1 \pm jy_1)$ and $(\pm x_2 \pm jy_2)$ be the two numbers, their product will be—

$$\pm (x_1 x_2 - y_1 y_2) \pm j(x_2 y_1 + x_2 y_1) = \pm X \pm jY.$$

*Reciprocal of a Complex Number.*—The reciprocal of a complex number has for its modulus the arithmetical reciprocal of the modulus of the number, and for its argument the negative value of the argument of the number. Thus, if $A \underline{/\beta}$ is the complex number, its reciprocal will be $\left(\frac{1}{A}\right) \overline{\backslash \beta} = \left(\frac{1}{A}\right) \underline{/-\beta}$.

If the complex number is say 10 $\underline{/20°}$, its reciprocal will be 0·1 $\overline{\backslash 20°}$.

If the number whose reciprocal is required, be analysed into components in the form $(\pm x \pm jy)$, its reciprocal will be $\frac{1}{\pm x \pm jy}$. This may be reduced to the original form, by multiplying both numerator and denominator by $(\pm x \mp jy)$. The result is $\frac{\pm x \mp jy}{x^2 + y^2} = \pm \left(\frac{x}{x^2 + y^2}\right) \mp j \left(\frac{y}{x^2 + y^2}\right)$.

*Division of Complex Numbers.*—Division of complex numbers can always be effected by taking the reciprocal of the divisor according to the last preceding rule, and then multiplying this reciprocal into the dividend. Thus if $A_1 \underline{/\beta_1}$ has to be divided by $A_2 \underline{/\beta_2}$ the quotient is $\left(\frac{A_1}{A_2}\right) \underline{/\beta_1 - \beta_2} = A \underline{/\beta}$. If $A_1 \underline{/\beta_1}$ and $A_2 \underline{/\beta_2}$ are OA and OB of Fig. 23 respectively, the quotient is $2 \underline{\backslash 105°}$.

*Powers and Roots of Complex Numbers.*—The $n$th power of a complex number is formed by the $n$th arithmetical power of the modulus, and multiplying the argument by $n$. That is—

$$(A \underline{/\beta})^n = A^n \underline{/n\beta} \quad . \quad . \quad . \quad \text{numeric } \angle \quad (127)$$

Similarly, the $n$th root of a complex number is formed by taking the $n$th arithmetical root of the modulus, and dividing the argument by $n$. That is—

$$\sqrt[n]{(A \underline{/\beta°})} = (A \underline{/\beta°})^{\frac{1}{n}} = A^{\frac{1}{n}} \underline{/\frac{\beta°}{n}} \quad . \quad \text{numeric } \angle \quad (128)$$

Summing up, we may say that two-dimensional arithmetic is performed by rules which degrade into those of ordinary or one-dimensional arithmetic when the arguments are all zero. When adding or subtracting complex numbers numerically, it is desirable to analyse them into components; but when multiplying and dividing them, or when taking powers and roots, it is desirable to express them in angular form.

*Trigonometrical Functions of Complex Angles.*—We have

already seen that in the case of a simple angle in generalized trigonometry (Figs. 1 and 2), the circular functions can be read from a circle diagram (Fig. 1), and the hyperbolic functions from corresponding elements of a hyperbola diagram (Fig. 2). When the angle to be dealt with is complex, or of the type $(x + jy)$ radians, both the circular and the hyperbolic functions can be derived from a mixed circle and hyperbola diagram.

*Circular Functions of a Complex Angle. Construction for sin* $(x \pm jy)$, Fig. 26.*—Take OA = 1 along the negative end of

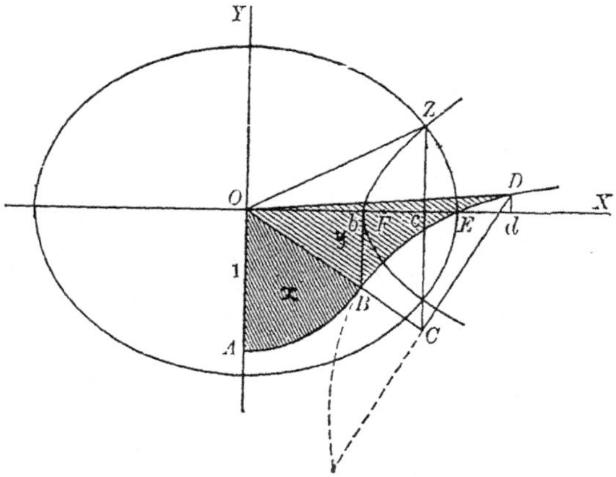


FIG. 26.—Construction for Sin $(x \pm jy)$ and Sin$^{-1}$ $(x \pm jy)$.

the Y-axis. From OA as initial line, mark off the circular angle $x$ and sector area AOB $= \frac{x}{2}$. From OB as initial line, mark off the hyperbolic angle $y$ and sector area BOD $= \frac{y}{2}$. Let C be the foot of the perpendicular from D on OB produced. Drop perpendiculars from C and D on the axis of reals OX, at $c$ and $d$ respectively. About $c$ as centre, rotate $cd$ positively through 90° to $cZ$. Then will the complex vector OZ $=$ O$c + jcd$ be the required circular sine of the complex angle $x + jy$ radians

* *See* Bibliography, 31.

In the case represented, $\sin(1 + j1) = 1{\cdot}299 + j0{\cdot}635 = 1{\cdot}446 \underline{/26{\cdot}05^\circ}$. As $y$ varies, Z moves along the hyperbola—

$$\frac{X^2}{\sin^2 x} - \frac{Y^2}{\cos^2 x} = 1 \quad . \quad . \quad . \quad . \quad \text{cm.}^2 \angle \quad (129)$$

as $x$ varies, Z moves along the ellipse—

$$\frac{X^2}{\cosh^2 y} + \frac{Y^2}{\sinh^2 y} = 1 . \quad . \quad . \quad . \quad \text{cm.}^2 \angle \quad (130)$$

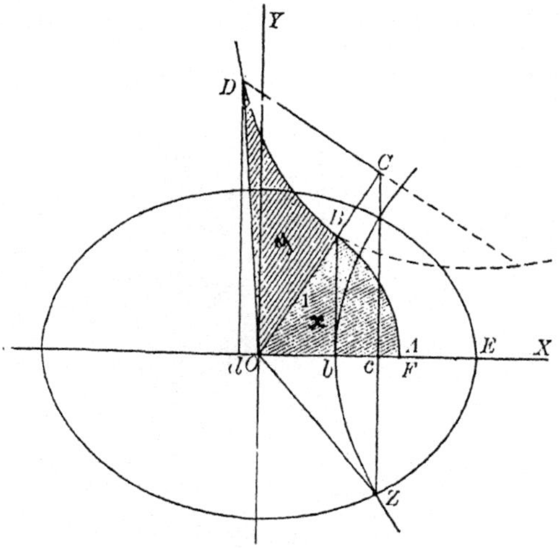


FIG. 27.—Construction for Cos $(x \pm jy)$ and Cos$^{-1}$ $(x \pm jy)$.

From the same figure we have also, if $Oc = u$ and $cZ = jv$

$$\sin^{-1} OZ = \sin^{-1}(u \pm jv) = \sin^{-1} Ob \pm j \cosh^{-1} OE$$

$$= \sin^{-1}\left\{\frac{\sqrt{(1+u)^2+v^2} - \sqrt{(1-u)^2+v^2}}{2}\right\} \pm$$

$$j\cosh^{-1}\left\{\frac{\sqrt{(1+u)^2+v^2} + \sqrt{(1-u)^2+v^2}}{2}\right\}. \quad \text{cm.} \angle \quad (131)$$

*Construction for Cos* $(x + jy)$. Fig. 27.—Take OA $= 1$ along the positive end of the axis of reals. From OA as initial line, mark off the circular angle $x$ or the sector area AOB $= \frac{x}{2}$, etc., precisely as in the preceding paragraph. The complex vector

OZ $= Oc + jcd$ thus obtained will be the required circular cosine of the complex angle $(x + jy)$ radians.

In the case represented—

$$\cos(1 + j1) = 0{\cdot}834 - j0{\cdot}989 = 1{\cdot}293 \quad \overline{49{\cdot}866}^\circ.$$

As $y$ varies, Z moves along the hyperbola—

$$\frac{X^2}{\cos^2 x} - \frac{Y^2}{\sin^2 x} = 1. \quad . \quad . \quad . \quad \text{cm.}^2 \angle \quad (132)$$

As $x$ varies, Z moves along the ellipse (130).

From the figure—

$$\cos^{-1} OZ = \cos^{-1}(u \pm jv) = \cos^{-1} Ob \mp \cosh^{-1} OE$$

$$= \cos^{-1}\left\{\frac{\sqrt{(1+u)^2+v^2} - \sqrt{(1-u)^2+v^2}}{2}\right\} \mp$$

$$j\cosh^{-1}\left\{\frac{\sqrt{(1+u)^2+v^2} + \sqrt{(1-u)^2+v^2}}{2}\right\}. \quad \text{cm.} \angle \quad (133)$$

*Hyperbolic Functions of a Complex Angle. Construction for Sinh* $(x \pm jy)$. Fig. 28.—Take OA $= 1$ along the positive end of the axis of reals. From OA as initial line, mark off the circular angle $y$ and sector area AOB $= \frac{y}{2}$. From OB as initial line, mark off the hyperbolic angle $x$ and sector area BOD $= \frac{x}{2}$. Let C be the foot of the perpendicular from D on OB produced. Drop perpendiculars from C and D on the axis of imaginary OY, at $c$ and $d$ respectively. About $c$ as centre, rotate $cd$ negatively, or clockwise, through 90° to $c$Z. Then will the complex vector OZ be the required hyperbolic sine of the complex angle $x + jy$ radians.

In the case presented in Fig. 28, $\sinh(1 + j1) = 0{\cdot}635 + j1{\cdot}2985 = 1{\cdot}446 \underline{/63{\cdot}95^\circ}$. As $x$ varies, Z moves along the hyperbola Z$bz$. As $y$ varies Z moves along the ellipse XE$xy$. Both the ellipse and the hyperbola have foci at F and $f$, points situated at unit distances from O on the Y axis.

From the same figure, if $cZ = u$, and $Oc = jv$, we have—

$$\sinh^{-1}(u \pm jv) = \cosh^{-1}\left\{\frac{\sqrt{(1+v)^2+u^2}+\sqrt{(1-v)^2+u^2}}{2}\right\} \pm$$
$$j\sin^{-1}\left\{\frac{\sqrt{(1+v)^2+u^2}-\sqrt{(1-v)^2+u^2}}{2}\right\}$$
$$= x \pm jy \quad . \quad . \quad . \quad . \quad . \quad . \quad \text{hyp. radians } \angle \quad (134)$$

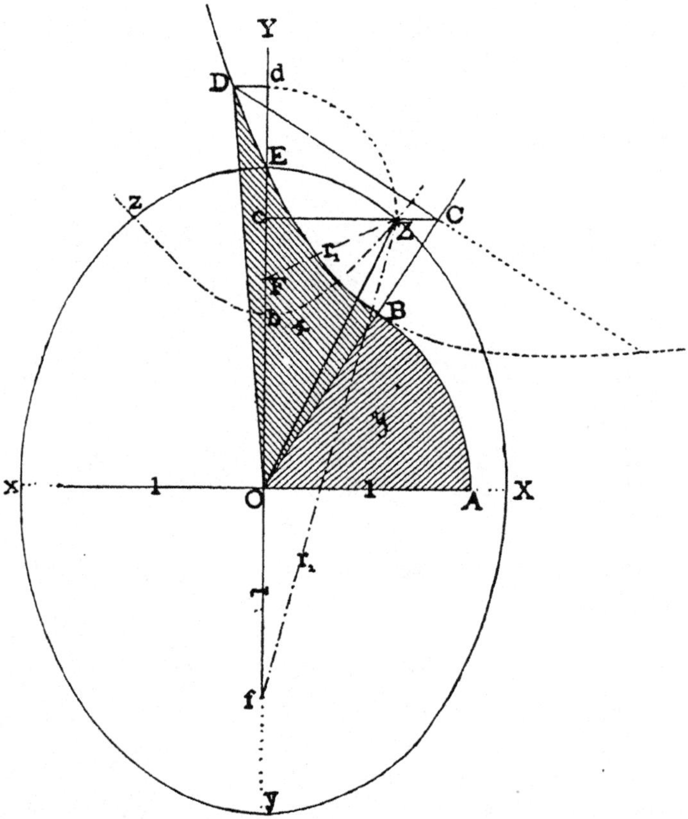


FIG. 28.—Graphical Constructions for Sinh $(x \pm jy)$ and Sinh$^{-1}$ $(x \pm jy)$.

*Construction for Cosh* $(x + jy)$. Fig. 29.—Take OA = 1 along the positive end of the axis of reals. From OA as initial line, mark off the circular angle $y$, or the sector area AOB $= \frac{y}{2}$.

From OB as initial line, mark off the hyperbolic angle $x$ and the sector area BOD $= \frac{x}{2}$. Let C be the foot of the perpendicular from D on OB produced. Drop perpendiculars from C and D on the axis of reals OX, at $c$ and $d$, respectively. About

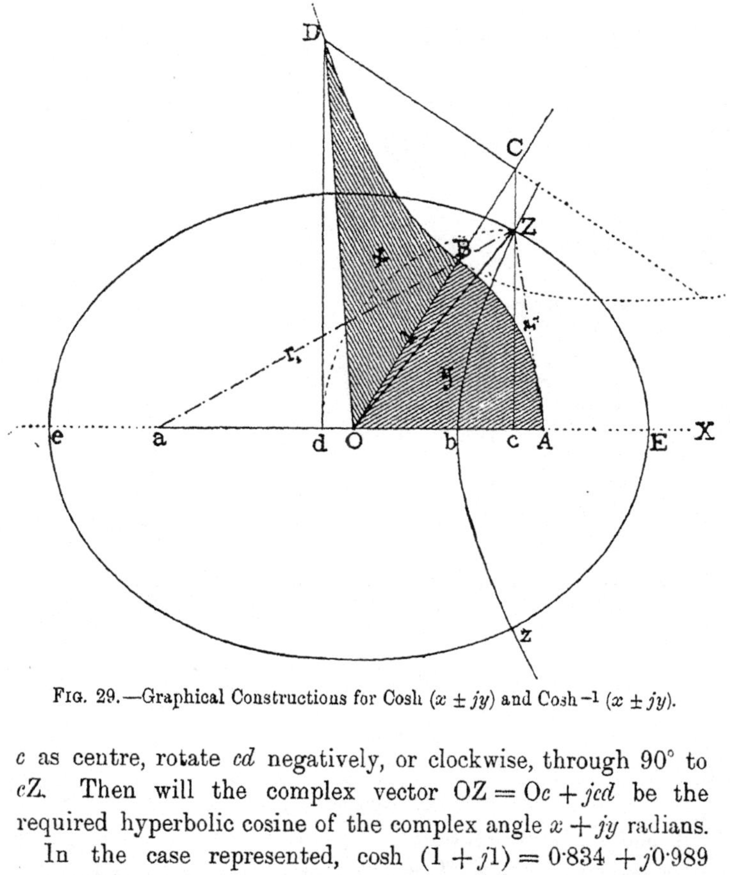


FIG. 29.—Graphical Constructions for Cosh $(x \pm jy)$ and Cosh$^{-1}$ $(x \pm jy)$.

$c$ as centre, rotate $cd$ negatively, or clockwise, through 90° to $c$Z. Then will the complex vector OZ $=$ O$c + jcd$ be the required hyperbolic cosine of the complex angle $x + jy$ radians.

In the case represented, cosh $(1 + j1) = 0{\cdot}834 + j0{\cdot}989$ $= 1{\cdot}293 \underline{/49{\cdot}866^\circ}$. As $x$ varies, Z moves along the hyperbola Z$bz$. As $y$ varies, Z moves along the ellipse EZ$ez$. Both ellipse and hyperbola have foci at $a$ and A, points on the real axis equally remote from O.

From the same figure, if $Oc = u$ and $cZ = jv$, we have—

$$\cosh^{-1}(u \pm jv) = \cosh^{-1}\left\{\frac{\sqrt{(1+u)^2+v^2}+\sqrt{(1-u)^2+v^2}}{2}\right\} \pm$$
$$j\cos^{-1}\left\{\frac{\sqrt{(1+u)^2+v^2}-\sqrt{(1-u)^2+v^2}}{2}\right\}$$
$$= x \pm jy \quad . \quad . \quad . \quad . \quad . \quad . \quad \text{hyp. radians } \angle \quad (135)$$

*Construction for Tanh* $(x + jy)$. Fig. 30.—Mark off on the axis of reals $xOX$, two points T and X, such that the former is distant by $\tanh x$ and the latter by $\coth x$ from the origin O. Find the point C midway between T and X.

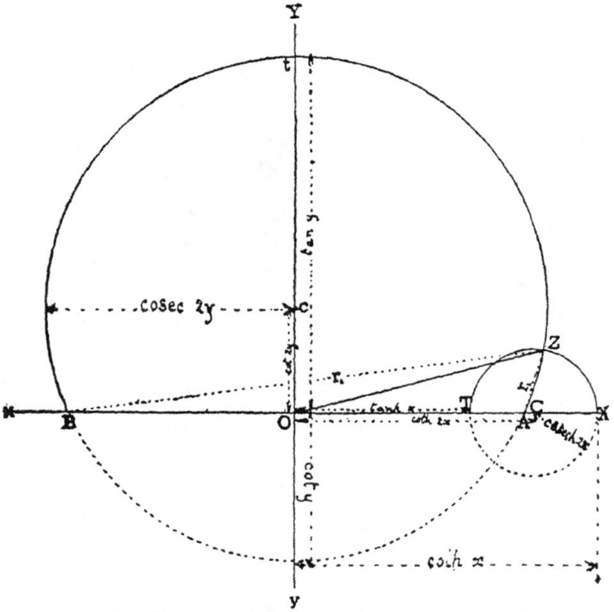


FIG. 30.—Graphical Constructions for Tanh $(x \pm jy)$ and Tanh$^{-1}$ $(x \pm jy)$.

This point will incidentally be distant $\coth 2x$ from O. With centre C and radius $CT = \operatorname{cosech} 2x$, draw the circle TXZ. Mark off on the axis of imaginaries $yOY$ two points $t$ and $y$, such that the former is distant by $\tan y$ and the latter $\cot y$ from the origin O. Find the point $c$ midway between $t$ and $y$. This point will incidentally be distant $\cot 2y$ from O. With centre $c$ and radius $ct = \operatorname{cosec} 2y$, draw the circle $ByAt$. This

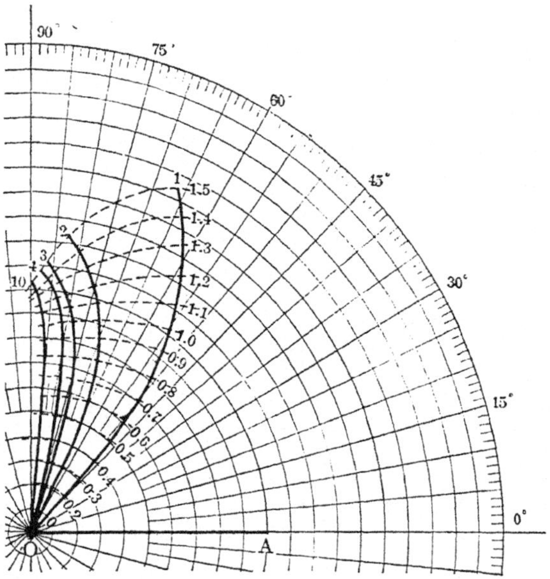


FIG. 31.—Loci of Sinh $\theta$ for the Imaginary-Real Ratios 1, 2, 3, 4 and 10.

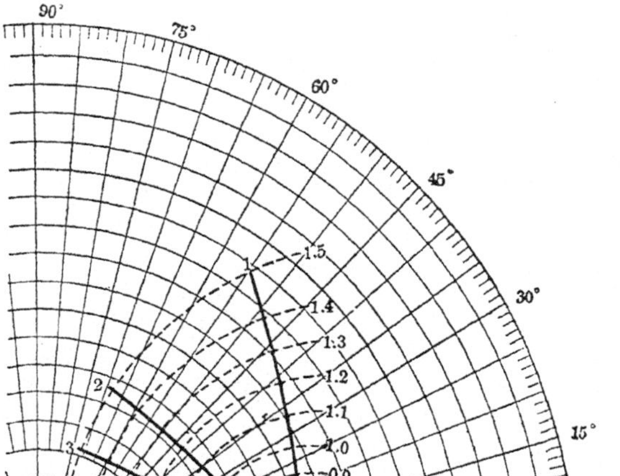


FIG. 32.—Loci of Cosh $\theta$ for the Imaginary-Real Ratios 1, 2, 3, 4 and 10.

circle will cut the axis of reals at two points A and B distant each one unit from O. It will also intersect the circle TXZ orthogonally in Z. Connect OZ. This vector OZ is the required hyperbolic tangent of the angle $x + jy$ radians.

In the case presented in Fig. 30, tanh $(1 + j1) = 1{\cdot}084 + j0{\cdot}2718 = 1{\cdot}118\underline{/14{\cdot}083^\circ}$.

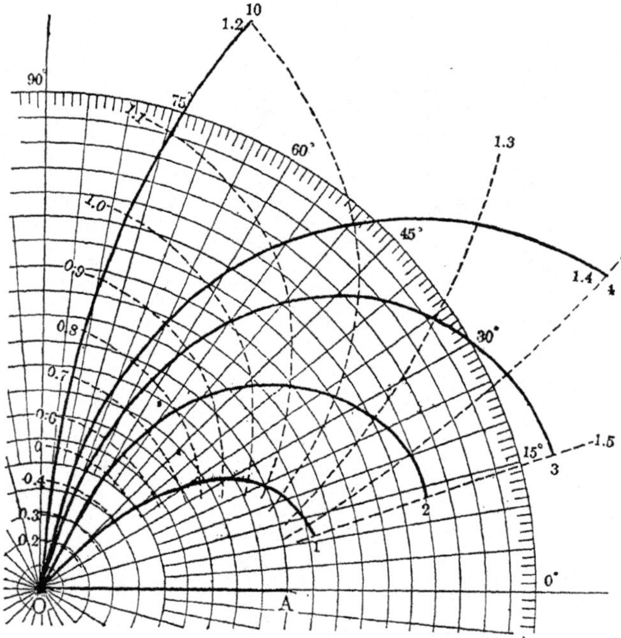


FIG. 33.—Loci of Tanh $\theta$ for the Imaginary-Real Ratios 1, 2, 3, 4 and 10.

As $x$ varies, Z moves along the circle A$t$B. As $y$ varies, Z moves along the circle TZX, performing one complete revolution for each $\pi$ units of increase in $y$.

From the same figure, if $OZ = u \pm jv$,

$$\tanh^{-1}(u \pm jv) = \tfrac{1}{2}\log_e \sqrt{\frac{(1+u)^2 + v^2}{(1-u)^2 + v^2}} \pm j\left\{\frac{\pi - \tan^{-1}\left(\frac{1+u}{\pm v}\right) - \tan^{-1}\left(\frac{1-u}{\pm v}\right)}{2}\right\}$$

$$= x \pm jy \quad . \quad . \quad . \quad . \text{ radians } \angle \qquad (136)$$

*Principal Formulas for Deriving the Hyperbolic Functions of Complex Angles.*—As distinguished from constructions for the

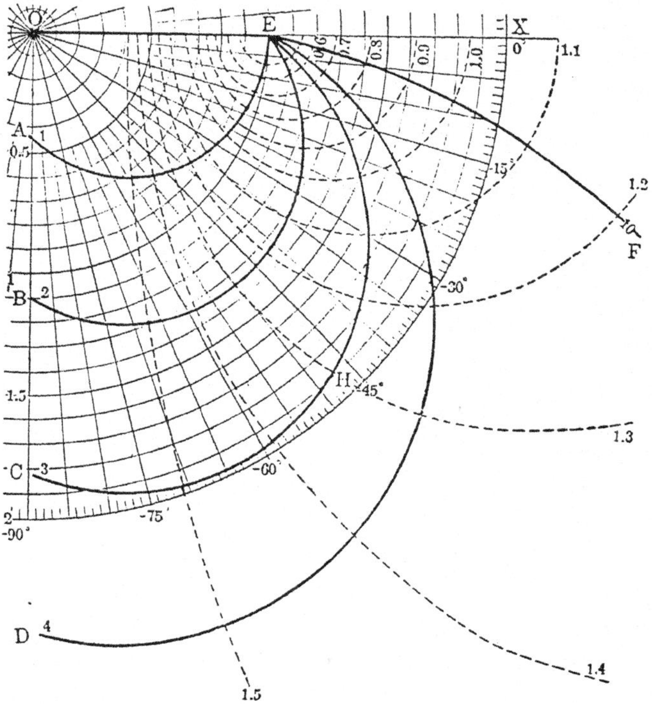


FIG. 34.—Loci of Sech $\theta$ for the Imaginary-Real Ratios 1, 2, 3, 4 and 10.

hyperbolic functions of complex angles, the following are among the most important formulas for computing them—

$$\sinh(x \pm jy) = \sinh x \cosh jy \pm \cosh x \sinh jy \quad . \quad . \quad . \quad (137)$$
$$= \sinh x \cos y \pm j \cosh x \sin y \quad . \quad . \quad . \quad . \quad (138)$$
$$= \sqrt{\sinh^2 x + \sin^2 y} \,/\underline{\pm \tan^{-1}(\coth x \tan y)} \quad (139)$$
$$= \sqrt{\cosh^2 x - \cos^2 y} \,/\underline{\pm \tan^{-1}(\coth x \tan y)} \quad (140)$$
$$\cosh(x \pm jy) = \cosh x \cosh jy \pm \sinh x \sinh jy \quad . \quad . \quad . \quad (141)$$
$$= \cosh x \cos y \pm j \sinh x \sin y) \quad . \quad . \quad . \quad . \quad (142)$$
$$= \sqrt{\cosh^2 x - \sin^2 y} \,/\underline{\pm \tan^{-1}(\tanh x \tan y)} \quad (143)$$
$$= \sqrt{\sinh^2 x + \cos^2 y} \,/\underline{\pm \tan^{-1}(\tanh x \tan y)} \quad (144)$$
$$\tanh(x \pm jy) = \frac{\sinh 2x}{\cosh 2x + \cos 2y} \pm j \frac{\sin 2y}{\cosh 2x + \cos 2y} \quad . \quad (145)$$

It is evident that the values of the complex hyperbolic functions may be computed and tabulated either for varying values of $x$ and $y$ * or for varying values of A and $\beta$, the modulus and argument of the angle expressed as a vector. For most electrotechnical purposes the latter are the more convenient.

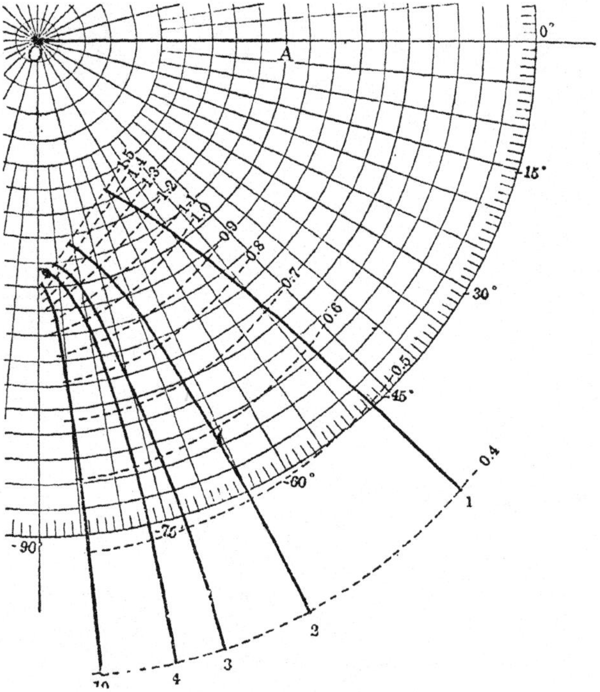


FIG. 35.—Loci of Cosech $\theta$ for the Imaginary-Real Ratios 1, 2, 3, 4 and 10.

Tables of these functions are still greatly needed. They have been published for the particular case of $\beta = 45°$.† Other tables are in course of preparation.

Since— $$\sinh\theta = \frac{1}{j}\sin j\theta \quad . \quad . \quad . \quad . \quad . \quad . \quad . \quad (146)$$

* Tables of sinh $(x + jy)$ and cosh $(x + jy)$ over a limited range have been published by Prof. James McMahon and by the General Electric Co., Schenectady (N.Y.). *See* Bibliography, 15 and 64.

† *See* Bibliography, 33 and 63.

$$\sin\theta = \frac{1}{j}\sinh j\theta \quad . \quad . \quad . \quad . \quad . \quad (147)$$

$$\cosh\theta = \cos j\theta \quad . \quad . \quad . \quad . \quad . \quad . \quad (148)$$

$$\cos\theta = \cosh j\theta \quad . \quad . \quad . \quad . \quad . \quad . \quad (149)$$

where $\theta$ is any angle, simple or complex, it follows that tables of complex hyperbolic functions can be used, with a little

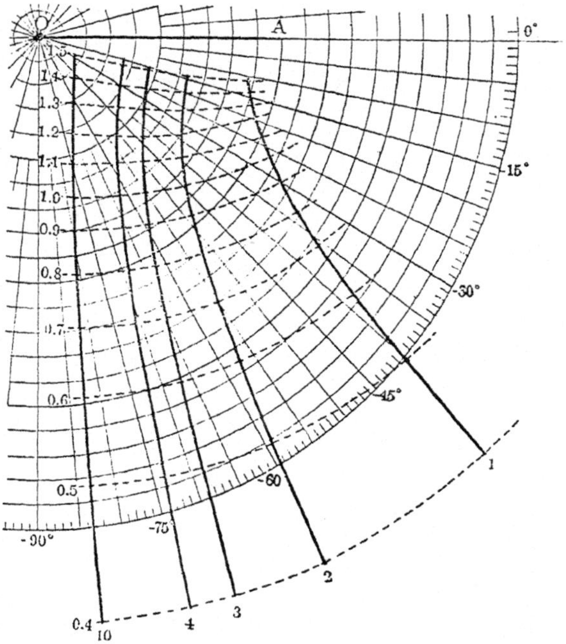


FIG. 36.—Loci of Coth $\theta$ for the Imaginary-Real Ratios 1, 2, 3, 4 and 10.

extra trouble, as tables of complex circular functions, and reciprocally.

Figs. 31 to 36 indicate the vector values of complex hyperbolic functions for the five values of $\beta$ whose tangents are 1, 2, 3, 4 and 10 respectively ($\beta = 45°$, $63{\cdot}43°$, $71{\cdot}57°$, $75{\cdot}97°$, and $84{\cdot}28°$) up to $A = 1{\cdot}5$. Thus taking Fig. 31 the hyp. sine of a complex angle $\theta = 1{\cdot}4\underline{/45°}$ or $1{\cdot}4\underline{/\tan^{-1}1}$ is found by following the heavy curve marked 1 to its intersection with the dotted line 1·4. The intersection marks the complex vector required. It is $1 \cdot 43 \underline{/63 \cdot 57^\circ}$.

Curve-sheets like those of Figs. 31 to 36 can be drawn on suitable scales for, say, each degree of $\beta$ and each step of 0·1 or less in modulus A. Such curves extend, theoretically, to infinity or cover the XY plane, not only once, but many times in succession, if A be taken large enough.

## CHAPTER VI

### THE PROCESS OF BUILDING-UP THE POTENTIAL AND CURRENT DISTRIBUTION IN A SIMPLE UNIFORM ALTERNATING-CURRENT LINE

In order to present the application of hyperbolic functions to the analysis of alternating-current lines from an alternative and illuminating view-point, we shall here discuss the simplest elements of electromagnetic wave motion over such lines during

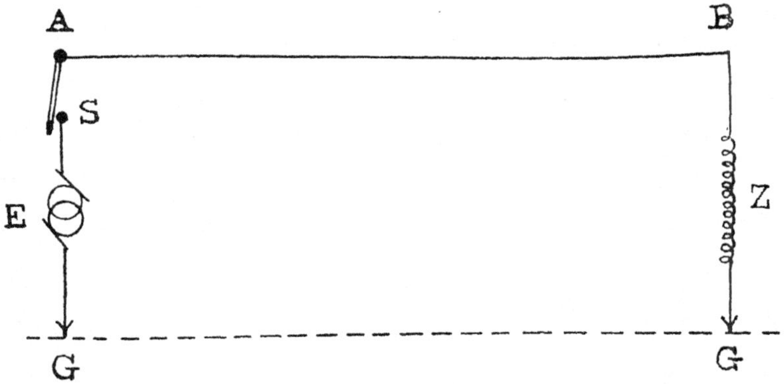


Fig 37.—Simple Alternating-Current Circuit.

the construction period which precedes the steady state. Since, however, the steady-state distribution is the object of our investigation, and the preceding unsteady state demands a like share of analysis for its investigation, we shall pass over the latter very briefly, and make certain assumptions as postulates which may be readily verified.

Fig. 37 presents the essential elements of a simple a.c. (alternating current) circuit. The line AB is uniform, and has uniformly distributed secondary constants; namely—

$r$ = linear resistance (ohms per wire-km.)
$l$ = „ inductance (henrys „ „ )
$g$ = „ conductance (mhos „ „ )
$c$ = „ capacitance (farads „ „ )
$f$ = the frequency of the impressed e.m.f. (cycles per second)
$\omega$ = the angular velocity (radians per second) $= 2\pi f$.
$z = (r + jl\omega)$ linear conductor impedance (ohms per wire-km.)
$y = (g + jc\omega)$ „ dielectric admittance (mhos „ „ )

$$a = \sqrt{z.y} = \sqrt{(r + jl\omega)(g + jc\omega)} \text{ . hyps. per wire-km.} \quad (150)$$

will then be the attenuation-constant or linear hyp. angle of

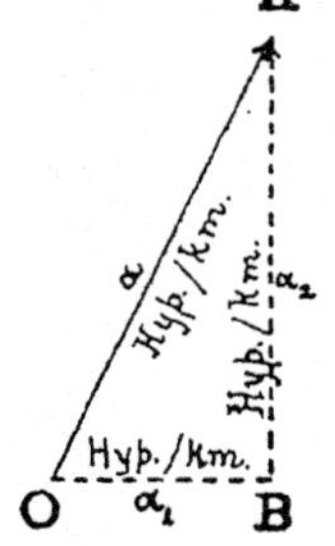


FIG. 38.—Analysis of the Vector Attenuation-Constant OA $= a$ into the real part OB $= a_1$ and the imaginary part BA $= a_2$.

the line. It is a vector possessing components $a_1 + ja_2$; so that (Fig. 38)—

$$a \,\underline{/\beta} = a_1 + ja_2 \text{ . . . hyps. per wire-km.} \quad (151)$$

The imaginary component $a_2$ of a hyperbolic angle $a$ has the properties of a circular angle. Again the surge-impedance of the line is—

$$z_0 = \sqrt{\frac{z}{y}} = \sqrt{\frac{r + jl\omega}{g + jc\omega}} \text{ . . . . . ohms } \angle \quad (152)$$

The hyperbolic angle of the line will be—

$$\theta \,\underline{/\beta} = \mathrm{L}a \,\underline{/\beta} = \mathrm{L}(a_1 + ja_2) = \sqrt{\mathrm{Z.Y}} = \theta_1 + j\theta_2 \text{. hyps.} \quad (153)$$

Where L is the line-length in km., Z is the total conductor

impedance of the line or $L(r + jl\omega) = Lz$ ohms and Y is the total dielectric admittance of the line, or $L(g + jc\omega) = Ly$ mhos. The real part $\theta_1$ of the complex angle is expressible in hyperbolic radians, and the imaginary part in circular radians.

It will be seen that (153) is merely a restatement of (19), (152) of (18), and (150) of (17) with two-dimensional signification. In other words, formulas (17), (18), and (19) for continuous-current circuits apply to the alternating-current case here to be discussed, when interpreted as involving complex resistances or impedances, and complex conductances or admittances. We use, therefore, $z$ for $r$, Z for R, $y$ for $g$, and Y for G, in order to emphasize the two-dimensional meaning; but we may use all the preceding continuous-current formulas unchanged, if we keep the vectorial interpretation before the mind.

The significance attached to the linear angle or vector attenuation-constant $\alpha$, is that any wave of potential or current running along the uniform line, shrinks or attenuates by the linear attenuation-factor $\varepsilon^{-\alpha}$ in each unit of length (miles or km.), for the particular a.c. frequency under consideration. If then, when starting, at end A, a wave of potential or current has a value taken as unity, it will, after having run 1 km., have dwindled to $\varepsilon^{-\alpha}$, after 2 km. to $\varepsilon^{-\alpha}.\varepsilon^{-\alpha}=\varepsilon^{-2\alpha}$, after $n$ km. to $\varepsilon^{-n\alpha}$, and just before arriving at B, to $\varepsilon^{-L\alpha} = \varepsilon^{-\theta}$, where $\theta$ is the line angle.

But $$\varepsilon^{-\alpha} = \varepsilon^{-(\alpha_1 + j\alpha_2)} = \varepsilon^{-\alpha_1} \times \varepsilon^{-j\alpha_2} \quad . \quad . \quad \frac{\text{numeric}}{\text{km.}} \angle \quad (154)$$

so that $\varepsilon^{-\alpha}$ is the product of two factors, namely, $\varepsilon^{-\alpha_1}$, which is a real numeric, and $\varepsilon^{-j\alpha_2}$, which is an angle, and may be written—

$$\varepsilon^{-j\alpha_2} = \underline{/-\alpha_2} = \overline{\backslash \alpha_2} \quad . \quad . \quad . \quad \text{radians} \quad (155)$$

Consequently, we may write—

$$\varepsilon^{-\alpha} = \varepsilon^{-\alpha_1}\overline{\backslash \alpha_2} \quad . \quad . \quad . \quad . \quad \frac{\text{numeric}}{\text{km.}} \angle \quad (156)$$

and— $$\varepsilon^{-\theta} = \varepsilon^{-\theta_1}\overline{\backslash \quad} = {}^{-L\alpha_1}\overline{\backslash L\alpha_2} \quad . \quad \text{numeric} \angle \quad (157)$$

The normal attenuation-factor of the line $\varepsilon^{-\theta}$, which is also the inevitable attenuation-factor for any one wave of potential or current running over the line, comprises a real component $\varepsilon^{-\theta_1}$, and an imaginary component $\varepsilon^{-j\theta_2}$. The former is the actual numerical attenuation-factor which applies to the amplitude of the wave; while the latter is an angle, and applies only to the phase of the wave. If, for the moment, we leave phase angles out of consideration, and consider only the shrinkage or attenuation of the waves running over the line, then we only consider the real component $\alpha_1$ of the attenuation-constant $\alpha$, and the real or hyperbolic component $\theta_1$ of the line complex angle $\theta$.

In practice, an alternating-current line, whether used for signal transmission, telegraphy, telephony, etc., may be provisionally regarded electrically as—

| | | | | |
|---|---|---|---|---|
| a very short line if | $\theta_1$ is less than | 0·1, | $\epsilon^{-\theta_1}$ being greater than | 0·90 |
| a short line if | ,, is between 0·1 and | 0·5, | $\epsilon^{-\theta_1}$ ,, ,, | 0·607 |
| a line of moderate length if | ,, ,, 0·5 ,, | 2·5, | $\epsilon^{-\theta_1}$ ,, ,, | 0·100 |
| a long line if | ,, ,, 2·5 ,, | 4·0, | $\epsilon^{-\theta_1}$ ,, ,, | 0·020 |
| a very long line if | ,, is over | 4·0, | $\epsilon^{-\theta_1}$ being less than | 0·020 |

it being understood that $\theta_1$ depends upon the frequency of alternation as well as upon the line-length and secondary constants. A given line when used, say, for power transmission, is ordinarily very short with respect to the fundamental frequency of the generator applied to it, but perhaps long to some higher harmonic frequency such as the 11th-frequency harmonic that the generator may produce.

The essential difference with respect to attenuation between the line-angle $\theta=\sqrt{ZY}$ of an a.c. line and that of a c.c. (continuous-current) line $\theta=\sqrt{R.G.}$, is that, in the former, only the real component $\theta \cos \beta$, or $\theta_1$, counts, while the latter is all real, and all counts. An a.c. line may have a complex angle of 50 hyperbolic radians, and yet, if the real component is only, say, 2 radians, the line is only of moderate length; whereas a c.c. line of 50 radians would be of enormous electrical length. As above defined, a very short line attenuates waves less than 10 %, so that they arrive at the distant end of the line with over 90 % of their amplitude at the home end. A short line attenuates less than 40 %, so that the wave attenuation-factor for one run over the line is over 0·6. A moderately long line reduces the wave attenuation-factor to something between 0·6 and 0·1. A long line may bring it down as low as 2 %, and a very long line to yet lower values. These will also be the normal attenuation-factors of the lines in their steady a.c. state, *i. e.* if the terminal-load impedance at B is equal to their surge-impedance $z_0$.

In Fig. 37, let the a.c. generator have negligible impedance, and produce a sinusoidal e.m.f., or pure sine wave. If the terminal-load impedance $Z_r$ is made infinite, the line is freed at B, if it is made zero the line is grounded at B. If it is made equal to $z_0$ the line is in the normal state as to attenuation, the attenuation-factor of the steady state being then the same as the attenuation-factor for any single wave in the unsteady state.

The physical significance of the surge-impedance $z_0$ is that, at any point, the line offers this impedance to an advancing wave of the frequency considered. That is, at any point—

$$e = iz_0 \quad . \quad . \quad . \quad . \quad . \quad . \quad . \quad . \quad \text{volts} \angle \quad (158)$$

where $e$ and $i$ are the instantaneous values of the potential and current strength at the point. Consequently, the surge-impedance of the line is not only the natural impedance which it offers everywhere to surges of the frequency considered, but it is also the initial impedance of the line at the sending end. It is, therefore, also the *initial sending-end impedance* of the line, as distinguished from the impedance which the line offers at the sending end in the steady state, when a number of waves are merged together.

The velocity with which any wave of the frequency $f$ runs over the line considered, is determined by the relation—

$$v = \frac{\omega}{a_2} \quad . \quad . \quad . \quad . \quad \text{km. per second} \quad (159)$$

When the linear conductor resistance $r$ and dielectric conductance $g$ are both made equal to zero in (150) we obtain—

$$a = ja_2 = j\omega\sqrt{cl} \quad . \quad . \quad . \quad \text{hyps. per km.} \quad (160)$$

and the velocity over such a line would become—

$$v = \frac{1}{\sqrt{cl}} \quad . \quad . \quad . \quad . \quad . \quad \text{km. per second} \quad (161)$$

which for a free uniform line, surrounded by air at all points, and ignoring inductance within the substance of the wire, is known to be the velocity $v$ of "light" in air, or 300,000 km. per second, owing to the fundamental relations between the inductance and capacity of such a wire. Owing, however, to the presence of effective linear resistance of all kinds that dissipate electromagnetic wave energy into heat in the conductor, and to effective linear conductance $g$, of all kinds that dissipate such energy into heat in the dielectric, the speed of the waves drops, even in an aerial line. Moreover, when the dielectric is a solid, the free limiting speed of waves is reduced inversely as the square root of the specific inductive capacity of the material; so that with solid-insulated conductors, the speed of wave-advance may be only a small fraction of 300,000 km./sec. On loaded lines, the wave speed, $v = \omega/a_2$, is artificially reduced still more.

Let us assume a line having at a certain frequency an attenuation-constant of $a = 0{\cdot}07675 + j0{\cdot}7854$ hyps. per km. Then, in running 1 km. over this line, a wave shrinks in amplitude by $\varepsilon^{-0{\cdot}07675}$ or 0·9259, *i.e.* to 92·59 %. It also shrinks in phase by 0·7854 radians ($\pi/4 = 45°$). This means that it loses this phase angle with respect to the phase of the wave then being delivered by the generator at A. When a wave of either potential or current is delivered to the line, that wave goes on with its phase unchanged with respect to its own parts. The crest of the wave remains a crest, and a zero-point on the wave remains a zero-point. But the generator keeps on changing its phase and the phase of the next wave that is emitted. Consequently, the wave which has been released and is travelling along the line is continually falling further and further behind the instantaneous phase of the generator end. It loses $\omega$ radians per second, and since the velocity of wave transmission is $v = \omega/\alpha_2$ km. per second, the advancing wave must lose $\omega/v = \alpha_2$ radians per km.

The magnitude and phase relation of a released wave of either potential or current as it runs over the first 10 km. of the circuit here considered, are indicated in Fig. 39. Assuming the initial magnitude and phase at start to be represented

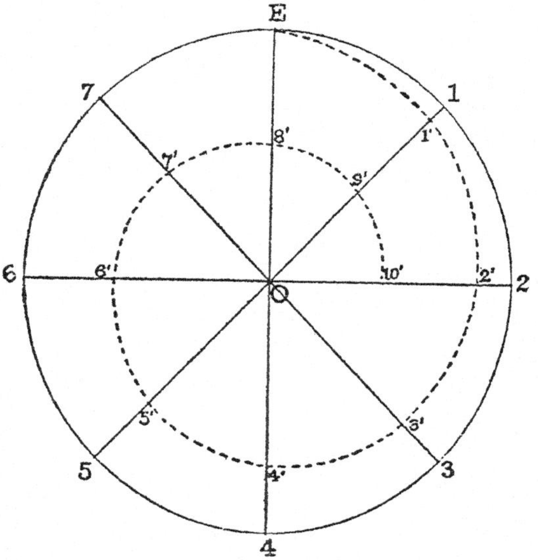


FIG. 39.—Diagram of Relative Magnitudes and Phases of Outgoing Wave over the Line of Fig. 37, for the first ten kilometers.

by the vector OE, these will have changed, after 1 km., to O, 1′, an amplitude of 0·926, and a phase lag of 45°. After 2 km. the vector has become O, 2′, with an amplitude of $(0{\cdot}926)^2 = 0{\cdot}857$, and a lag of $2(45°) = 90°$. After running 10 km. the amplitude vector is O, 10′, with an amplitude of 0·464 and a phase lag of 450°, corresponding to the attenuation-factor $\varepsilon^{-10(\alpha_1 + j\alpha_2)} = \varepsilon^{-10\alpha}$. Another view of the condition is indicated in Fig. 40, where it will be seen that at the point BB′, 30 km. from A, the wave has fallen $3\frac{3}{4}$ complete cycles or 1350° in space-phase behind the generator at A, and the wave then being emitted. The length of the wave on the line must in all cases be—

$$\lambda = \frac{2\pi}{\alpha_2} \quad . \quad . \quad . \quad . \quad . \quad . \quad . \quad \text{km.} \quad (162)$$

a relation which holds either for the unsteady or steady state.

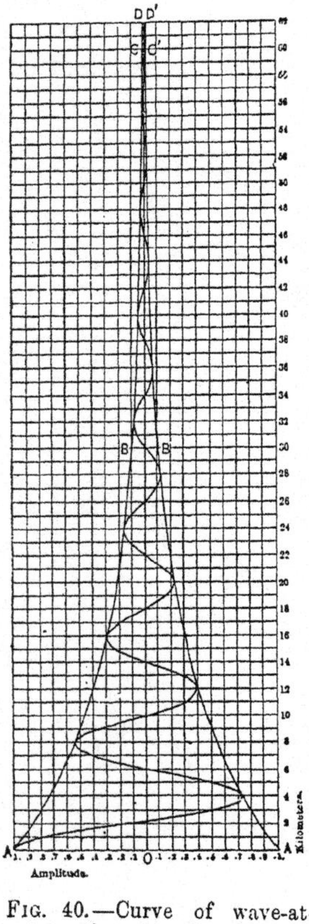


FIG. 40.—Curve of wave-attenuation for the first 62 kilometers of a circuit of attenuation-constant $\alpha = 0{\cdot}07675 + j0{\cdot}7854$ per kilometer.

The wave-length may be regarded as the distance through which a wave must run in order to lose 1 cycle, $2\pi$ radians, or 360° with respect to the generator phase.

The distance $L_\epsilon$ to which a wave must run over a uniform line in order that its amplitude may shrink to $^1/\epsilon$th part (1/2·71828 or 0·3679) will evidently be such that $L_\epsilon \alpha_1 = 1$ or

$$L_\epsilon = 1/\alpha_1 \quad . \quad . \quad . \quad . \quad \text{km.} \quad (163)$$

Similarly the distance $L_{\frac{1}{2}}$ to which a wave must run in order to be attenuated down to $\frac{1}{2}$ amplitude or lose 50 per cent. will be—

$$L_{\frac{1}{2}} = \frac{0{\cdot}69315}{\alpha_1} \quad . \quad . \quad \text{km.} \quad (164)$$

In the case considered the waves would fall to $\frac{1}{2}$ in 9·03 km. and to $^1/\epsilon$th in 13·03 km.

A mechanical model might be constructed to illustrate the preceding principles, in the manner indicated in Fig. 41.* A wooden shaft 00′ is mounted in a long wooden box so as to rotate in bearings at opposite ends, and so as to be rotated by the external handle H. The shutter *ss*, or lid of the box, is arranged to slide in a groove at the top, and is geared with the

* This idea has evidently no novelty. *See* Bibliography, 34 and 44.

handle in such a manner that starting closed with the handle at the top, it slides, to open the lid, by one wave-length for each turn of the handle. Radial pins are then permanently inserted in the shaft according to the plan of Figs. 39 and 40: *i.e.* so spaced angularly as to correspond to the circular component $a_2$ of the linear complex angle or attenuation-constant, and so spaced longitudinally as to correspond to the hyperbolic component $a_1$. If now light falls vertically upon the closed lid of the box, and the handle H is slowly turned, the lid begins to open and the vertically falling shadow of

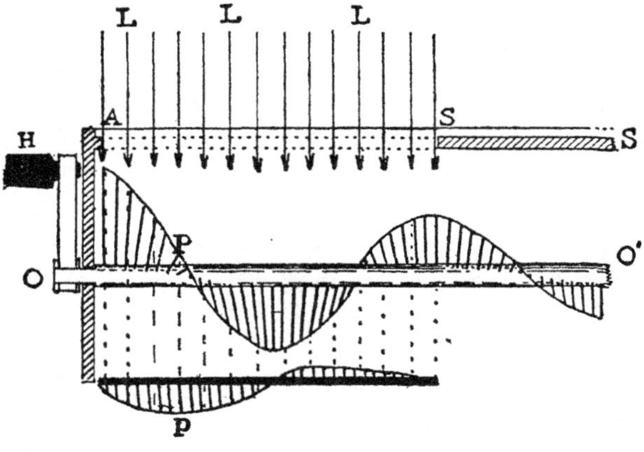


FIG. 41.—Diagram of Model for exhibiting the propagation of the first outgoing wave-train along a uniform line.

each pin begins to execute a simple harmonic motion on the floor of the box. The phase and amplitude of this shadow-motion at any instant and distance along the model correspond to the phase and amplitude of the wave considered in motion over the line, assuming that no reflections occur.

*Terminal Reflections.*—In what follows it will be assumed that when a potential wave, or wave of electric flux, comes to the open end of a freed line, it is reflected backwards without loss of amplitude, of time, or of velocity, and without any change of sign. When, however, it comes to a grounded end, it is reflected back with reversed sign, or 180° change in phase

Also, when a current wave, or wave of magnetic flux, comes to a grounded end, it is reflected back without any change of sign; but when it comes to a freed end, it is reflected back with a reversal of sign, or 180° change in phase. The presence of the impedanceless generator E in the circuit, Fig. 37, does not change these conditions of reflection. The only action of the generator, which is significant in this discussion, is that it continually generates an electric disturbance at the home end, and seeks to send electromagnetic waves over the line.

*Initial Disturbances.*—If we close the generator switch S at the peak of a positive impulse, a wave will immediately be urged along the line under the instantaneously applied full e.m.f. Along with the normal wave, there will however be an exponentially decaying wave which dies out faster than the accompanying normal wave element. The first few waves in the advancing train will therefore not be sinusoidal; although they will tend to recover the sinusoidal form as they advance. If, however, we close the generator switch at the moment when the generated e.m.f. is passing through zero, there will be, in the simplest condition, no abrupt disturbance to the system, and the outgoing waves are all sinusoidal. We shall assume that the switch is closed at the proper instant to avoid initial disturbance, as is always theoretically possible.

*Line Freed at the Distant End.*—If the aerial line-length is, say, just 300 km., a wave would traverse it in one millisecond, if there were no retardation due to attrition in conductor and dielectric ($r=o$, $g=o$). We may suppose, for simplicity, that the line-length is so chosen that with the actual velocity $\omega/\alpha_2$, the time of one passage, or *traverse-time*, is just one millisecond.

We now close the switch at A, and the first potential wave of the entering wave-train runs along the line undergoing attenuation as it runs. Nothing happens at B until after the lapse of one millisecond. Then the leading wave arrives, and instantly retreats backwards, as though the return were a prolongation of the wire that happened to be bent back parallel to the actual line. The amplitude of the wave at the start from A was E volts. On arriving at B it had become $E\varepsilon^{-\theta}$, where $\theta$ is the complex line-angle. But the returning wave after reflection also has an amplitude of $E\varepsilon_{-\theta}$; so that after the wave strikes the distant end B, it immediately doubles up, or produces an amplitude of 2 $E\varepsilon^{-\theta}$ volts, which continues to undergo simple harmonic variation at B. When the head-wave gets back to A, it goes to ground, and returns reflected on the line as $-E\varepsilon^{-2\theta}$ volts. It runs back again to B, getting there in three milliseconds from the start. Its condition is now $-2\ E\varepsilon^{-3\theta}$, allowing for the doubling on reflection. It returns to A after the total lapse of four milliseconds, in the condition $-E\varepsilon^{-4\theta}$. It goes to ground and comes back instantly reversed, or as $E\varepsilon^{-4\theta}$, running back to B, where it arrives after a total lapse of five milliseconds, in the condition $E\varepsilon^{-}{}_{5\theta}$, when it doubles up on itself and returns again to A. This state of to-and-fro activity continues theoretically for eternity; but in any practical case the residue is insignificant after a comparatively few traverses and milliseconds, owing to the continual attrition and attenuation. It is interesting to watch the process of accumulation at B.

We have after the total lapse of 1 millisecond $2\ E\varepsilon^{-\theta}$ volts $\angle$

| | | | | | |
|---|---|---|---|---|---|
| " | " | " | 3 milliseconds | $-2\ E\varepsilon^{-3\theta}$ | " |
| " | " | " | 5 " | $+2\ E\varepsilon^{-5\theta}$ | " |
| " | " | " | 7 " | $-2\ E\varepsilon^{-7\theta}$ | " |

and so on. Each new term is added on to the general accumulation, and since each term is a vector, the addition must be made vectorially. The total accumulation at B becomes then—

$$E_B = 2E(\varepsilon^{-\theta} - \varepsilon^{-3\theta} + \varepsilon^{-5\theta} - \varepsilon^{-7\theta} + \ldots .) \text{ . volts } \angle \quad (165)$$

$$= 2E\varepsilon^{-\theta}(1 - \varepsilon^{-2\theta} + \varepsilon^{-4\theta} - \varepsilon^{-6\theta} + \ldots .) . \quad \text{"} \quad (166)$$

$$= 2E\varepsilon^{-\theta}\frac{1}{1+\varepsilon^{-2\theta}} \ldots \ldots \quad \text{"} \quad (167)$$

$$= \frac{2E}{\varepsilon^{\theta}+\varepsilon^{-\theta}} = \frac{E}{\cosh\theta} = E \operatorname{sech}\theta \ \ldots . \quad \text{"} \quad (168)$$

a result which agrees with (22) when that formula is interpreted vectorially. That is, the hyperbolic function fiua potential at B in the steady state is due to the superposition of successive jumps of potential, each smaller than its predecessor, which arrive at different amplitudes and phases according to a simple exponential law.

As an example, we may take the case of a submarine cable having a linear resistance $r = 10$ ohms per naut, $l = o$; $c = \frac{1}{3} \times 10^{-6}$ farad per naut, $g = o$, $f = 100$ cycles per second, $\omega = 628{\cdot}3$ radians per second. If an e.m.f. of 100 volts maximum cyclic amplitude is applied without initial disturbance at A, required to find the potential at the distant free end when L, the length of the cable, is 24·906 nauts. Here—

$$z = 10\,\underline{/0^\circ},\ y = 2{\cdot}094 \times 10^{-4}\,\underline{/90^\circ},\ a = \sqrt{20{\cdot}94\,\underline{|90^\circ} \times 10^{-4}}$$
$$= 0{\cdot}04576\,\underline{/45^\circ} = 0{\cdot}03236 + j0{\cdot}03236 \text{ hyps. per naut};$$
$$z_0 = 218\,\overline{\backslash 45^\circ} \text{ ohms};\ \theta = 0{\cdot}8059 + j0{\cdot}8059 = 1{\cdot}14\,\underline{/45^\circ} \text{ hyp.}$$

The amplitude of the first wave reaching the end B is $100\ \varepsilon^{-0{\cdot}8059}\,\overline{\backslash 0{\cdot}8059 \text{ radian}} = 100\ \varepsilon^{-0{\cdot}8059}\,\overline{\backslash 46{\cdot}2^\circ} = 44{\cdot}67\,\overline{\backslash 46{\cdot}2^\circ}$. This doubles on arrival and becomes $89{\cdot}34\,\overline{\backslash 46{\cdot}2^\circ}$, as indicated at O1 in Fig. 42. That is, the first potential wave would build up to 89.34 volts amplitude in a simple harmonic motion, and if no other waves arrived, the voltage at B would be $89{\cdot}34/\sqrt{2}$ volts by static voltmeter. The phase of this voltage would be 46·2° behind that of the generator at A. The next return of the leading wave would be in the condition $-100\ \varepsilon^{-2{\cdot}4177}\,\overline{\backslash 2{\cdot}4177} = -8{\cdot}9\,\overline{/138{\cdot}5^\circ} = +8{\cdot}9\,\underline{/41{\cdot}5^\circ}$. The doubling up of this makes a rise $17{\cdot}8\,\underline{/41{\cdot}5^\circ}$ volts as indicated in the figure at O2. The total harmonic e.m.f. at B is now the vector sum of O1 and O2 or $O2^1$, and if no other reflected waves arrived, the potential would perform this harmonic motion at B with a frequency of 100 cycles per second. The third return of the leading wave is in the condition $+100\ \varepsilon^{-4{\cdot}0295}\,\overline{\backslash 4{\cdot}0295} = 1{\cdot}78\,\overline{/230{\cdot}9^\circ}$, which contributes on doubling, the element $O3 = 3{\cdot}56\,\overline{/230{\cdot}9^\circ}$. At the fourth return the condition of the leading wave is $-100\ \varepsilon^{-5{\cdot}641}\,\overline{\backslash 5{\cdot}641} = -0{\cdot}356\,\overline{/323{\cdot}3^\circ}$, which contributes $0{\cdot}712\,\overline{/143{\cdot}3^\circ}$ volts. The vector addition of all these components is $OB = 88{\cdot}17 \overline{\backslash 35{\cdot}0^\circ}$. But by formula (22) the voltage in the steady state at B is $E \operatorname{sech} \theta = 100 \operatorname{sech} (1{\cdot}14 \underline{/45^\circ}) = 100 \times 0{\cdot}882 \overline{\backslash 35{\cdot}0^\circ} = 88{\cdot}2 \overline{\backslash 35{\cdot}0^\circ}$. Consequently, neglecting all terms after the

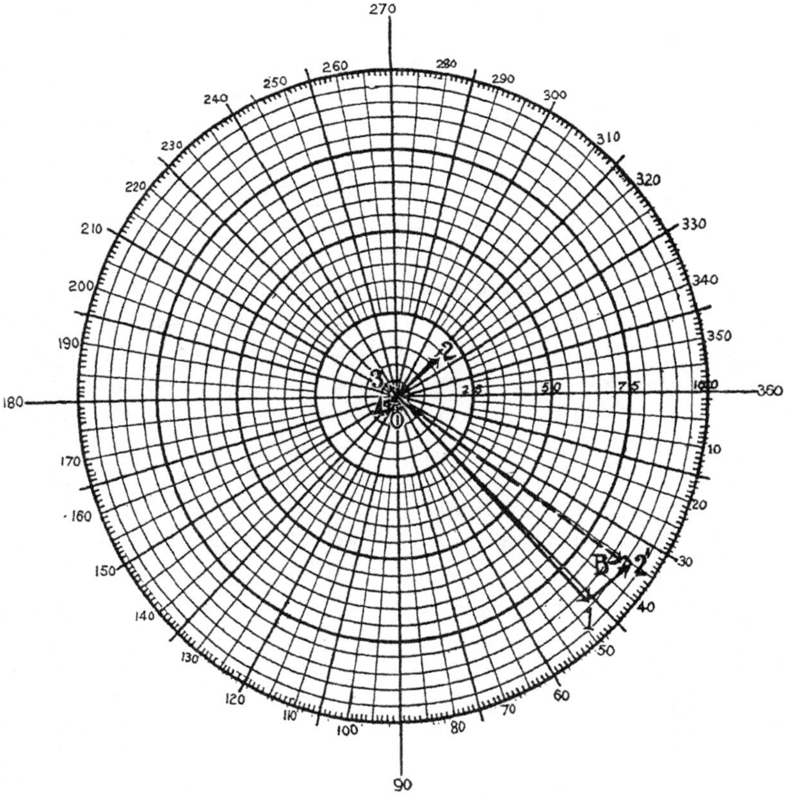


FIG. 42.—Successive Leaps of the Potential at free end of a particular alternating-current line during the constructive stage prior to the steady state.

fourth, the total increment of voltage agrees satisfactorily with the value determined by the steady-state formula.

In the case considered, $\theta_2$ the $j$ component of the hyp. angle of the line, *i.e.* the circular angle component, is 0·8059 radians, or a little more than 45° (46·16°). Consequently the successive increments of voltage, which come at phase intervals of $2\theta_2$, are nearly perpendicular to each other. If, however, the $j$ component of the line angle happens to approximate one right angle, or any odd number of right angles, the successive increments will be nearly in the same phase, so that the final distant end voltage will build up; or sech $\theta$ will be greater than $\varepsilon^{-\theta}$. This will be particularly the case when $\theta_2$ is just 90°, or the line has one quarter wave-length. If the attenuation is small, the successive increments do not rapidly dwindle, and being in the same phase, they may build up to a voltage greatly in excess of that impressed by the generator on the sending end of the line. This effect is called the *Ferranti-effect.* While it must occur whenever the line is an odd number of quarter wave-lengths, it will only be noticeable as an actual increase of potential towards the distant free end when the real component $\theta_1$ of the line angle is distinctly less than the imaginary component $\theta_2$, *i. e.* when the attenuation is relatively small. No rise of potential can occur towards the free distant end of a line in the steady state when $\theta_1$ is greater than $\theta_2$. On the other hand when $\theta_1$ is made sufficiently small, and $\theta_2=\pi/2$, the Ferranti-effect of potential rise can theoretically be made indefinitely great, and practically the free-end potential can be many times the impressed potential at the home end. A small terminal load usually suffices to destroy the effect. (See Chapter VII.)

If on the other hand the line has half-wave length, or any integral multiple thereof ($\theta_2 = n\pi$), where $n$ is any real integer, the successive increments of potential during the constructive state arrive in opposite phases; so that the distant end potential does not tend to build up, even with low attenuation.

By a similar summation of outgoing and reflected waves at A, we should find the total amount to $E_A$ volts, or merely that impressed by the generator; because each reflection from ground at A takes the negative sign and cancels the effect of the arriving potential wave.

Similarly, if we take the current-wave at B, with the line

free: on the first arrival, the amplitude is $I_0 \varepsilon^{\theta}$ amperes, where—

$$I_0 = \frac{E_A}{z_0} \quad . \quad . \quad . \quad . \quad . \quad \text{amperes} \angle \quad (169)$$

is the initial outgoing current at A. But the reflected current-wave from the open end at B is immediately $-I_0 \varepsilon^{-\theta}$, which cancels the arrival; so that the resultant rise of current is nil. The same action occurs at each return of the current-wave. Hence the current remains at zero throughout the steady state, as is, of course, the inevitable condition at an open end.

Again, with the distant end B still free, let us trace the building-up of current-waves at A. The outgoing wave, as we have seen, is $I_0$. On first return from B it has become $-I_0 \varepsilon^{-2\theta}$, which on being reflected from ground at A, takes no change of sign, and so doubles the increment to $-2I_0 \varepsilon^{-2\theta}$. On second return from B it is $+I_0 \varepsilon^{-4\theta}$, which likewise doubles at A. Continuing this process, the summation at A is—

$$I_A = I_0 - 2I_0 \varepsilon^{-2\theta} + 2I_0 \varepsilon^{-4\theta} - 2I_0 \varepsilon^{-6\theta} + \ldots \text{amperes} \angle \quad (170)$$

$$= I_0 \{1 - 2\varepsilon^{-2\theta} (1 - \varepsilon^{-2\theta} + \varepsilon^{-4\theta} - \ldots \quad \text{,,} \quad \text{,,} \quad (171)$$

$$= I_0 \left\{1 - \frac{2\varepsilon^{-2\theta}}{1 + \varepsilon^{-2\theta}}\right\} \quad . \quad . \quad . \quad . \quad . \quad . \quad \text{,,} \quad \text{,,} \quad (172)$$

$$= I_0 \left\{1 - \frac{2}{\varepsilon^{2\theta} + 1}\right\} = I_0 \left\{\frac{\varepsilon^{\theta} - \varepsilon^{-\theta}}{\varepsilon^{\theta} + \varepsilon^{-\theta}}\right\} = I_0 / \coth \theta$$

$$\text{amperes} \angle \quad (173)$$

$$= \frac{E_A}{z_0 \coth \theta} \quad . \quad . \quad . \quad . \quad . \quad . \quad . \quad . \quad . \quad . \quad \text{,,} \quad \text{,,} \quad (174)$$

which agrees with (21) when that formula is interpreted in complex numbers.

*Line Grounded at Distant End.*—If we ground the line at B, the series of reflected current-waves returning to the sending end A is the same as in the preceding case (170) except there is no change of sign in the successive elements. The summation of current at A is then—

$$I_A = I_0 + 2I_0\varepsilon^{-2\theta} + 2I_0\varepsilon^{-4\theta} + 2I_0\varepsilon^{-6\theta} + \dots \text{ amperes } \angle \quad (175)$$

$$= I_0\{1 + 2\varepsilon^{-2\theta}(1 + \varepsilon^{-2\theta} + \varepsilon^{-4\theta} + \dots \quad \text{"} \quad \text{"} \quad (176)$$

$$= I_0\left\{1 + \frac{2\varepsilon^{-2\theta}}{1 - \varepsilon^{-2\theta}}\right\} = I_0\left\{\frac{1 + \varepsilon^{-2\theta}}{1 - \varepsilon^{-2\theta}}\right\}$$

$$= I_0 \coth\theta \quad \dots \quad \text{"} \quad \text{"} \quad (177)$$

$$= \frac{E_A}{z_0 \tanh\theta} \quad \dots \quad \text{"} \quad \text{"} \quad (178)$$

which agrees with (24) when that formula is interpreted in complex numbers.

Again, if we ground the line at B, and sum the current-waves arriving at B, their conditions are $I_0\varepsilon^{-\theta}$, $I_0\varepsilon^{-3\theta}$, $I_0\varepsilon^{-5\theta}$, etc. Being reflected at B without change of sign, they contribute doubled increments at B. Hence the summation at B is—

$$I_B = 2(I_0\varepsilon^{-\theta} + I_0\varepsilon^{-3\theta} + I_0\varepsilon^{-5\theta} + \dots \text{ amperes } \angle \quad (179)$$

$$= 2I_0\varepsilon^{-\theta}(1 + \varepsilon^{-2\theta} + \varepsilon^{-4\theta} + \dots \quad \text{"} \quad \text{"} \quad (180)$$

$$= \frac{2I_0\varepsilon^{-\theta}}{1 - \varepsilon^{-2\theta}} = I_0\left(\frac{2}{\varepsilon^{\theta} - \varepsilon^{-\theta}}\right) = \frac{I_0}{\sinh\theta} \quad \dots \quad \text{"} \quad \text{"} \quad (181)$$

$$= \frac{E_A}{z_0 \sinh\theta} \quad \dots \quad \text{"} \quad \text{"} \quad (182)$$

which agrees with (25) when that formula is interpreted in complex numbers.

With the line grounded at B, the potential waves cancel their arrivals on each reflection and at each end of the line; so that the summation of potential at the sending end is always $E_A$, and at the receiving end always zero.

If now the line is grounded at B through an impedance $z_r$ ohms, any current-wave on arriving at B is split into two parts; namely, a transmitted part which is absorbed to ground without further reflection, and a reflected part which goes back as though from an open end. Let $m$ be the fraction of the wave that is transmitted or the transmission coefficient, and $1-m$ the fraction that is reflected or the reflection coefficient. It was shown by Heaviside that in the symbols here used—

$$m = \frac{2z_0}{z_0 + z_r} \quad . \quad . \text{ numeric } \angle \quad (183)$$

$$1 - m = \frac{z_r - z_0}{z_0 + z_r} \quad . \quad . \quad \text{"} \quad \text{"} \quad (184)$$

the reflected current-wave retreats with its sign reversed, or as with a coefficient $m-1$.

The current-wave to ground at B on first arrival is $mI_0\varepsilon^{-\theta}$, and $(m-1)I_0\varepsilon^{-\theta}$ goes back to A. It reaches A in the condition $(m-1)I_0\varepsilon^{-2\theta}$, and is reflected back to B without change of sign. It arrives at B for the second time in the condition $(m-1)I_0\varepsilon^{-3\theta}$. Of this $m\,(m-1)I_0\varepsilon^{-3\theta}$ is absorbed to ground, and the remainder $(m-1)^2I_0\varepsilon^{-3\theta}$ retreats to A. The final summation current absorbed to ground at B is—

$$I_B = mI_0\varepsilon^{-\theta} + m\,(m-1)I_0\varepsilon^{-3\theta} + m\,(m-1)^2I_0\varepsilon^{-5\theta} + \ldots \quad \text{amperes} \angle \quad (185)$$

$$= mI_0\varepsilon^{-\theta}\{1+(m-1)\varepsilon^{-2\theta}+(m-1)^2\varepsilon^{-4\theta}+ \ldots \quad \text{"} \quad \text{"} \quad (186)$$

$$= \frac{mI_0\varepsilon^{-\theta}}{1-(m-1)\varepsilon^{-2\theta}} = \frac{mI_0}{\varepsilon^{\theta}-(m-1)\varepsilon^{-\theta}}$$

$$= \frac{I_0}{\dfrac{\varepsilon^{\theta}}{m} - \left(\dfrac{m-1}{m}\right)\varepsilon^{-\theta}} \quad \text{"} \quad \text{"} \quad (187)$$

$$= \frac{I_0}{\left(\dfrac{2-m}{m}\right)\cosh\theta + \sinh\theta} = \frac{I_0}{\dfrac{z_r}{z_0}\cosh\theta + \sinh\theta}$$

$$= \frac{I_0 z_0}{z_0\sinh\theta + z_r\cosh\theta} \;. \quad \text{amperes} \angle \quad (188)$$

$$= \frac{E_A}{z_0\sinh\theta + z_r\cosh\theta} \quad \ldots\ldots\ldots \quad \text{"} \quad \text{"} \quad (189)$$

which agrees with (59) when that formula is interpreted in complex numbers.

It can readily be seen that the potential at B in the steady state is $I_B z_r$ volts, and that potential reflections at A cancel off.

We might similarly sum the potential or current waves at any intermediate point on the line and derive formulas (37) to (42).

## CHAPTER VII

### THE APPLICATION OF HYPERBOLIC FUNCTIONS TO ALTERNATING-CURRENT POWER-TRANSMISSION LINES.

ALTERNATING-CURRENT power-transmission lines differ from alternating-current power-distribution lines in having only terminal loads applied to them, as distinguished from a number, usually a large number, of intermediate distributed loads. They are ordinarily of the three-phase type, as indicated in Fig. 43, and consist of three line conductors. The system may be regarded as being divisible into three independent single-phase lines, AB, A′B′, A″B″, each operated, at star voltage, to

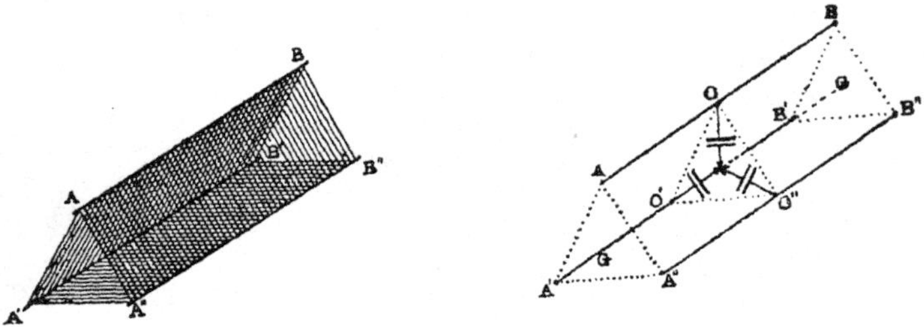

FIG. 43.—Set of Three-phase Transmission Wires. Nominal circuit of three-phase set of wires. Three equal condensers connected in star between mid-length points O, O′, O″.

ground, or neutral-potential surface. If the three wires are symmetrically disposed, the three individual single-phase lines will have equal wire inductances, and also equal wire capacities, acting like a star group of condensers, with zero potential at the neutral point (Fig. 43). The corresponding conditions for a two-wire system are indicated in Figs. 44, 45 and 46. If the geometrical disposition of the three wires in the system of Fig. 43 is dissymmetrical, the three individual inductances and capacities of the independent single-phase wires will be unequal, but can ordinarily be computed from the geometrical data.

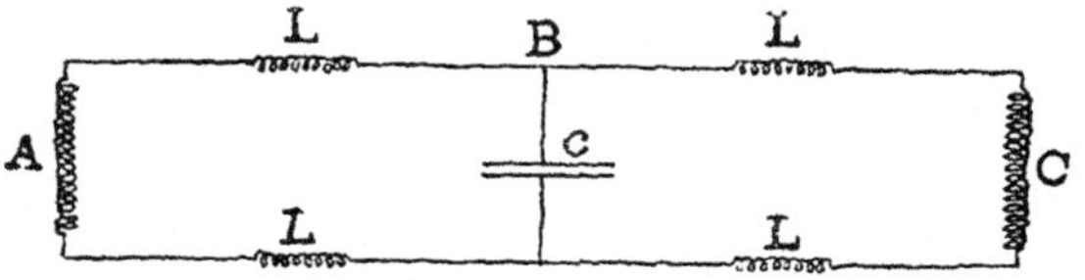


FIG. 44.—Diagram of an Alternating-current Circuit with the entire Line Capacity centered at B, and the resistance and inductance divided between the four choking-coils L, L, L, L.

Every alternating-current power-transmission system may therefore be analysed into a group of parallel wires, each operating independently to ground potential. Strictly speaking,

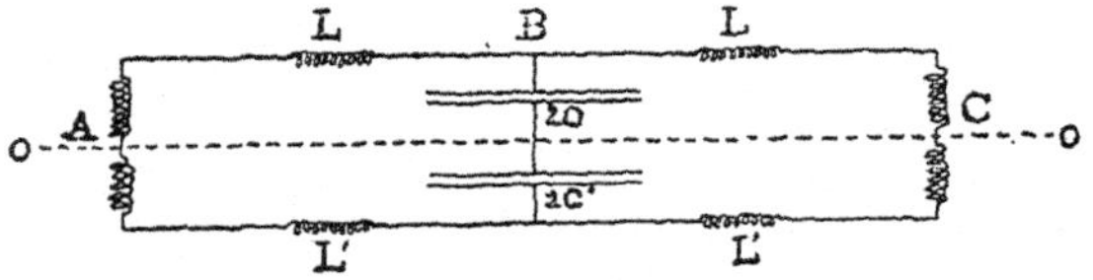


FIG. 45.—Division of the Circuit of Fig. 44 into two equal and symmetrical portions about the neutral mid-plane OO of zero potential.

the capacity of each line is uniformly distributed, and it is the recognition of this condition that introduces hyperbolic functions as the natural key to the true behaviour of such lines

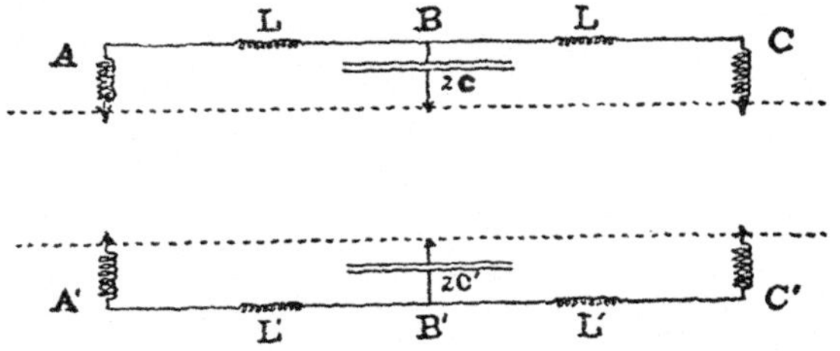


FIG. 46.—Analysis of the Double-wire Circuit of Fig. 44 into two equivalent single-wire circuits A, B, C and A′, B′, C′, each having twice the condenser capacity of the circuit in *Fig.* 44.

in the steady state; but, as a first approximation, which is sufficiently good for all but very long lines, at ordinary operating frequencies, the capacity may be either lumped into a single condenser at the middle of the line, as in Figs. 45 to 47, thus forming the nominal *T* of the line; or, as is usually more convenient for the purposes of computation, all the capacity may be collected into two equal condensers, and these applied one at each end of the wire, as shown in Fig. 48, thus forming the nominal *Π* of the line. This method has been called the

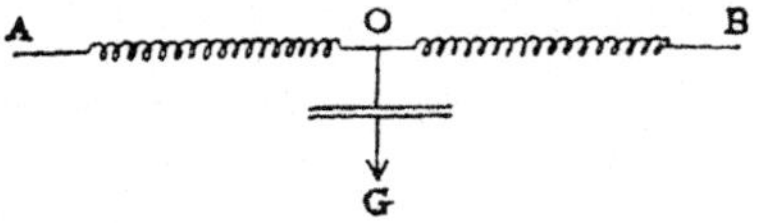


FIG. 47.—T-Conductor Equivalent to an alternating-current transmission wire.

"split condenser" method of analysing approximately the electrical behaviour of a transmission line.* One wire AB of a transmission system is represented, in Fig. 49, as operated to neutral potential. The wire has a vector impedance $R + jX = Z \angle \beta$ ohms. At the receiving end B, is a motor M, or other load, of known magnitude and power-factor, at a given voltage $E_B$. At the sending end A, a single-phase generator G,

FIG. 48.—Π-Conductor Equivalent to an alternating-current transmission wire.

delivers such a voltage $E_A$ as will maintain the given voltage $E_B$ at B. Half of the capacity of the wire to neutral surface is applied as a condenser at A, and the other half at B. Each such condenser offers an admittance of $Y_A = Y_B$ mhos. The admittance $Y_B$ receives a current $I'_B$ from the voltage $E_B$; while $Y_A$ similarly receives a current $I'_A$ from the voltage $E_A$. The line current I has the same strength at all points between A and B. It is the vector sum of the load current $I_B$ and the condenser current $I'_B$. All voltages are r.m.s. vector star-

* "Calculation of the High-Tension Line," by P. B. Thomas: *Trans. Am. Inst. Electrical Engineers*, Part I, vol. xxviii, pp. 641–686, June 1909.

voltages, in the case of a three-phase system, and all currents are r.m.s. vector amperes. We take the phase of the voltage $E_B$ as standard, and refer all other voltages to this phase.

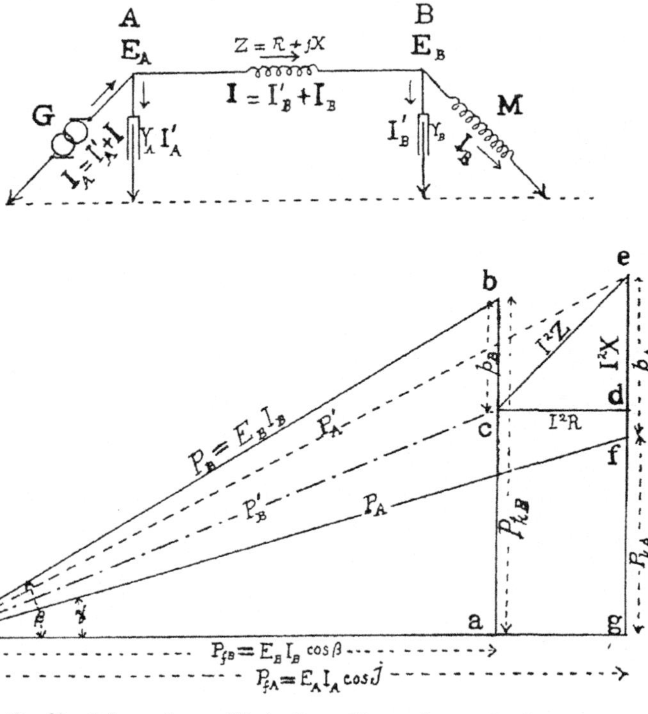


FIG. 49.—Circuit Connections and Vector Power Diagram for one wire of a power-transmission system. Nominal Π.

The diagram O $a$ $b$ $g$ $e$, Fig. 49, is the stationary vector power diagram relating to this wire. Let the vector O$b$ represent, to scale, the power load delivered at B in the branch M to the phase of current $I_B$ as standard. Denote this load by—

$$P_B = E_B \,.\, I_B \angle\beta \quad . \quad . \quad \text{watts} \angle \quad (190)$$
$$\text{or volt-amperes} \angle$$

Then the horizontal component $Oa$ of this load is the effective power delivered; or—

$$P_{jB} = E_B \,.\, I_B \,.\, \cos\beta \quad . \quad . \quad \text{watts (191)}$$

and the vertical component $ab$ is the reactive power; or—

$$P_{kB} = jE_B \,.\, I_B \,.\, \sin\beta \quad . \quad . \, j \text{ watts (192)}$$

The power-factor of the load is $\cos\beta$.

Since the receiving end of the line is assumed to be maintained at a steady star-voltage $E_B$, the charging current in the admittance $Y_B$ is—

$$I'_B = jE_B \,.\, Y_B \quad . \quad . \quad \text{amperes} \angle \text{ (193)}$$

The power absorbed by this admittance is—

$$p_B = E_B I'_B = jE^2_B \,.\, Y_B \quad . \quad j \text{ watts (194)}$$

This is $+j$ reactive power, to voltage standard phase, but must be reckoned as $-j$ reactive power with respect to current standard phase. It is, therefore, measured along $ba$, and is indicated at $bc$. The power supplied at the B end of the line including the admittance $Y_B$, or half-line condenser, is $Oc$ watts or volt-amperes. In other words, part of the reactive power in the load is supplied by the end-condenser $Y_B$.

The line current is the vector-sum of the load current $I_B$ and the condenser current $I'_B$; or—

$$I = I_B + I'_B \quad . \quad . \quad . \quad . \quad . \quad . \quad . \quad \text{r.m.s. amperes} \angle \text{ (195)}$$

The power expended in the line is—

$$I^2Z = I^2R + jI^2X \quad . \quad \text{watts or volt-amperes} \angle \text{ (196)}$$

where the phase of the line current must be taken as standard. This power is represented by the vector $ce$, $cd$ being the effective or dissipated power, and $de$ the reactive or non-dissipated power. Consequently, the power supplied to the line at A, beyond the condenser $Y_A$, is represented, to scale, by the vector $Oe = P'_A$.

The power developed in the condenser admittance $Y_A$ is $-jE_A^2Y_A$ watts, or $p_A$ on the diagram. For this item, we find the value of $E_A$; namely—

$$E_A = E_B + IZ \quad . \quad . \quad . \quad \text{volts} \angle \text{ (196)}$$

Finally, the power supplied by the generator G to the end A of the single wire considered, including the condenser $Y_A$, is $Of = P_A$ vector watts or volt-amperes; of which the horizontal component $Og = P_{fA} = E_A I_A \cos \gamma$ is the effective component, and $P_{kA} = gf$ is the reactive component. The power-factor at the generator is $\cos \gamma$. The electrical efficiency of the line is $Oa/Og$.

The simplest method of taking distributed capacity into account in such a problem is to substitute, by formulas (77) and (78), the equivalent $\Pi$ for the nominal $\Pi$.

We may take the example of a three-phase transmission line having a length L = 250 km. (155·34 English statute miles), consisting of three No. 000 A.W.G. copper wires, 1·041 cm. in diam. (0·41″), supported symmetrically, on pole insulators, at a uniform interaxial distance of 193 cm. (72″). The following values of linear resistance, inductance and capacitance are taken for each of these three wires, to neutral potential surface—

$r = 0{\cdot}206$ ohm / wire km. $= 0{\cdot}33$ ohm / wire mile.
$l = 1{\cdot}22229$ millihenry / wire km. $= 1{\cdot}967$ millihenry / wire mile.
$c = 0{\cdot}0094828$ microfarad / wire km. $= 0{\cdot}01526$ microfarad / wire mile.

The linear leakance is taken as negligible. The frequency of operation is $f = 25$ cycles per second; or $\omega = 157{\cdot}08$ radians per second.

With the above linear secondary constants we obtain for the total line constants—

$$Lr = R = 51{\cdot}5 \text{ ohms}, \qquad Ll = 0{\cdot}30557 \text{ henry},$$
$$Lc = C = 2{\cdot}3707 \text{ microfarads}.$$

We may assume that the star voltage at the receiving end of the line is 50 kilovolts r.m.s. or 86·6 kv. between any pair of the three wires.

At the above fundamental frequency, the linear reactance of each wire will be $jx = jl\omega = j1{\cdot}22229 \times 0{\cdot}15708 = j0{\cdot}191996$ ohms per km. The total line reactance $jLx = j47{\cdot}999$ ohms.

The linear dielectric admittance $jc\omega = j1{\cdot}48955 \times 10^{-6}$ mho per km., and the total dielectric admittance $jY = jLc\omega = j3{\cdot}72390 \times 10^{-4}$ mho. The total wire impedance is therefore $Z = 51{\cdot}5 + j47{\cdot}999 = 70{\cdot}40 \underline{/42^\circ\ 59'\ 05''}$ ohms.

Fig. 43 shows, at AB, the nominal $\Pi$ of this line for the above-mentioned frequency and conditions. It consists of a line impedance with half the capacity susceptance at each end; *i.e.* $j1{\cdot}86194 \times 10^{-4}$ mho corresponding to a reactance of $5370{\cdot}74 \overline{|90^\circ}$ ohms.

The hyperbolic angle $\theta$ subtended by the line will be, by (19), $\theta = \sqrt{ZY} = \sqrt{70{\cdot}40\underline{/42^\circ\ 59'\ 05''} \times 3{\cdot}72390 \times 10^{-4}\overline{|90^\circ}}$ $= \sqrt{0{\cdot}0262167\underline{/132^\circ\ \ 59'\ \ 05''}} = 0{\cdot}161915\underline{/66^\circ\ \ 29'\ \ 32''}$ $= 0{\cdot}064583 + j0{\cdot}148476$ hyp. The nominal $\Pi$ of this line is presented in Fig. 50, at ABGG, for the frequency of 25 ~. The architrave impedance is the line impedance Z above mentioned, and half of the dielectric admittance is placed in each pillar.

In order to form the equivalent $\Pi$ of the line, we require to form and apply the ratios $\frac{\sinh\theta}{\theta}$ and $\frac{\tanh\frac{\theta}{2}}{\frac{\theta}{2}}$.* By the help of (139) and (145) we find—

$$\sinh\theta = 0{\cdot}161431\ \underline{/66^\circ\ 40'\ 32''}$$

$$\text{and } \tanh\frac{\theta}{2} = 0{\cdot}0810865\ \underline{/66^\circ\ 23'\ 17''};$$

consequently—

$$k_{\rho''} = \frac{\sinh\theta}{\theta} = 0{\cdot}99703\ \underline{/0^\circ\ 11'\ 00''}$$

$$k_{g''} = \frac{\tanh\frac{\theta}{2}}{\frac{\theta}{2}} = 1{\cdot}0016\ \overline{\backslash 0^\circ\ 06'\ 15''}.$$

* Tables of $\frac{\sinh\theta}{\theta}$ and $\frac{\tanh(\theta/2)}{\theta/2}$ with five significant digits have been computed by the writer for each degree between 60° and 90° of argument and each 0·1 of modulus in $\theta$ up to 0·5. These tables are shortly to be published (Bibliography, 73).

That is, the correction-factor which transmutes the nominal into the equivalent $\Pi$ for this 250 km. line differs from unity

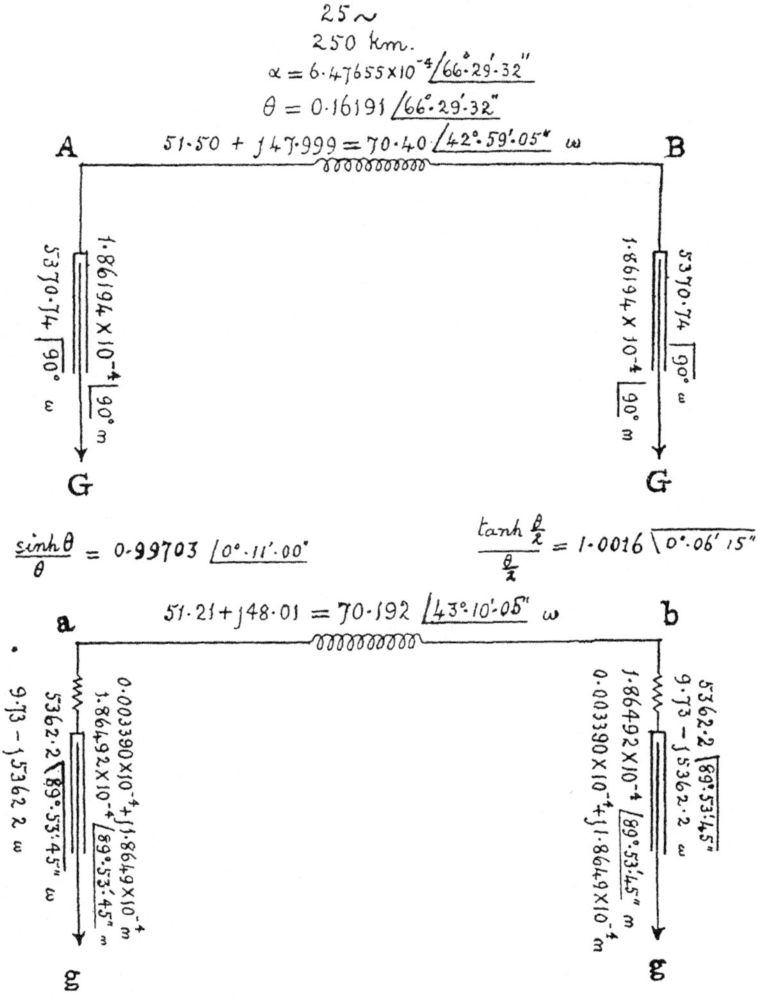


FIG. 50.—Nominal and Equivalent $\Pi$ for the particular transmission line at 25 ~.

by only 0·3 per cent. for the architrave and 0·16 per cent. for the pillars. In other words, the correction for distributed capacity is negligible for such a line.

In Fig. 50, when we multiply the architrave impedance AB of the nominal $\Pi$ by 0·99703 $\underline{/0^\circ\ 11'\ 00''}$, we obtain the architrave impedance 70·192 $\underline{/43^\circ\ 10'\ 05''}$ ohms, $a$, $b$, of the equivalent $\Pi$. This means that the line behaves in the steady state, at this frequency, as though its conductor resistance were reduced from 51·50 to 51·21 ohms, and its inductive reactance increased from 47·999 to 48·01 ohms. Similarly, multiplying the pillar admittance AG or BG, of the nominal $\Pi$, by 1·0016 $\overline{\backslash 0^\circ\ 06'\ 15''}$, we obtain the pillar admittances $1{\cdot}86492 \times 10^{-4}$ $\underline{/89^\circ\ 53'\ 45''}$ mho, $a, g$, or $b, g$, of the equivalent $\Pi$.* This is equivalent to assuming that either a certain small resistance (9·73 ohms) is inserted in series with a slightly increased condenser (1·1873 $\mu f$); or, that a non-inductive leak of 0·339 micromho, has been applied to each condenser in shunt. We shall retain the latter conception for convenience.

The circuit-connections and the vector power diagram, for one wire of the line considered, are given in Fig. 51, under an assumed load of 4000 kw. (4 megawatts) of effective power (12 megawatts for the entire three-phase system), at a power factor of 0·8. The apparent or resultant power delivered at B is, therefore, $P_B = 5{\cdot}0\ \underline{/36^\circ\ 52'\ 12''}$ megawatts, to standard current phase, and the inductively reactive power $j3$ megawatts. The current received through the load, under 50 kilovolts at B, is thus $80 - j60 = 100\ \overline{\backslash 36^\circ\ 52'\ 12''}$ amperes to B-voltage phase. That is, the load current lags by this angle behind the voltage at the receiving end of the line. The current in the leaky condenser at B is $0{\cdot}01695 + j9{\cdot}3245$ amperes, carrying a power of $0{\cdot}8475 + j466{\cdot}23$ kw. to B-voltage phase, or $0{\cdot}8475 - j466{\cdot}23$ kw. with respect to current phase. The total power delivered at B, including the terminal condenser, is thus: $4{\cdot}000848 + j2{\cdot}53377$ megawatts. All pressures and currents are expressed in r.m.s. values.

The current in the line is $80{\cdot}017 - j50{\cdot}676$ amperes to B-

* It will be understood that the degree of arithmetical precision aimed at in these examples, for the sake of thoroughness, is much greater than is necessary in ordinary transmission-engineering computation.

voltage phase $= 94{\cdot}714 \setminus 32^\circ\ 20'\ 47''$. The IZ drop in the line is $94{\cdot}714 \setminus 32^\circ\ 20'\ 47'' \times 70{\cdot}192\ /43^\circ\ 10'\ 05'' = 6648\ /10^\circ\ 49'\ 18''$ $= 6530 + j1248$ volts. The $I^2Z$ power is $(94{\cdot}714\ /0^\circ)^2 \times 70{\cdot}192\ /43^\circ\ 10'\ 05'' = 629686\ /43^\circ\ 10'\ 05''$ watts $= 0{\cdot}6297$

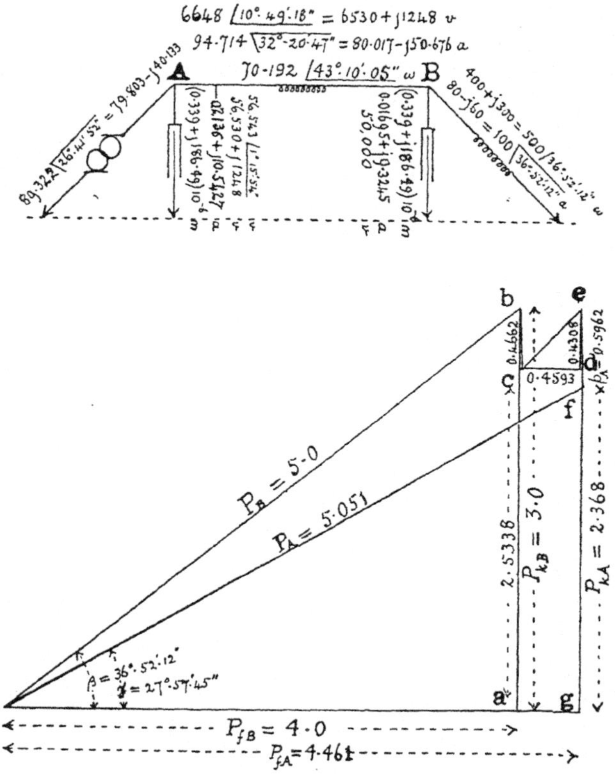


FIG. 51.—Circuit Connections and Power Vector Diagram for one wire of the particular transmission line at 25 ∼.

$/43^\circ\ 10'\ 05''$ megawatt $= 0{\cdot}45926 + j0{\cdot}43080$. In this power computation, the current must be taken as of standard phase, as all power products $P = E\,I$ watts $\angle$, require one of the vectors E or I to be taken at standard phase, or zero argument.

The voltage at A is the vector sum of the B-voltage and

the IZ line drop. It amounts to $56{,}530 + j1248 = 56543 \underline{/1^\circ\ 15'\ 54''}$ volts. At this voltage, the current in the leaky condenser at A is $56543 \underline{/1^\circ\ 15'\ 54''} \times 1{\cdot}86492 \times 10^{-4} \underline{/89^\circ\ 53'\ 45''} = 10{\cdot}5449 \underline{/91^\circ\ 09'\ 39''}$ amp. $= -0{\cdot}2136 + j10{\cdot}5427$. The power delivered to this condenser is $56543 \underline{/0^\circ} \times 10{\cdot}5449 \underline{/89\ 53'\ 45''} = 596{,}247 \underline{/89^\circ\ 53'\ 45''}$ watts $=$ $1084 + j596{,}240$ watts with reference to voltage phase, or $1084 - j596{,}240$ watts with reference to current phase. Adding this power vectorially on the diagram, by the vector *ef*, we arrive at *f*, and O*f* is the vector power delivered by the generator at $A = 5{\cdot}051 \underline{/27^\circ\ 57'\ 45''}$ megawatts $= 4{\cdot}461 + j2{\cdot}368$ megawatts at a power-factor of 0·8833.

The current delivered by the generator is the vector sum of the line current and the A-leak current; or $79{\cdot}803 - j40{\cdot}133 = 89{\cdot}322 \overline{\backslash 26^\circ\ 41'\ 52''}$ amperes, under a pressure of $56{,}543 \underline{/1^\circ\ 15'\ 54''}$ volts, with a power of $89.322 \underline{/0^\circ} \times 56{,}543 \underline{/27^\circ\ 57'\ 46''} = 5{,}051{,}000 \underline{/27^\circ\ 57'\ 46''}$ watts (current phase) which checks the preceding result.

The efficiency of the line under this load is $4{\cdot}0/4{\cdot}461 = 0{\cdot}8967$.

It is evident that the power diagram and line computations would be only slightly modified, in this case, if we employed the nominal $\Pi$ of the line, instead of the equivalent $\Pi$. The computations would also be simplified; because the leaks at A and B are pure reactances or $j$-quantities in the nominal $\Pi$, and pertain to pure condensers; whereas, in the equivalent $\Pi$, they are complex quantities, or pertain to leaky condensers.

*Relation of Line Angle to Length and Frequency.*—The hyperbolic angle $\theta$ subtended by a uniform alternating-current line manifestly increases with the length of the line. If the line had no conductor resistance, or dielectric leakance, the angle would be, by (150)—

$$\theta = jL\omega\sqrt{lc} \quad . \quad . \quad \text{hyp. radians} \quad (197)$$

$$= L\omega\sqrt{lc} \quad . \quad \text{circular radians} \quad (198)$$

which shows that it increases directly in proportion to the frequency. To a first approximation, therefore, the line angle increases with the frequency, provided that the dissipative linear constants of the line ($g$ and $r$) are relatively small, and this is true for power-transmission lines. As a rough rule, we may say that 1000 km. of wire such as is used in power-transmission, operated at a frequency of 50 ~, has a line-angle of about 1 hyp. at an argument usually between 60° and 80°. Consequently, to the same low degree of precision, the modulus of the hyp. angle subtended by a wire of L km. operated at a frequency of $f$ ~ is roughly—

$$\theta \backsimeq \frac{\mathrm{L}f}{50{,}000} \quad . \quad . \quad . \quad \text{numeric} \quad (199)$$

Thus, although the hyperbolic angle subtended by a transmission line, at its fundamental working frequency, may be so small that there is very little difference between its nominal and its equivalent $T$ or $\Pi$, yet steadily increasing differences will develop with the ascending harmonics in the impressed e.m.f., if such harmonics are present.

In a properly constructed three-phase system it is well known that no harmonic frequencies can exist of three times, or of $3n$ times, the fundamental frequency. Such harmonics as exist must be of 5, 7, 11, etc. times the fundamental frequency. In Fig. 52, the nominal and equivalent $\Pi$'s are presented for the quintuple frequency of 125 ~ in the case of the 250-km. line already considered. The nominal $\Pi$ of the line differs only from the nominal $\Pi$ at 25 ~, in having five times the inductive reactance in the architrave, and five times the condenser susceptance in the pillars. The hyperbolic angle of the line is $\theta = \sqrt{245{\cdot}458 \underline{/77° \, 53' \, 19''} \times 1{\cdot}86194 \times 10^{-3} \underline{|90°}} = 0{\cdot}676039 \underline{/83° \, 56' \, 40''} = 0{\cdot}071318 + j0{\cdot}672267$ hyp. The correcting ratios for this angle are $k_{p_{\prime\prime}} = 0{\cdot}92722 \underline{/0° \, 56' \, 40''}$, and $k_{g_{\prime\prime}} = 1{\cdot}03893 \overline{\backslash 0° \, 28' \, 50''}$. Applying these factors to the architrave and pillars of the nominal $\Pi$ respectively, we obtain $227{\cdot}595 \underline{/78° \, 49' \, 59''}$ ohms for the architrave, and $9{\cdot}6721 \times 10^{-4} \underline{/89° \, 31' \, 10''}$ mho for the pillars of the equivalent $\Pi$.

The corresponding conditions for the septuple-frequency

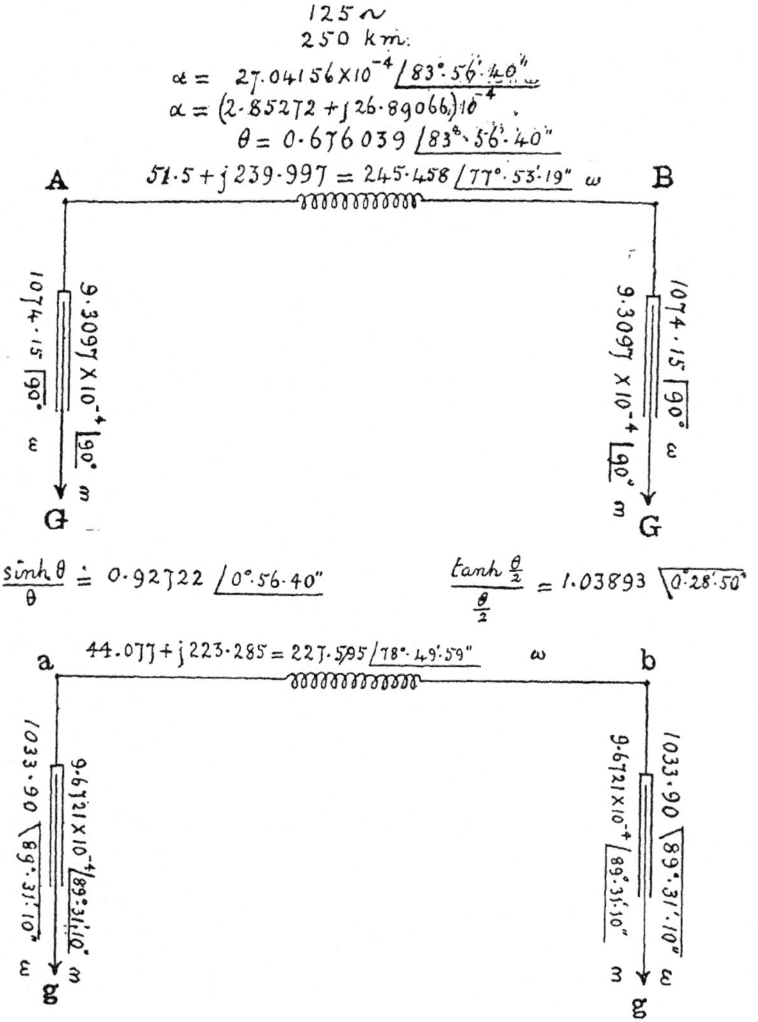


FIG. 52.—Nominal and Equivalent Π for the particular transmission line at 125~.

harmonic 175~ are indicated in Fig. 53. The hyperbolic angle has reached 0·941311 /85° 38′ 34″ hyp. The correcting

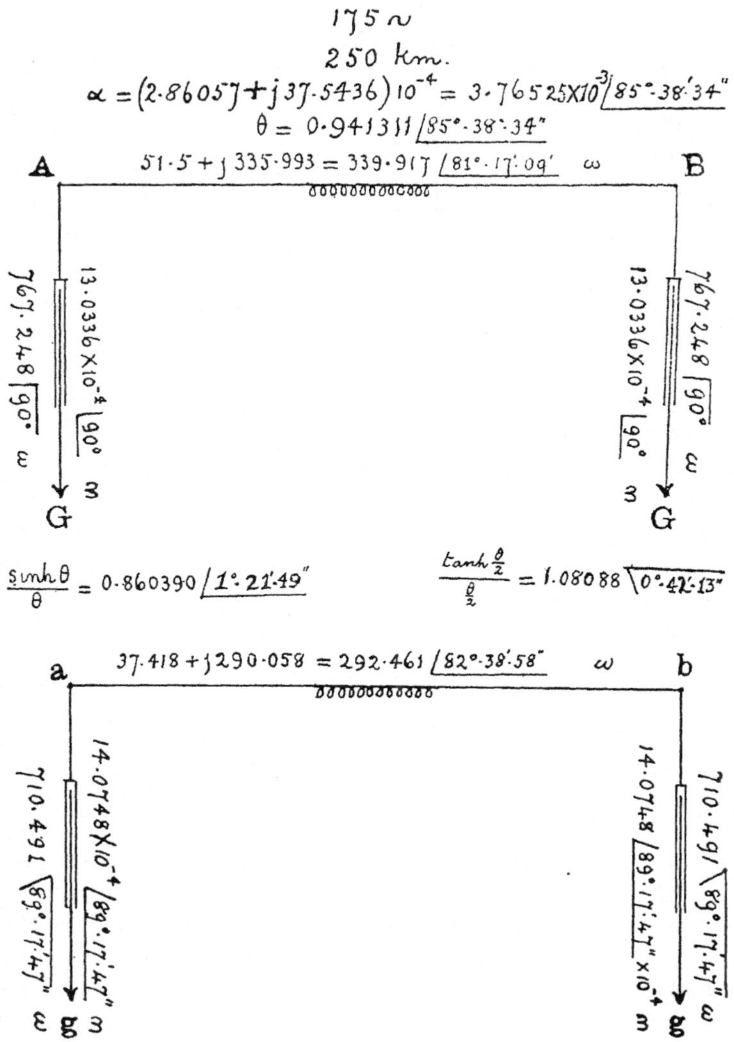


FIG. 53.—Nominal and Equivalent Π for the particular transmission line at 175 ~.

factor $k_{p_{,,}}$ has a modulus of 0·86039, and $k_{g_{,,}}$ 1·08088. The equivalent $\Pi$ has not only less line reactance, but also less line resistance than the nominal $\Pi$, which means that a given current strength delivered over the line dissipates less power in transmission, by reason of the uniform distribution of capacity, compared with that which would be dissipated with the capacity in two terminal lumps.

*Graphical Method of Combining Harmonic Maximum or R.M.S. Values of Voltage or Current into a Resultant Maximum or R.M.S. Value.*—If the impressed e.m.f. at the generating end of the line is impure, and the magnitudes of the various harmonics are known, then, in order to determine completely the distribution of voltage and current over the line, taking distributed capacity into account, it is necessary to compute the equivalent $\Pi$ or $T$ of the line for each harmonic frequency, as well as for the fundamental, to ascertain the impedance which the load at B offers to each frequency respectively, and to compute the voltage-current distribution for each frequency independently, in the manner indicated at A B, Fig. 51. Finally, knowing the components of r.m.s. voltage, or current, at any point in the system, the resultant r.m.s. value is found by the process of "crab addition," or the successive addition of components, each added perpendicularly to the last resultant. Thus, in Fig. 54, let OA be to scale, the r.m.s. value of a fundamental frequency component of voltage, or of current, at a given point in the system, $A_1B$, $B_1C$, $C_1D$, other-frequency r.m.s. components of the voltage, or of the current, at the same point, in any order of selection. In practice $A_1B$ might be a 5th harmonic (quintuple-frequency harmonic), $B_1C$ a 7th harmonic, and $C_1D$ a 13th harmonic, and so on, but the proposition applies equally well to components of any different frequencies, whether harmonic or not. Then, no matter what the relative phases of the different components may be, the resultant r.m.s. value of all the components may be found graphically by adding them rectangularly, and successively, in any order. Thus, taking OA as 1·0, representing say 1000 volts r.m.s. as the voltmeter value of a fundamental e.m.f., associated with a 5th harmonic of 0·333, or 333 volts r.m.s., also with a 7th harmonic of 0·111, or

111 volts r.m.s., also with an 11th harmonic of 0·037 or 37 volts r.m.s., the resultant of all would be Od = 1·0605, or 1060·5 volts r.m.s.

The same reasoning and process evidently applies if each component is expressed in terms of its maximum cyclic, instead

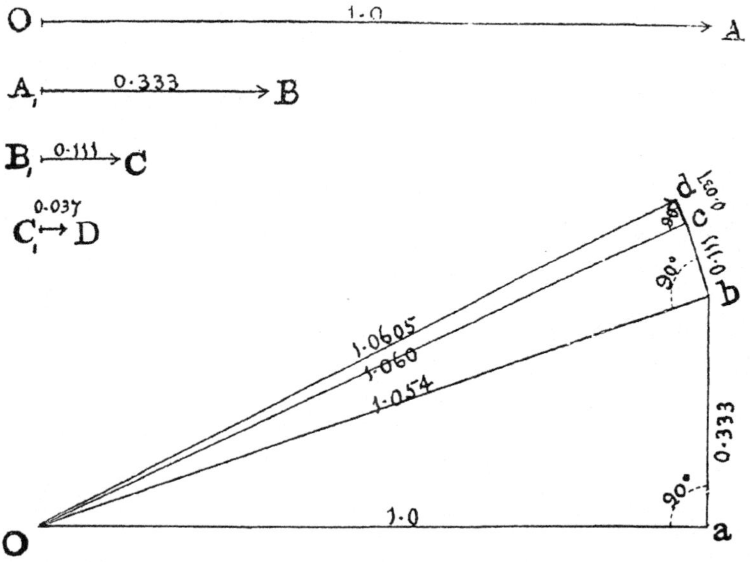


FIG. 54.—Composition of Fundamental and Harmonic Frequency R.M.S. components of voltage or current into a resultant R.M.S. value by the process of "crab addition" or perpendicular summation.

of its r.m.s. value. In such a case, the resultant would also be a maximum cyclic value.

Moreover, if each harmonic component is analysed into two quadrature sub-components, of the sine and cosine type respectively, such sub-components, although of the same frequency, may be included correctly in the rectangular summation process. In other words, quadrature sub-components of any harmonic component of voltage, or current, act, in this particular, as though they had different frequencies. Expressing the same proposition algebraically, if a complex harmonic quantity be analysed into the Fourier series—

$$A + B \sin \omega t + C \sin 3 \omega t + D \sin 5 \omega t + E \sin 7 \omega t +$$
$$+ b \cos \omega t + c \cos 3 \omega t + d \cos 5 \omega t + e \cos 7 \omega t +$$

where the constants A B C, etc., are maximum cyclic values. Then the resultant modulus is well known to be—

$$\sqrt{A^2 + B^2 + b^2 + C^2 + c^2 + D^2 + d^2 + E^2 + e^2 +}$$

which is obviously the value given geometrically by the rectangular summation process. Moreover, since the maximum cyclic value of any single harmonic component is $\sqrt{2}$ times its r.m.s. value, the proposition must be capable of application either to maximum cyclic, or to r.m.s. values, throughout.

Summing up, then, the conclusions reached in this chapter we may say that power-transmission lines of the greatest lengths in ordinary industrial service to-day, operated at ordinary frequencies, do not require correction for distributed electrostatic capacity, if analysed on the basis of the nominal $\Pi$, or split-condenser method, unless great precision is required; because the hyperbolic angle subtended by such lines is usually less than 0·5 in modulus. If, however, high harmonics have to be taken into consideration, the hyperbolic angle subtended by the line may be over 1 hyp., and the correction for distributed capacity in such cases may be material.

In every case where a correction for distributed capacity is required, the simplest method of effecting it is to substitute the equivalent for the nominal $\Pi$ of the line, at the frequency considered.

*Ferranti-Effect.*—The property of an alternating-current power-transmission line to develop a higher voltage at the receiving end than at the sending end, or to possess a negative drop of potential, has been called the Ferranti-effect, having been first pointed out in connection with the Deptford–London power transmission cables in 1890.*

* "Capacity and Self-Induction in Alternate Current Working," by Gisbert Kapp, *The Electrician*, Dec. 19, 1890, p. 197, and Dec. 26, 1890, p. 229. "On the Rise of Electromotive Force observed in the Deptford Mains," by R. T. Glazebrook, *The Electrician*, Dec. 26, 1890, pp. 232–233.

The Ferranti-effect is observable on such lines only at or near no-load. It usually disappears with a very small load on the line. It is commonly supposed to depend upon the influence of distributed electrostatic capacity in the line; but it is produced by the charging current of the line passing through the inductive reactance of the wire, and this charging current may be regarded as due to the total line capacity lumped in a single condenser at the middle point, as in the nominal *T*; or lumped in two split condensers, one at each end, as in the nominal *Π*. The only influence exerted by the distribution of capacity on the Ferranti-effect is that due in detail to the substitution of the equivalent *T* or *Π* for the nominal *T* or *Π*.

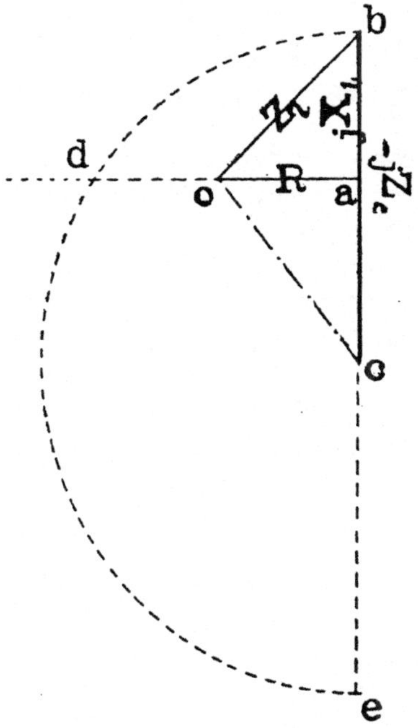


FIG. 55.—Vector Diagram indicating the Limitation of the Ferranti-effect as determined by the line-resistance.

Referring to the nominal *Π* of such a line-wire as is represented at AB, in Fig. 49, let $E_A$ be the r.m.s. vector voltage impressed at A, in the steady state. Let the end B of the line be freed, so that there is no load on the line, and let $E_B$ be the vector r.m.s. voltage developed at B. Then, if $Z = R + jX$ is the vector impedance of the line and $Z_c = -jX_c$ the vector impedance of the semi-line condenser at B, we have, as in the continuous-current circuit—

$$E_B = E_A \frac{Z_c}{Z + Z_c} \quad . \quad . \quad . \quad \text{volts} \angle \quad (200)$$

That is, the B-voltage is the A-voltage multiplied by the vector fraction $\frac{Z_c}{Z + Z_c}$. In Fig. 55, let *ob* represent, to a scale of ohms, the line nominal impedance Z, as the vector sum of the line resistance R and reactance $jX$. Let *bc* be the impedance $Z_c$ to the same scale, of the semi-line condenser at B. Then *oc* will be the vector sum $Z + Z_c$. Consequently, in (200), the B-voltage is the A-voltage multiplied by the ratio $\frac{bc}{oc}$. This ratio will always be greater than unity, if the condenser impedance *bc* is greater than the line reactance *ab*, provided that the line resistance *ca* is less than *da*, the point *d* being on the circle drawn with centre *c* and radius $cb = jZ_c$ ohms. In practice, the line reactance $jX$ is always small compared with the semi-line condenser reactance $jZ_c$; so that the B-voltage in the steady state must exceed the A-voltage at no load; *i.e.* the Ferranti-effect must occur on any normal well insulated line, unless the line resistance R is more than the critical value *da*; or algebraically, unless—

$$R \nless \sqrt{X(2Z_c - X)} \quad . \quad . \quad . \quad \text{ohms} \quad (201)$$

Thus, in the case presented by the nominal $\Pi$, AB, of Fig. 50, where $R = 51{\cdot}50$, $X = 48$, $Z_c = 5370{\cdot}7$ ohms, the Ferranti-effect occurs, and must occur, until R is not less than $\sqrt{48\,(10741{\cdot}4 - 48)} = 714{\cdot}8$ ohms. The equivalent $\Pi$ of this line (*ab*, Fig. 50) differs so little from the nominal $\Pi$, that this deduction is scarcely affected by the uniform distribution of electrostatic capacitance.

Since the linear resistance of all power-transmission lines must be kept relatively low, their line resistances are, in practice, well below the above critical value; so that they exhibit the Ferranti-effect, at no load, almost invariably. But the magnitude of the effect is ordinarily very small, although it increases when the length of the line is increased. Thus, if we repeat the diagram of Fig. 55 to the scale pertaining to the nominal $\Pi$ of Fig. 50, and the 250 km. power wire operated at 25 ∼, we obtain the diagram of Fig. 56, where *ob* is the line-wire impedance of $51{\cdot}5 + j48$ ohms, and *bc* the split-condenser impedance of $-j5370{\cdot}74$ ohms. The vector *oc* is therefore $5322{\cdot}99 \overline{\backslash 89° 26' 44''}$ ohms and the Ferranti-effect factor—

$$\frac{bc}{oc} = \frac{5370{\cdot}74 \overline{|90°}}{5322{\cdot}99 \overline{\backslash 89° 26' 44''}} = 1{\cdot}00897 \overline{\backslash 0° 33' 16''}$$

which means that the voltage at the B-end of the line exceeds the impressed voltage at the A-end by 0·897 per cent., and lags in phase by 0° 33′ 16″. The substitution of the equivalent *Π* for the nominal *Π* of the wire barely affects this result.

If, however, we increase the length of the line; or if, leaving the line-length unchanged, we increase the frequency of operation; then the line reactance *ab* increases, while the capacity reactance *bc* diminishes, so that the Ferranti-effect factor increases. At the same time, the change from the nominal to the equivalent *Π* becomes more marked. Consequently, the Ferranti-effect, which is insignificant on ordinary aerial lines, at ordinary low frequencies, may become very large at extraordinary lengths of line, and especially at upper harmonic frequencies.

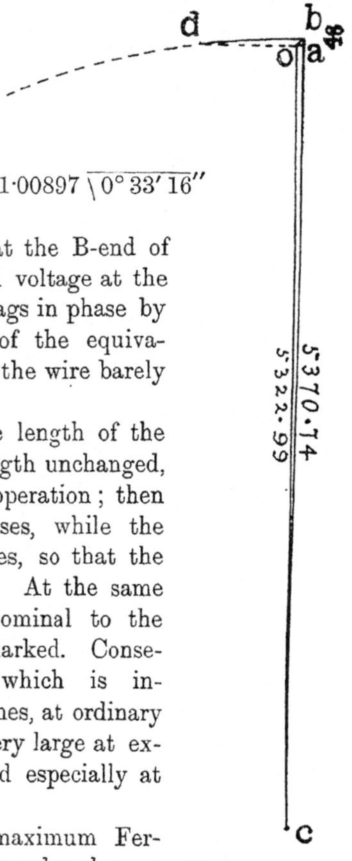


FIG. 56.—Vector Diagram indicating the magnitude of the Ferranti-effect in the case of the particular line at no load.

In order to determine the maximum Ferranti-effect factor that can be produced on a given line, it is expedient to refer to Fig. 42, which shows the successive vector additions to the receiving, or B-end, free voltage, as built up by reflections during the unsteady state It is evident from an inspection of that diagram, that in order to build up the maximum B-voltage it is necessary that there shall not only be small attenuation on the line; so that the successive modulus additions may be large, and numerous; but also that they should arrive in the same phase-direction. This means that the argument of the angle $\theta$ shall be $\beta = \pi/2$ radians, or 90°; because the phases of the successive increments are $-2\beta$ apart. But we have seen that this requires the $j$-component of the hyperbolic line angle $\theta$ to be $\pi/2$, and the wave-length $\lambda$ of the line to be just four times the length of the line. In other words, the line must be a quarter-wave in length. When a line is operated at such a frequency as makes it a quarter wave-length, then the successive voltage increments in the preliminary unsteady state fall vectorially into line with each other, and build up the maximum Ferranti-effect multiplier that the attenuation over the line will permit. On the contrary, if the frequency of operation and line-length are such that the line has half a wave-length, then the successive increments of voltage in the unsteady state arrive in alternate directions and produce a minimum Ferranti-effect factor. The same proposition applies with reduced force to all quarter and all half wave-lengths, as the length of line is increased.

In the case of the 250-km. power-transmission line above considered, it can be readily found that the frequency of $f = 293{\cdot}424 \sim$, with $\omega = 1843{\cdot}64$ radians per second, brings the line into resonance at one quarter wave-length. For the line angle at this frequency is $\theta = 0{\cdot}071650 + j1{\cdot}5708 = 0{\cdot}071650 + j\,\pi/2 = 1{\cdot}5724338 \underline{/87° 23' 18''}$ hyps. The attenuation-constant is also $\alpha = (2{\cdot}86599 + j62{\cdot}832)\,10^{-4}$ hyp. per km. The wave-length is then by (162) $\lambda = 2\pi/(62{\cdot}832 \times 10^{-4} = 1000$ km. and the line-length is just one-fourth of this. The velocity of propagation is also, by (159) $v = \omega/\alpha_2 = 1843{\cdot}64/6{\cdot}2832 \times 10^{-3} = 293424$ km/sec. This frequency is 11·737 times the fundamental frequency of operation, and is, therefore, not exactly a harmonic frequency; but if the fundamental frequency were increased from 25 to 26·675 $\sim$ or by 6·7 per cent., it would then become exactly the 11th harmonic frequency. Under these special circumstances of an 11th harmonic frequency, we should expect to derive the maximum Ferranti-effect possible on this type of line-wire.

The nominal and equivalent *Π* of the line-wire for this resonant frequency are indicated in Fig. 57. It will be noticed that the equivalent line resistance has fallen from 51·5 to 16·435 ohms, ignoring "skin-effect" or extra resistance due to imperfect conductor penetration, which begins to be appreciable at this frequency for the degree of precision under consideration. The line reactance has fallen to $j360{\cdot}31$ ohms; while the semi-line condenser reactance has fallen to $9{\cdot}355 - j359{\cdot}64$ ohms; leaving in circuit a total impedance of $25{\cdot}790 + j0{\cdot}67 = 25{\cdot}7985 \underline{/1^\circ\, 29'\, 24''}$ ohms.

The vector-diagram of this case is presented in Fig. 58. The Ferranti-effect factor is the ratio $\frac{bc}{oc} = \frac{359{\cdot}77 \overline{\;88^\circ\, 30'\, 36''}}{25{\cdot}7985 \underline{/1^\circ\, 29'\, 24''}} =$ $13{\cdot}945 \overline{|90^\circ}$, which means that the B-voltage at the receiving end lags 90° behind the impressed voltage at A, and is 13·945 times as large. This checks the result of formula (22); because $\cosh 1{\cdot}57243 \underline{/87^\circ\, 23'\, 18''} = 0{\cdot}0717109 \underline{|90^\circ}$, and the voltage at the distant free end of the line is $E_A / 0{\cdot}0717109 \underline{90^\circ} = 13{\cdot}945\, E_A \overline{|90^\circ}$.

No rise of voltage nearly so great as 13·945-fold has yet been reported upon any actual transmission line. The conditions are, however, very special, since a 250-km. line, perfectly insulated, is assumed to be operated at the relatively high frequency of 293·424 ~. Nevertheless, tests made in the laboratory with an artificial power-transmission line at a similar frequency have produced a resonant rise of potential of the same order, in good agreement with the results obtained by hyperbolic formulas, and if an actual line of the length and conditions here considered failed to develop so large a Ferranti-effect factor, it would be owing to attenuation and energy dissipation due to extraneous causes omitted from the preceding calculations, such as imperfect insulation, dielectric loss, or the like.

In practice, on actual long lines, no such large Ferranti-effect

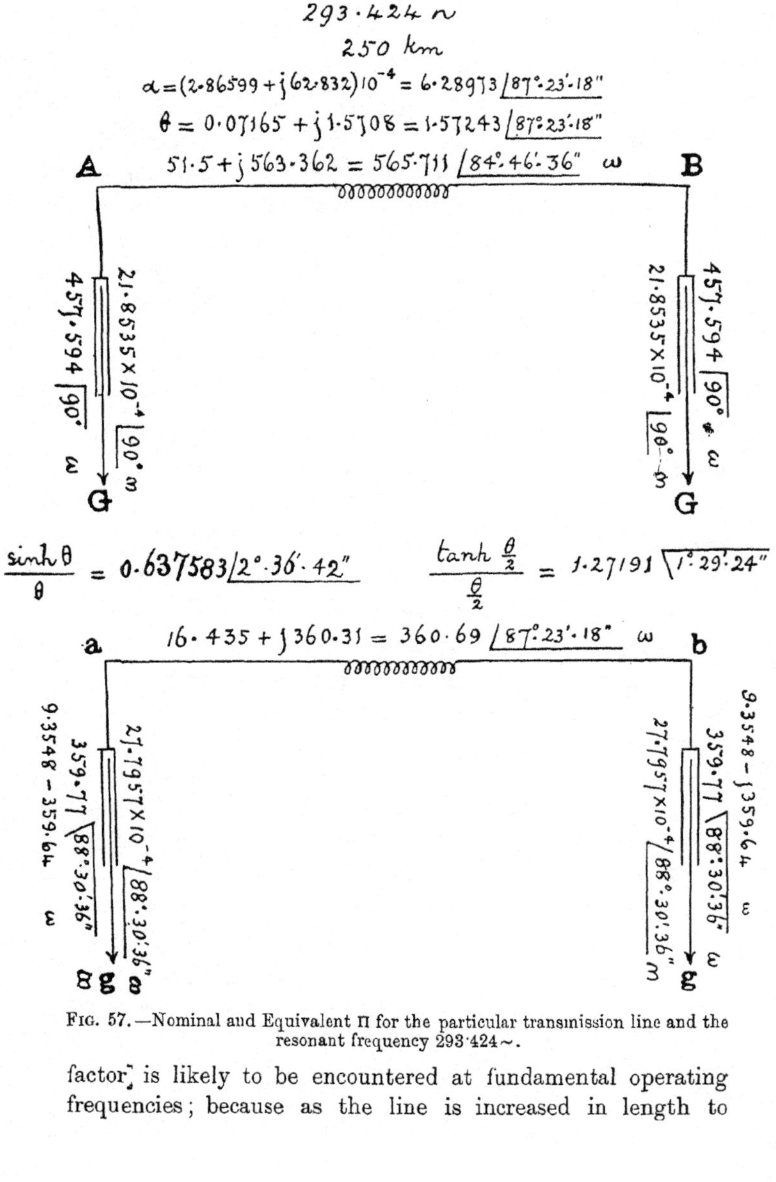


FIG. 57.—Nominal and Equivalent Π for the particular transmission line and the resonant frequency 293·424∼.

factor, is likely to be encountered at fundamental operating frequencies; because as the line is increased in length to develop quarter-wave length, the line resistance increases, and so adds to the base *oc* of the vector-diagram, Fig. 58. Moreover, even if the line had such a length as corresponded to the resonant condition, the effect disappears very rapidly as load is applied at B; because the large resonant rise clearly depends upon a somewhat sensitive adjustment of positive line-reactance in opposition to a nearly equal negative condenser-reactance in the pillar of the equivalent *Π*. Shunting the pillar by a load would tend to destroy the balance, to take the resonant load off the generator, and to keep down the excessive voltage at B. It would seem, therefore, that the danger of resonance could be avoided by keeping the line always under load, either at the distant end, or at intermediate points, or at both. Corona losses along the line would also probably assist automatically in keeping down the excessive potential.

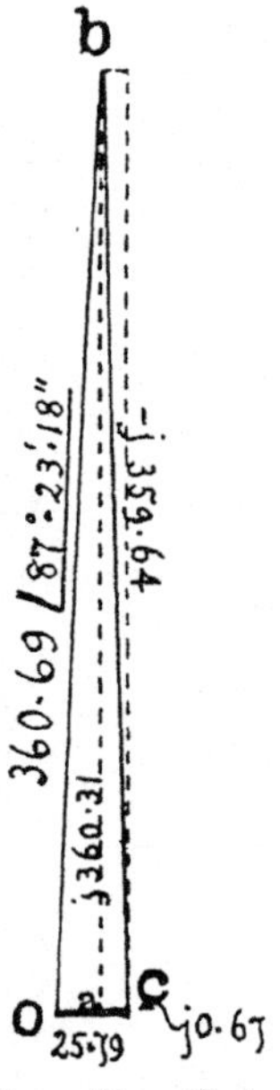


Fig. 58. — Vector Diagram indicating the Ferranti-effect in the case of the particular line at no load and resonance.

But if the generator delivered voltage to the 250-km. line at a fundamental frequency of 26·675 ∼, and happened to possess an appreciable 11th harmonic, say 5 per cent. of the amplitude of the fundamental; then, with no load on the line, there would be a distant-end harmonic component of about 70 per cent. of the fundamental, and the resultant B-voltage would be $\sqrt{1+0{\cdot}7^2} = 1{\cdot}22$ times the A-voltage, or a Ferranti-effect of 22 per cent. due almost wholly to harmonic resonance. This effect would speedily disappear with the application of load. Therefore, when an aerial line of such a length as 250 km., operated at as low a frequency as 25 ∼, displays an evident Ferranti-effect, the presence of an upper harmonic nearly in resonance with the line is to be suspected.

For convenience of reference, in connection with the hyperbolic theory of transmission lines, the data concerning the particular 250-km. wire above considered are collected in the following Tables—

*Table of Fundamental Data.*

| Frequency $f$ ~ | Angular Velocity $\omega$ rad/sec. | Hyp. Angle $\theta$ Hyp. Radians. Modulus | Argument. | Surge-Impedance $z_0$ ohms ∠ Modulus | Argument. | Transmission Velocity $v$ km/sec. | Attenuation-Constant $\alpha_1$ Hyp/km. | $\alpha_2$ Rad/km. |
|---|---|---|---|---|---|---|---|---|
| | | | | | | | $\times 10^{-4}$ | $\times 10^{-4}$ |
| 25 | 157·08 | 0·161914 | /66° 29′ 32″ | 434·80 | \23° 30′ 28″ | 264,487 | 2·58332 | 5·93904 |
| 125 | 785·398 | 0·676039 | /83° 56′ 40″ | 363·083 | \6° 03′ 20″ | 292,070 | 2·85272 | 26·89066 |
| 175 | 1099·556 | 0·941311 | /85° 38′ 34″ | 361·110 | \4° 21′ 26″ | 292,875 | 2·86057 | 37·54362 |
| 293·424 | 1843·635 | 1·572434 | /87° 23′ 18″ | 359·768 | \2° 36′ 42″ | 293,424 | 2·86599 | 62·832 |

*Table of Secondary Data.*

| Frequency. $f$ ~ | Sinh $\theta$. Modulus | Argument. | Cosh $\theta$. Modulus | Argument. | Tanh $\frac{\theta}{2}$. Modulus | Argument. | Ferranti-effect Factor.* Modulus | Argmnt. |
|---|---|---|---|---|---|---|---|---|
| 25 | 0·161431 | /66° 40′ 32″ | 0·9911077 | /0° 33′ 10″ | 0·0810865 | /66° 23′ 17″ | 1·00897 | \0° 33′ |
| 125 | 0·626835 | /84° 53′ 20″ | 0·78566 | /3° 14′ 37″ | 0·351178 | /83° 27′ 55″ | 1·2728 | \3° 15′ |
| 175 | 0·809895 | /87° 00′ 23″ | 0·595244 | /5° 34′ 00″ | 0·5082545 | /84° 56′ 21″ | 1·6800 | \5° 34′ |
| 293·424 | 1·002557 | \|90° | 0·0717109 | \|90° | 1·000 | /85° 53′ 54″ | 13·945 | \|90° |

*Table of Correcting Factors.*

| Frequency. $f$ ~ | $\frac{\text{Sinh } \theta}{\theta}$ | $\frac{\text{Tanh } (\theta/2)}{(\theta/2)}$ |
|---|---|---|
| 25 | 0·99703 /0° 11′ 00″ | 1·0016 \0° 06′ 15″ |
| 125 | 0·92722 /0° 56′ 40″ | 1·03893 \0° 28′ 50″ |
| 175 | 0·86039 /1° 21′ 49″ | 1·08088 \0° 42′ 13″ |
| 293·424 | 0·63758 /2° 36′ 42″ | 1·27191 \1° 29′ 24″ |

* In the case of an actual line in California, 248 km. long, with a total single-wire resistance of 50 ohms, a wire capacity to neutral surface of 2·2 microfarads and a single-wire inductance of 0·323 henry, operated at 60 cycles per second, by an alternator giving a fairly pure sine wave, the observed Ferranti-effect factor was 1·238, as reported by G. Faccioli ("Electric Line Oscillations": *Proc. Am. Inst. Elect. Engrs.*, July 1911, pp. 1621–1668). As the fundamental Ferranti-effect factor on this nominal Π would only be 1·053, the assistance of harmonics is suggested.

*Equivalent T of One Line Wire at Different Frequencies.*

| Frequency. $f$ ~ | Total Wire Impedance. Ohms. | Equivalent Wire Inductance. Henrys. | Staff Impedance. Ohms. | Equivalent Capacity. Microfarads. |
|---|---|---|---|---|
| 0 | 51·50 + *j*0<br>51·50 /0° | 0·30557 | − ∞ | 2·3707 |
| 25 | 51·669 + *j*47·952<br>70·5128 /42° 52′ 50″ | 0·30564 | − 8·649 − *j*2693·3<br>2693·3 /90° 11′ 03″ | 2·3636 |
| 125 | 55·594 + *j*248·88<br>255·014 /77° 24′ 21″ | 0·31688 | − 9·545 − *j*579·15<br>579·23 /90° 56′ 39″ | 2·1985 |
| 175 | 60·065 + *j*362·124<br>367·072 /80° 34′ 56″ | 0·32934 | − 10·610 − *j*445·75<br>445·873 /91° 21′ 49″ | 2·0403 |
| 293·424 | 84·114 + *j*714·594<br>719·528 /83° 17′ 02″ | 0·38760 | − 16·351 − *j*358·48<br>358·849 /92° 36′ 42″ | 1·5131 |

*Equivalent Π of One Line Wire at Different Frequencies.*

| Frequency. $f$ ~ | Total Wire Impedance. Ohms. | Equivalent Wire Inductance. Henrys. | Each Pillar Impedance. Ohms. | Equivalent Line Capacity. Microfarads. |
|---|---|---|---|---|
| 0 | 51·5 + *j*0<br>51·50 /0° | 0·30557 | ∞ | 2·3707 |
| 25 | 51·21 + *j*48·01<br>70·192 /43° 10′ 05″ | 0·30564 | 9·726 − *j*5362·2<br>5362·2 \89° 53′ 45″ | 2·3745 |
| 125 | 44·077 + *j*223·285<br>227·595 /78° 49′ 59″ | 0·28430 | 8·67 − *j*1033·86<br>1033·90 \89° 31′ 10″ | 2·4631 |
| 175 | 37·418 + *j*290·058<br>292·461 /82° 38′ 58″ | 0·26380 | 8·87 − *j*710·44<br>710·491 \89° 17′ 47″ | 2·5603 |
| 293·424 | 16·435 + *j*360·313<br>360·69 /87° 23′ 18″ | 0·19544 | 9·355 − *j*359·65<br>359·77 \88° 30′ 36″ | 3·0103 |

*Ferranti-Effect in Wireless Aerials.*—A very marked Ferranti-effect on a line of very short mechanical length may be found in a wireless aerial conductor free at the top and grounded at the base through a high-frequency alternator. If the frequency of the alternator can be raised to the point at which the aerial is a quarter-wave line, the Ferranti-effect may become very large.

# CHAPTER VIII

## THE APPLICATION OF HYPERBOLIC FUNCTIONS TO WIRE TELEPHONY

HYPERBOLIC functions find a wide field of usefulness in the problems of telephone engineering. This is for the reason that telephone circuits are long, and subtend relatively large hyperbolic angles. Consequently, the equivalent $T$ or $\Pi$ of such a circuit is markedly different from the nominal $T$ or $\Pi$, so different, in some cases, as occasionally to astonish the computer, who would be likely to discredit the results of some calculations, were it not that no discrepancy has yet been detected between hyperbolic theory and actual measurements on telephone circuits; although there is still a very large unexplored territory in both the theory and the measurement of wire telephony.

According to the present theory, a simple telephone circuit consists of a pair of uniform wires, with a generating apparatus at one end, and a receiving apparatus at the other. The transmitter diaphragm at the sending end is thrown into complex vibrations, by the air vibrations incident upon it, as excited by the vocal organs of the speaker. The vocal tones are known to be very varied and complex. They vary in pitch range from about 100 to 10,000 cycles per second, or higher, the ordinary range of acoustic sensibility to pitch being between about 16 and 20,000 cycles per second. It seems that the telephone diaphragm, when executing forced vibrations, in obedience to incident vocal vibrations, responds much more powerfully to some frequencies than to others; so that the resultant vibration of the diaphragm differs considerably, in detail, from the resultant vibration of the air near the diaphragm. Moreover, a simple harmonic pressure delivered by the transmitter diaphragm to the carbon granules in the adjacent microphone may, perhaps, set up a simple harmonic variation of electric resistance in the same; but a simple harmonic variation of an otherwise uniform resistance in a circuit cannot produce a simple harmonic variation in the current flowing through that circuit. It seems therefore probable that some distortion is produced in the transmitted vibrations, both by the mechanical and electrical constraints of the microphone transmitter. Nevertheless, the electromagnetic waves emitted from the transmitting apparatus, after undergoing further electrical distortion along the line, and probably yet further electromechanical distortion in exciting the receiver diaphragm, are still able to convey interpretable sounds to the listener's ear, owing to the intelligent appreciation—more or less trained by habit—of the listener. By reason of this intelligent automatic selection and interpretation, it is fortunately possible to dispense with transmitted sounds of frequency much above 2000 cycles per second.

*Fundamental Assumptions.*—The theory of telephony here presented starts with, and assumes, the terminal sending and receiving apparatus as standard, and deals, in the main, with the electrical phenomena of the line connecting them. Future developments of the theory will probably extend to the characteristics and phenomena of the terminal apparatus.

It is also assumed in the present theory that the essential electrical phenomena in transmission of vocal electromagnetic frequencies along lines are steady-state phenomena. It is assumed, in other words, that whenever a syllable is caught and interpreted by the listener's ear, it has lasted long enough to develop a reasonable number of vibrations, and of underlying electromagnetic waves; so that there will have been a sufficient number of such waves to permit the steady state for that frequency to be approached within the degree of precision required. Thus, a single complete vibration in a component complex vocal vibration entering into a syllable would, it is assumed, be insufficient either to set up the steady electric state for that frequency in the circuit, or to affect the listener's ear intelligibly; but a wave-train of, say, a dozen such vibrations, lasting in all perhaps, at most, only one-tenth of a second, would not only permit a sufficiently close approximation to the steady state for that frequency, but also to affect the listener's ear. Direct experimental demonstration for this important postulate is lacking; but indirect experimental support exists in the sense that our hyperbolic formulas apply accurately to the steady state of alternating-current circuits, and that no reliable discrepancy has yet been brought to light between the acoustic transmission of telephone circuits and the conclusions derived from the hyperbolic theory. Much remains to be investigated in this direction.

On the assumption that telephonic wave-transmission over line conductors is essentially a steady-state phenomenon, it follows that the relative phases of the various frequencies as arriving at the receiver must ordinarily be very different from that existing when leaving the transmitter. According to Helmholtz,* the ear can analyse the complex tone into its constituents independently of their phase relations; whereas, if the syllables contained only a wave or two of some important tone, the phase relation of the accompanying tones might be significant.

*Attenuation-Factor.*—All electromagnetic waves running over a long telephone line are subject to weakening or attenuation; but owing to the sensitiveness of the normal human ear, and to the high degree of perfection of the normal receiving instrument, a large amount of attenuation can be permitted. If we define the ratio of the amplitude of a single-frequency alternating voltage or current, at a given receiving point, to the corresponding amplitude of that voltage or current at a given sending point, as the *attenuation-factor*, or *attenuation-coefficient* between those points; then it appears that an attenuation-factor of 0·05 may be permitted in the most important currents without seriously affecting commercial telephonic transmission over a long line, and that an attenuation-factor of 0·01 may correspondingly occur over such a line before experts may

* *Sensations of Tones*, by Von Helmholtz, chap. vi.

be unable to communicate telephonically with the standard apparatus.

*Distortionless Circuits, and Distortion Ratio.*—When the attenuation-factor of a circuit is the same for all the frequencies of current transmitted telephonically, then the circuit is said to be *distortionless.* In general, however, and especially on cabled conductors, the attenuation is more marked on higher than on lower frequencies; so that the attenuation factor may be only 0·05 at 800 ~ but 0·01, or less, at 2000 ~. This disparity of attenuation over a line is called electrical distortion, as distinguished from electro-mechanical distortion existing in the terminal apparatus. The effect of distortion is manifestly to alter the acoustic character of the sound-waves as repeated in the receiver. The quality or timbre of the voice is altered.

The ratio of the attenuation-factor at a certain higher frequency to that at a particular lower frequency of reference, may be called the *distortion ratio* for that circuit and upper frequency. A certain amount of distortion, and distortion ratio, can be tolerated by the ear, and always exists in wire telephony, over any but the shortest lines; but when the distortion ratio at 2000 ~ falls below a certain value, the telephonic service becomes unsatisfactory, even with good and clear enunciation on the part of the speaker. The voice becomes "drummy" and indistinct. Consequently, the limiting length of a certain type of line over which conversation can be commercially carried on depends not only on the attenuation at the most important telephonic frequency of reference, but also on the distortion ratio between this and the highest important frequency. A lower attenuation factor can be permitted on the standard frequency of reference, if the distortion ratio at the upper end of the essential telephone register is prevented from falling too low.

*Hyperbolic Angles of Telephone Lines.*—The hyperbolic angle of a telephone line, disconnected from its terminal apparatus, is by (19) and (153), for a given frequency, a complex quantity, with a real and imaginary component. The real component is to be intrepreted as a real hyperbolic angle, and the imaginary component as an imaginary hyperbolic angle, or as a real circular angle. The real or hyperbolic component affects the attenuation factor of the line, and the imaginary component the phase of the arriving waves. Telephone-line hyperbolic angles range up to 50 hyps. or more, in modulus.

*Attenuation-Constants.*—The linear hyperbolic angle, or hyperbolic angle per unit-length of a telephone line, is, for a given frequency, a similar complex quantity $a\underline{/\beta} = a_1 + ja_2$, of which the real hyperbolic component $a_1$, affects the attenuation of the waves of that frequency, and the circular component $a_2$ the phase. A wave of current, starting with unit amplitude over the line, from a given point, becomes attenuated, after 1 kilometer, to the amplitude $\varepsilon^{-a} = \varepsilon^{-(a_1+ja_2)} = \varepsilon^{-a_1} \cdot \varepsilon^{-ja_2} = \varepsilon^{-a_1}\overline{\backslash a_2}$. That is, it has shrunk in amplitude, from 1 to $\varepsilon^{-a_1}$, and has retarded in phase $a_2$ radians with respect to the phase of waves starting at the instant of arrival. For this reason, the linear hyperbolic angle of the line is called the *attenuation-constant* of the line. We have already seen that the attenuation-constant has the same numerical value in hyps. per mile or km., whether we take the linear resistance, inductance, capacitance, and leakance per wire-km. or per loop-km. In what follows, we shall use the wire-km. constants consistently.

*Normal and Actual Attenuation-Factors.*—If a line-wire were indefinitely long, the attenuation-factor over a given length L km. would be—

$$k_L = \varepsilon^{-La_1} \quad . \quad . \quad . \quad \text{numeric } \angle \quad (202)$$

But if the line, instead of continuing indefinitely beyond L, stops and connects to terminal apparatus, the reflection of the waves at this terminal load alters, in general, the attenuation-factor. If, however, the impedance of the terminal load happens to be the same as the surge-impedance $z_0$ of the line, then the attenuation-factor of the line is the same as is given in (202) for a length L of an indefinitely continuing line. This value may therefore be called the *normal attenuation-factor* of the length L, to distinguish it from the *actual attenuation-factor* in the presence of a particular terminal load.

Thus, we have by (59) the amplitude of current received through the terminal apparatus—

$$I_{mB} = \frac{E_{mA}}{z_o \sinh L\alpha + z_r \cosh L\alpha} \text{ max. cyclic amperes } \angle \quad (202a)$$

where $E_{mA}$ is the maximum cyclic e.m.f. of the simple harmonic frequency considered, impressed on the sending end of the line at A, $z_o$ is the vector surge-impedance of the line, L is the length of line in km., $\alpha$ the vector attenuation-constant, $L\alpha$ the hyperbolic angle $\theta$ of the line, $z_r$ the vector impedance of the terminal apparatus to ground potential at B; or half the vector impedance of the terminal apparatus between the two wires of the loop line. The equation must be worked out by the rules of two-dimensional arithmetic, or plane-vectors, as explained in Chapter V.

The current at the sending end of the line is also by (56)

$$I_{mA} = \frac{E_{mA}}{z_o \tanh (L\alpha + \theta')} \quad . \quad \text{max. cyclic amperes } \angle \quad (202b)$$

where $\theta'$ is the auxiliary hyperbolic angle subtended by the receiving apparatus to ground when connected to this particular type of line, such that—

$$\tanh \theta' = \frac{z_r}{z_o} \quad . \quad . \quad \text{numeric } \angle \quad (202c)$$

The actual attenuation-factor of the line at this frequency is therefore—

$$k_L = \frac{I_{mB}}{I_{mA}} = \frac{z_o \tanh (L\alpha + \theta')}{z_o \sinh L\alpha + z_r \cosh L\alpha} \quad . \quad \text{numeric } \angle \quad (202d)$$

If, however, the impedance $z_o$ of the receiving apparatus to ground, happens to be identical with $z_o$, the surge-impedance of the line, we obtain—

$$k_L = \frac{I_{mB}}{I_{mA}} = \frac{\tanh (L\alpha + \theta')}{\sinh L\alpha + \cosh L\alpha} = \frac{\tanh (L\alpha + \theta')}{\varepsilon^{L\alpha}} \quad \text{numeric } \angle \quad (202e)$$

At the same time, however, $\tanh \theta' = 1$ or $\theta' = \infty$. Consequently, $\tanh (L\alpha + \theta') = 1$, and we conclude that—

$$k_L = \frac{1}{\varepsilon^{L\alpha}} = \varepsilon^{-L\alpha} = \varepsilon^{-L\alpha_1} \backslash\overline{\alpha_2} \quad . \quad . \quad \text{numeric } \angle \quad (202f)$$

or the actual attenuation-factor becomes the normal attenuation-factor.

*Values of Attenuation-Constants.*—If we consider an aerial telephone line consisting of a pair of No. 10 A.W.G. (American Wire Gauge) copper wires, of diameter 0·2589 cm. (0·1019″), interaxially separated by one foot (30·48 cm.), we may take the following linear constants—

| | |
|---|---|
| $r'' = 10{\cdot}6$ ohms per loop mile. | $r = 3{\cdot}293$ ohms per wire km. |
| $l'' = 3{\cdot}676 \times 10^{-3}$ henrys per loop mile. | $l = 1{\cdot}142 \times 10^{-3}$ henry per wire km. |
| $c'' = 0{\cdot}8018 \times 10^{-8}$ farad per loop mile. | $c = 0{\cdot}9964 \times 10^{-8}$ farad per wire km. |
| $g'' = 0.$ | $g = 0.$ |

Then using formula (150), we have for the attenuation-constant at various frequencies up to 15920 ∼, the data in Table I.

*Table I*

For single-line copper wires No. 10 A.W.G. 0·2589 cm. diam. at interaxial distance of 30·48 cms. $r = 3{\cdot}293$, $l = 0{\cdot}001142$, $g = 0, c = 0{\cdot}9964 \times 10^{-8}$ kilometer units.

| $f$ Cycles per Second. | $\omega$ Radians per Second. | $\alpha$ Vector Attenuation-constant hyp. per Kilometer. | | $\alpha_1$ Real Attenuation-constant hyp. per Kilometer. | $\alpha_2$ Radians per Kilometer. | $\lambda$ Kilometers. | $v$ Kilometers per Second. | % Free Air Velocity. | L¼ Kilometers. |
|---|---|---|---|---|---|---|---|---|---|
| 7·96 | 50 | 0·001281 | 45° 30′ | 0·000898 | 0·0009134 | 6880 | 54,750 | 18·25 | 772·2 |
| 15·92 | 100 | 0·001812 | 46° | 0·001259 | 0·001303 | 4822 | 76,750 | 25·6 | 550·5 |
| 39·80 | 250 | 0·002869 | 47° 29′ | 0·001939 | 0·002114 | 2972 | 118,300 | 39·4 | 357·4 |
| 79·60 | 500 | 0·004080 | 49° 55′ | 0·002627 | 0·003122 | 2012 | 160,154 | 53·4 | 263·8 |
| 159·2 | 1,000 | 0·005892 | 54° 34′ | 0·003417 | 0·00480 | 1309 | 208,330 | 69·4 | 202·8 |
| 200 | 1,257 | 0·006706 | 56° 47′ | 0 003673 | 0·00561 | 1120 | 224,000 | 74·7 | 188·7 |
| 398 | 2,500 | 0·01042 | 65° 28′ | 0·004326 | 0·00948 | 662·8 | 263,700 | 87·9 | 160·2 |
| 796 | 5,000 | 0·01812 | 75° 01′ | 0·004684 | 0·01751 | 358·8 | 285,600 | 95·5 | 148·0 |
| 1,592 | 10,000 | 0·03442 | 81° 57′ | 0·004819 | 0·03408 | 184·3 | 293,430 | 97·8 | 143·8 |
| 5,920 | 100,000 | 0·3374 | 89° 11′ | 0·004859 | 0·3373 | 18·62 | 296,470 | 98·8 | 142·6 |

The Table shows that the real attenuation-constant $\alpha_1$ increases very slowly, beyond the frequency of 400 ∼. Thus, taking the frequency of 796 ∼ ($\omega = 5000$), as the reference frequency, the real attenuation-constant $\alpha_1$ is 0·004684 hyps.

per km., so that at this frequency a wave would diminish in amplitude by $\varepsilon^{-0.004684}$, or 0·47 %, after running one km. At the frequency of 1592 ~ ($\omega = 10{,}000$) or one octave higher, the real attenuation-constant has only increased to 0·004819; so that the normal distortion-ratio is only $\varepsilon^{-0.000135}$ for 1 km. and this octave.

The wave-length $\lambda$ is obtained by formula (162), and becomes shorter the higher the frequency. The velocity of propagation $v$ is obtained by formula (159). As the frequency increases, it approaches the velocity of light in air ($3 \times 10^5$ km./sec.). It falls short of that value; first, because there is some internal inductance within the substance of the wire, and this constitutes a load distributed along the line. Only when a wire has no internal inductance can the velocity of propagation attain that of light in the dielectric; and, second, owing to loss of energy into the substance of the wire, the speed of propagation falls short of the speed of disturbances in the external medium. Only when there is no loss of energy either in the conductor or in the dielectric, can the velocity $v$ attain that of a disturbance in the medium.

The last column gives the semi-amplitude range, or the distance in km. to which the waves can run, at each frequency, before being normally attenuated to one half of their amplitude at the start, as obtained from the equation—

$$\varepsilon^{-x\alpha_1} = 0{\cdot}5 \quad . \quad . \quad . \quad . \quad \text{numeric (203)}$$

or since— $\varepsilon^{-0.69315} = 0{\cdot}5$; $x = 0{\cdot}69315/\alpha_1$ km.

In comparison with the above results, let us consider the attenuation-constant at different frequencies of cabled copper wires, No. 19 A.W.G., 0·0912 cm. in diameter (0·03589″), paper insulated in twisted pairs, with the following linear constants—

| | |
|---|---|
| $r'' = 90$ ohms per loop mile. | $r = 27{\cdot}96$ ohms per wire km. |
| $c'' = 0{\cdot}08 \times 10^{-6}$ farad per loop mile. | $c = 0{\cdot}994 \times 10^{-7}$ farad per wire km. |
| $l'' = 1{\cdot}126 \times 10^{-3}$ henry per loop mile. | $l = 0{\cdot}35 \times 10^{-3}$ henry per wire km. |
| $g'' = 0.$ | $g = 0.$ |

The results are given in the following Table—

*Table II*

For single-line copper wires in twisted pair cables No. 19 A.W.G. 0·0912 cm. diam.

| $f$<br>Cycles per Second. | $\omega$<br>Radians per Second. | $\alpha$<br>Attenuation-constant hyp. per Kilometer. | | $\alpha_1$<br>Real Attenuation-constant hyp. per Kilometer. | $\alpha_2$<br>Radians per Kilometer. | $\lambda$<br>Kilometers. | $v$<br>Kilometers per Second. | %<br>Free Air Velocity. | $L\frac{1}{2}$<br>Kilometers. |
|---|---|---|---|---|---|---|---|---|---|
| 9·95 | 62·5 | 0·01318 | 45° 0′ | 0·00932 | 0·00932 | 674·1 | 6,706 | 2·2 | 74·38 |
| 19·9 | 125 | 0·01862 | 45° 0′ | 0·01317 | 0·01317 | 477·1 | 9,492 | 3·2 | 52·63 |
| 39·8 | 250 | 0·02636 | 45° 06′ | 0·01790 | 0·01796 | 349·8 | 13,920 | 4·6 | 38·72 |
| 79·6 | 500 | 0·03724 | 45° 10′ | 0·02626 | 0·02641 | 237·9 | 18,930 | 6·3 | 26·39 |
| 159·2 | 1,000 | 0·05272 | 45° 22′ | 0·03704 | 0·03752 | 167·5 | 26,650 | 8·9 | 18·72 |
| 397·9 | 2,500 | 0·08335 | 45° 54′ | 0·0580 | 0·05986 | 105·0 | 41,760 | 13·9 | 11·95 |
| 796 | 5,000 | 0·1180 | 46° 47′ | 0·08079 | 0·08602 | 73·04 | 58,130 | 19·4 | 8·58 |
| 1592 | 10,000 | 0·1672 | 48° 34′ | 0·1106 | 0·1253 | 50·14 | 79,810 | 26·6 | 6·26 |

It will be seen that the real component of the attenuation-constant increases with the frequency. Thus, at $\omega = 5000$, $\alpha_1 = 0{\cdot}08079$, and at $\omega = 10{,}000$, $\alpha_1 = 0{\cdot}1106$. The normal linear distortion-ratio or the distortion-ratio for 1 km., and this octave, is therefore, $\varepsilon^{-(0{\cdot}1106 - 0{\cdot}0808)} = \varepsilon^{-0{\cdot}0298}$. After passing over 50 km. (31·07 miles) the distortion-ratio would be $\varepsilon^{-1{\cdot}49} = 0{\cdot}2254$, so that the amplitude of the 10,000 rad-per-sec. waves would be 22·5 per cent. of the amplitude of the 5000 rad-per-sec. waves, assuming normal attenuation, and that they started with the same initial amplitude. This is, in the present state of information, an approximate limit to the distortion-ratio for this octave in satisfactory commercial telephony. That is, if the normal distortion-ratio falls below $1/\varepsilon^{1{\cdot}5}$ for this octave, the articulation is unsatisfactory. Further measurements are needed, however, in this direction.

It should be pointed out that the real attenuation-constant is somewhat larger in the cable of Table II than is obtainable in practice; for the reason that the linear capacity is taken as 0·08 microfarad per loop mile (0·05 $\mu f$ per loop km.), whereas in good practice, such cables are made with a linear capacity of 0·072 $\mu f$ per loop-mile (0·046 $\mu f$ per loop km.) corresponding to 0·092 $\mu f$ per wire-km. or $c = 0{\cdot}92 \times 10^{-7}$. On the other hand, however, the most recent measurements reported in the United States give an effective leakance of $g = 1{\cdot}73 \times 10^{-}$ mho per loop mile ($3{\cdot}46 \times 10^{-6}$ mho per wire mile, $=2{\cdot}15 \times 10^{-6}$ mho per wire km.), and in Great Britain * $5 \times 10^{-6}$ mho per loop mile ($3{\cdot}1 \times 10^{-6}$ mho per loop km) $= 10^{-5}$ mho per wire mile; or $g = 6{\cdot}2 \times 10^{-6}$ mho per wire km. This effective leakance is probably due to dielectric hysteresis aided by the presence of residual moisture, rather than to leakage conductance. In either case, however, it represents loss of energy in the dielectric, increasing as the length, and as the square of the voltage. This small effective leakance tends to increase the real attenuation-constant $a_1$, by reducing the argument of $a$. It is known that a relatively very small amount of moisture resident in the paper dielectric of a lead-covered cable will bring about an increase both of capacitance and leakance. Moreover, paper absorbs moisture so readily that it is difficult to secure and maintain the cable insulating material moisture-free. The changes in $a_1$ effected by the above amendments are, however, relatively small.

In comparisons of electric or acoustic properties of lines, telephone engineers frequently employ as their standard a telephone cable of the following constants. Copper wires No. 19 A.W.G. paper-and-air insulated in twisted pairs, dry, within a leaden sheath. Diameter 0·03589″ (0·0912 cm.).

| | |
|---|---|
| $r''$ 88 ohms per loop mile | = 54·68 ohms per loop km. |
| $l''$ $10^{-3}$ henry per loop mile | = $0{\cdot}6213 \times 10^{-3}$ henry per loop km. |
| $g''$ $5 \times 10^{-6}$ mho per loop mile | = $3{\cdot}107 \times 10^{-6}$ mho per loop km. |
| $c''$ $0{\cdot}54 \times 10^{-7}$ farad per loop mile | = $0{\cdot}3355 \times 10^{-7}$ farad per loop km. |

Referring these to the wire kilometer we have—

$r=27{\cdot}34$; $l=0{\cdot}3107\times10^{-3}$; $g=6{\cdot}214\times10^{-6}$; $c=0{\cdot}6711\times10^{-7}$.

With the preceding data we obtain the following values—

* For these data, the author is indebted to the courtesy of the Engineering Departments of the American Telegraph & Telephone Co., and of the National Telephone Co. Ltd., of Great Britain.

*Table III*

For single-line copper wires in twisted-pair "standard" cable.

| $f$ | $\omega$ | $\alpha$ | | $\alpha_1$ | $\alpha_2$ | $\lambda$ | $v$ | % | $L_{\frac{1}{2}}$ |
|---|---|---|---|---|---|---|---|---|---|
| Cycles per Second. | Radians per Second. | Attenuation-constant hyp. per Kilometer. | | Real Attenuation-constant. | Radians per Kilometer. | Wave-length. Kilometers. | Kilometers per Second. | Free Air. | Semi-amplitude Range. Kilometers. |
| | | Modulus. | Argument. | | | | | | |
| 99·5 | 625 | 0·03405 | $\underline{/40° 54' 44''}$ | 0·02573 | 0·02230 | 281·8 | 28,029 | 9·3 | 26·94 |
| 199 | 1,250 | 0·04796 | $\underline{/43° 17' 19''}$ | 0·03491 | 0·03288 | 191·1 | 38,013 | 12·7 | 19·86 |
| 398 | 2,500 | 0·06776 | $\underline{/44° 48' 03''}$ | 0·04803 | 0·04775 | 131·6 | 52,357 | 17·5 | 14·42 |
| 796 | 5,000 | 0·09586 | $\underline{/46° 05' 44''}$ | 0·06647 | 0·06907 | 90·97 | 72,395 | 24·1 | 10·43 |
| 1592 | 10,000 | 0·13590 | $\underline{/47° 58' 35''}$ | 0·09098 | 0·10096 | 62·24 | 99,051 | 33 | 7·62 |

The following is an example from Table III of the method of computation. It assumes $\omega = 5000$, $r = 27{\cdot}34$, $l = 0{\cdot}3107 \times 10^{-3}$, $g = 6{\cdot}214 \times 10^{-6}$, and $c = 0{\cdot}6711 \times 10^{-7}$.

$$\alpha = \sqrt{(27{\cdot}34 + j1{\cdot}5535)(6{\cdot}214 + j335{\cdot}5)10^{-6}}.$$
$$= \sqrt{(27{\cdot}384 \,\underline{/3° 15' 08''})(335{\cdot}56 \times 10^{-6} \,\underline{/88° 56' 20''}}.$$
$$= \sqrt{9188{\cdot}91 \times 10^{-6} \,\underline{\backslash 92° 11' 28''}} = 95{\cdot}859 \times 10^{-3} \,\underline{/46° 05' 44''}.$$
$$= 0{\cdot}095859 \,\underline{/46° 05' 44''} = 0{\cdot}066474 + j0{\cdot}069066 \text{ hyp. per km.*}$$

$$\lambda = \frac{2\pi}{\alpha_2} = \frac{6{\cdot}2832}{0{\cdot}069066} = 90{\cdot}974 \text{ km.}$$

$$v = \frac{\omega}{\alpha_2} = \frac{5000}{0{\cdot}069066} = 72{,}395 \,\frac{\text{km.}}{\text{sec.}}$$

$$L_{\frac{1}{2}} = \frac{0{\cdot}69315}{\alpha_1} = \frac{0{\cdot}69315}{0{\cdot}066474} = 10{\cdot}427 \text{ km.}$$

It will be seen, from an examination of Table III, that the normal linear distortion-ratios of this line are in four successive octaves, $\varepsilon^{-0{\cdot}009}$, $\varepsilon^{-0{\cdot}013}$, $\varepsilon^{-0{\cdot}018}$, and $\varepsilon^{-0{\cdot}024}$, showing that the distortion increases as the frequency is increased.

* These arithmetical steps, ordinarily taken to a lower degree of numerical precision, are expedited by certain tables of interconversion between a complex quantity $(a + jb)$ and its plane vector $\rho\underline{/\beta}$, which have been completed by Herr Gáti Béla; but not yet published. They are a type of Tables resembling what are called in the Science of Navigation "Traverse Tables."

The attenuation-constant per English statute mile of cable is always obtainable from the attenuation-constant per kilometer, by multiplying each of the components $\alpha_1$ and $\alpha_2$, by 1·60933.

The vector attenuation-constants recorded in Tables I and II are shown graphically in Fig. 59. The origin is at O. The

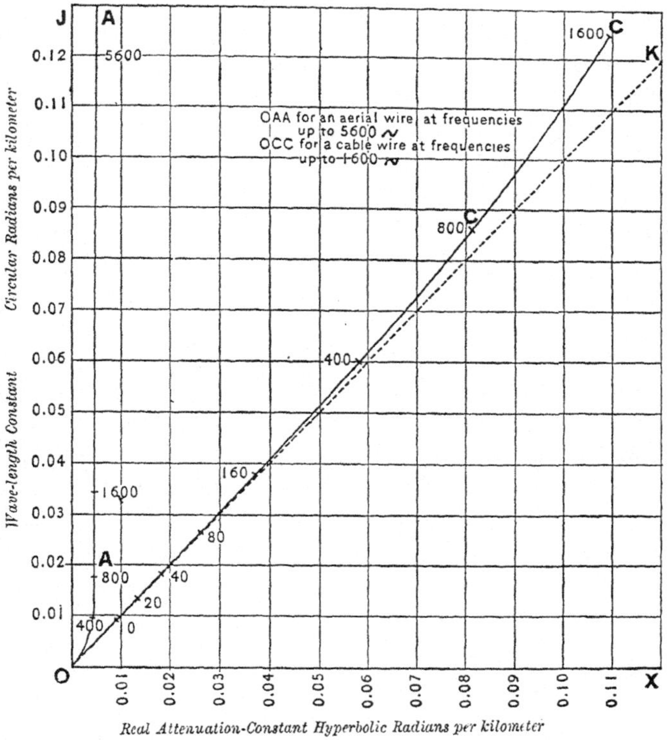


FIG. 59.—Curves showing the Loci of Vector Attenuation-Constants.

curve OCC indicates the locus of the attenuation-constant for the cabled wires of Table II, and OAA the locus of that for the aerial wires of Table I. The real components $\alpha_1$ extend along OX, and the circular components $\alpha_2$ along OJ. The dotted line OK at 45° with each axis, marks the locus of the attenuation-constants in lines of such small linear inductance, with respect to capacity, that the former may be neglected. The small amount of linear inductance in the cabled wires of Table II causes the locus to bend upwards, and to leave the dotted line perceptibly at the frequency of 160 ~ ; while the relatively greater linear inductance and smaller linear capacity of the aerial wires in Table I cause the locus OAA to bend up sharply, and to become nearly parallel to the $j$ axis. The real attenuation-constant $a_1$ is, therefore, roughly the same for all frequencies on the aerial line above 400 ~ ; whereas it continually increases with the frequency in the cabled wires. Thus, not only is the electric distortion at telephone frequencies much more marked on cabled wires than on aerial wires, but the attenuation is also much greater at all telephonic frequencies on ordinary sizes of cabled wires than on ordinary sizes of aerial wires, so that about 50 km. of such cable, as is referred to in Table II, would be approximately the commercial limiting telephonic range, as against about 670 km. of such aerial wire as is referred to in Table I.

*Particular Values of the Attenuation-Constant.*—In the particular case when $l = 0$ and $g = 0$, that is, when both the linear inductance and leakance are negligible, a case closely approached by well insulated cabled wires, the argument of the conductor impedance, $\beta_1 = 0$, and the argument of the dielectric admittance $\beta_2 = 90°$; so that the argument of the attenuation-constant is 45°, or may be called a *semi-imaginary* quantity, having as large a real as imaginary component. In such a case—

$$a_1 = \frac{a}{\sqrt{2}} = \sqrt{\frac{cr\omega}{2}} \quad . \quad . \quad \text{hyps. per km. (204)}$$

Here the real attenuation-constant $a_1$ increases as the square root of the frequency.

When a considerable amount of linear inductance exists, as in aerial lines, the real attenuation-constant $a_1$ tends to a limiting value as the frequency increases. This value is—

$$a_1 = \frac{\frac{r}{2}}{\sqrt{\frac{l}{c}}} = \frac{\frac{r}{2}}{z_{00}} \quad . \quad . \quad \text{hyps. per km. (205)}$$

where $z_{oo}$ is the limiting value of the surge-impedance in (152), or $\sqrt{l/c}$ ohms. Thus, for the aerial line-wire referred to in Table I, $z_o$ tends to the limit 328·6 $\underline{/0°}$ ohms, and $\frac{r}{2} = 1·647$ ohms per semikilometer. Consequently $a_1$ tends to the limit 1·647/338·6 = 0·004864 hyps. per km., as is indicated in Table I.

When an appreciable amount of linear leakance exists, an approximate value for $a_1$ can be obtained by adding a correction factor to (205), thus—

$$a_1 = \frac{\frac{r}{2}}{\sqrt{\frac{l}{c}}}\left(1 + \frac{gl}{cr}\right) \quad . \quad . \quad \text{hyps. per km. (206)}$$

Otherwise, the full formula for the real attenuation-constant is—

$$a_1 = \sqrt{\tfrac{1}{2}\left\{\sqrt{(r^2 + l^2\omega^2)(g^2 + c^2\omega^2)} + (gr - lc\omega^2)\right\}}$$
hyps. per km. (207)

and the imaginary component, or wave-length constant, is—

$$a_2 = \sqrt{\tfrac{1}{2}\left\{\sqrt{(r^2 + l^2\omega^2)(g^2 + c^2\omega^2)} - (gr - lc\omega^2)\right\}}$$
radians per km. (208)

As a general rule, it is easier and more convenient to employ the vector formula (150), which is also readily remembered, than to employ the cumbersome non-vector, or scalar, formulas (207) and (208). Moreover, the results obtained with the scalar formulas are more liable to be vitiated by slight arithmetical errors than if the vector formula (150) is used.

*Surge-Impedances of Telephone Lines.*—The surge-impedance $z_o$ of a telephone line varies considerably with the frequency, as is shown by formula (152). When $r$, the linear conductance of the conductor, and $g$, the linear leakance of the dielectic, are relatively very small; or, in any case, when the angular velocity is high, the surge-impedance approximates to the value—

$$z_{oo} = \sqrt{\frac{l}{c}} \quad . \quad . \quad . \quad . \quad . \quad \text{ohms (209)}$$

and is therefore greater in aerial lines than in cabled lines. This is the impedance * which a line tends to offer to its own surges in the unsteady state. The limiting value of (209) is a pure resistance, or has no reactive component; so that in the limiting case there is no difference in phase between a wave of e.m.f. and its accompanying wave of current travelling along the line. If there be relatively greater conductor reactance, $l\omega$, than dielectric susceptance $c\omega$; so that the argument of the conductor impedance $\beta_1 = \tan^{-1}\left(\frac{l\omega}{r}\right)$, exceeds the argument of the dielectric admittance $\beta_2 = \tan^{-1}\left(\frac{c\omega}{g}\right)$, then the actual surge-impedance has a positive argument $\frac{\beta_1 - \beta_2}{2}$, and the line at and beyond any point behaves like a positive reactance or induction coil. In practice, however, the reverse is the case, and the dielectric admittance argument $\beta_2$ always exceeds the conductor impedance argument $\beta_1$; so that the argument of the surge-impedance is $-\left(\frac{\beta_2 - \beta_1}{2}\right)$, a negative angle, and the line behaves as a condenser associated with a resistance. If the surge-impedance of the line had a positive argument, it would mean that any wave of current travelling along the line would stow away more energy of magnetic form in the dielectric than is stored electrically; while, in practice, the fact that the surge-impedance of either unloaded aerial or unloaded cabled conductors has a negative argument indicates that any wave travelling over the line stows away more energy of electric form in the dielectric than is stored magnetically.

Table IV gives the surge-impedance of the aerial line above considered in Table I for various frequencies between 8 and 16,000 cycles per second, both as a vector, and as a complex number, of ohms per wire. If we consider the surge-impedance of the circuit, we know that it will be twice the surge-impedance per wire (36).

* Dr. Steinmetz has recently suggested the name "natural impedance" for this important quantity. See *Journal of the Franklin Institute*, July 1911, "Electric Transients," by Charles P. Steinmetz, p. 46.

*Table IV*

Initial Sending-end Impedance per Single Wire for a pair of No. 10 A.W.G. Copper Wires (0·1019″ or 0·2589 cm.) inter-axially separated 1 ft. (30·48 cms.) at different impressed frequencies.

| $f$ Cycles per Second. | $\omega$ Radians per Second. | $z_0$ Vector Ohms. Modulus. | Argument. | $z_0$ Complex Quantity. Ohms. | |
|---|---|---|---|---|---|
| 7·95 | 50 | 2570· | − 44° 30′ | 1833· | − $j$ 1801· |
| 15·9 | 100 | 1818· | − 44° 00′ | 1308· | − $j$ 1216· |
| 39·8 | 250 | 1152· | − 42° 32′ | 849· | − $j$ 778·8 |
| 79·6 | 500 | 819·0 | − 40° 05′ | 626·7 | − $j$ 527·4 |
| 159·2 | 1,000 | 591·4 | − 35° 26′ | 481·9 | − $j$ 342·9 |
| 397·9 | 2,500 | 418·4 | − 24° 32′ | 380·6 | − $j$ 173·7 |
| 795·8 | 5,000 | 363·7 | − 14° 59′ | 351·3 | − $j$ 94·02 |
| 1,592· | 10,000 | 345·4 | − 8° 03′ | 342·0 | − $j$ 48·36 |
| 7,958· | 50,000 | 338·8 | − 1° 39′ | 338·6 | − $j$ 9·76 |
| 15,920· | 100,000 | 338·7 | − 0° 50′ | 338·5 | − $j$ 4·93 |

At the infra-telephonic frequency of 7·95 ∼ the surge-impedance of this wire is $2570 \diagdown\overline{44^\circ 30'}$ ohms, and it diminishes to $338{\cdot}7 \diagdown\overline{0^\circ 50'}$ ohms at 15,920 ∼; but undergoes very little change after reaching 800 ∼. The limiting value (209) is, in fact, $338{\cdot}5 \diagdown\overline{0^\circ}$ ohms.

The wire surge-impedances of the cable circuit considered in Table II are also presented for various frequencies in Table V. At 10 ∼ the impedance commences at $2122 \diagdown\overline{45^\circ}$ ohms, and at 15,916 ∼ it has fallen to $67{\cdot}12 \diagdown\overline{19^\circ 19'}$. It is evident that when a first wave of e.m.f. with local amplitude $e$ volts moves along this line, at a frequency of say 796 ∼, it propels a wave of current whose local amplitude is—

$$i = \frac{e}{z_0} = \frac{e}{237{\cdot}4 \diagdown\overline{43^\circ 13'}} = 0{\cdot}004212e \, \underline{/43^\circ 13'} \quad \text{ampere} \ \angle \qquad (210)$$

or a current which leads the local voltage by 43° 13′. The power developed in urging the current forward will be $ei$ cos 43° 13′ watts, and the reactive power developed in storing energy in the dielectric will be $ei$ sin 43° 13′ watts.

*Table V*

Initial Sending-end Impedance per Single Wire for a twisted pair of No. 19 A.W.G. copper wires paper-covered in cable.

| $f$ Cycles per Second. | $\omega$ Radians per Second. | $z_0$ Vector. Ohms. | | $z_0$ Complex Quantity. Ohms. | |
|---|---|---|---|---|---|
| 9·95 | 62·5 | 2122· | − 45° | 1500· | $-j$ 1500· |
| 19·9 | 125· | 1500· | − 44° 58′ | 1061· | $-j$ 1060· |
| 39·8 | 250· | 1061· | − 44° 55′ | 751·2 | $-j$ 749·1 |
| 79·6 | 500· | 750·1 | − 44° 50′ | 531·9 | $-j$ 528·9 |
| 159·2 | 1,000· | 530·4 | − 44° 39′ | 377·5 | $-j$ 372·8 |
| 397·9 | 2,500· | 335·4 | − 44° 07′ | 240·8 | $-j$ 233·5 |
| 795·8 | 5,000· | 237·4 | − 43° 13′ | 173· | $-j$ 162·6 |
| 1,591·6 | 10,000· | 168·4 | − 41° 26′ | 126·3 | $-j$ 111·4 |
| 15,916· | 100,000· | 67·12 | − 19° 19′ | 63·35 | $-j$ 22·2 |

Table VI gives the surge-impedance of "standard" telephone cable, referred to in Table III, for five frequencies—

*Table VI*

Wire Surge-Impedance of "Standard" Telephone Cable.

| $f$ Frequency. Cycles per Second. | $\omega$ Angular Velocity. Radians per Second. | $z_0$ Vector. Ohms. |
|---|---|---|
| 99·5 | 625 | 803·0 $\overline{\backslash 40° 39′ 38″}$ |
| 199 | 1,250 | 570·1 $\overline{\backslash 42° 28′ 29″}$ |
| 398 | 2,500 | 403·6 $\overline{\backslash 43° 10′ 25″}$ |
| 796 | 5,000 | 285·65 $\overline{\backslash 42° 50′ 36″}$ |
| 1592 | 10,000 | 202·47 $\overline{\backslash 46° 29′ 35″}$ |

The graphs of the surge-impedances in Tables IV and V are given, in Fig. 60, at A A A′ for the aerial wire, and at C C C′ for the cabled wire. The abscissas represent the effective resistance, and the negative ordinates the condensive reactance at each frequency.

*Initial Sending-end Impedance.*—It will be evident from what has preceded, in Chapter II, that the surge-impedance $z_0$ of a line is also the initial impedance of that line at the sending end. If the line is indefinitely long; or, is finite, but connected to ground through a terminal impedance of $z_0$ ohms; then no reflected waves return from that end to modify the strength of the outgoing waves; so that the initial sending-end impedance $z_0$ remains, for all time, the final sending-end impedance. In general, however, the reflection of current waves from the

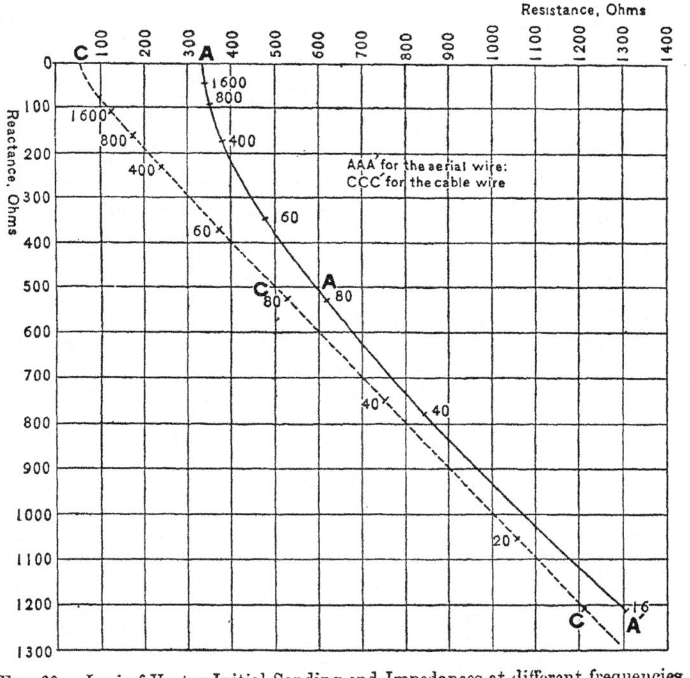


FIG. 60.—Loci of Vector Initial Sending-end Impedances at different frequencies.

distant end alters the outgoing current during the unsteady state; so that the final sending-end impedance becomes changed from $z_0$ to $z_0$ tanh $\theta$, if the line, of angle $\theta$, is directly grounded at the distant end, and from $z_0$ to $z_0$ tanh $(\theta + \theta')$, if the line is grounded through an impedance which subtends with the line an auxiliary hyperbolic angle $\theta'$.

*Transmission over an Indefinitely Long Line.*—We have already seen [(15) (16)] that the pressure and current along a line of uniform electric constants in the steady state are subject to the following conditions—

$$E_P = E_A \cosh L_1\alpha - I_A z_o \sinh L_1\alpha \quad . \quad \text{r.m.s. volts} \angle \quad (211)$$

and $$I_P = I_A \cosh L_1\alpha - \frac{E_A}{z_o}\sinh L_1\alpha \quad \text{r.m.s. amperes} \angle \quad (212)$$

where $E_P$ and $I_P$ are respectively the voltage and current at the point considered, $L_1$ km. (or miles) from the sending end; while $E_A$ and $I_A$ are the impressed r.m.s. voltage and the entering r.m.s. current at the sending end of the line. $z_o$ is the vector surge-impedance and $\alpha$ the vector attenuation-constant of the line at the frequency considered; so that $L_1\alpha$ is the hyperbolic angle of the line, as measured from the sending end to the point considered. When the line is of such great length that the returning waves reflected from the distant end of the line during the unsteady state can be ignored, the final sending-end impedance remains equal to the initial sending-end impedance $z_o$, and the final outgoing current in the steady state remains the same as the initial outgoing current; viz.—

$$I_A = \frac{E_A}{z_o} \quad . \quad \text{r.m.s. amperes} \angle \quad (213)$$

Consequently, for such long lines (211) and (212) become—

$$E_P = E_A(\cosh L_1\alpha - \sinh L_1\alpha) = E_A\varepsilon^{-L_1\alpha}$$
$$= E_A\varepsilon^{-L_1\alpha_1} \backslash \overline{L_1\alpha_2} \quad . \quad \text{r.m.s. volts} \angle \quad (214)$$

$$I_P = \frac{E_A}{z_o}(\cosh L_1\alpha - \sinh L_1\alpha) = \frac{E_A\varepsilon^{-L_1\alpha}}{z_o}$$
$$= \frac{E_A\varepsilon^{-L_1\alpha_1}}{z_o} \backslash \overline{L_1\alpha_2} \quad \text{r.m.s. amperes} \angle \quad (215)$$

Thus the r.m.s. voltage and current, at a distance $L_1$ km. from the sending end, are the normally attenuated values of the initial values impressed at the sending end. The ratio of the voltage and current remains constant at $z_o$ ohms for all points along the line.

As an example, we may consider a circuit length of 100 miles (160·9 km.) of the twisted pair cable of No. 19 A.W.G. copper wires already referred to, subjected to an impressed e.m.f.

of 4 volts, at a frequency of 796 cycles per second. This will correspond to 2 volts on each wire. Table II gives the attenuation-constant at $0{\cdot}118 \underline{/46° 47'} = 0{\cdot}08079 + j0{\cdot}08602$ hyp. per km.; while Table V give the initial sending-end impedance at $237{\cdot}4 \overline{\backslash 43° 13'}$ ohms. The initially outgoing current on each wire will therefore be $\frac{2}{237{\cdot}4 \overline{\backslash 43° 13'}} = 0{\cdot}008425 \underline{/43° 13'}$ amperes; or 8·425 milliamperes, leading the impressed e.m.f. by 43° 13′, or nearly one-eighth of a cycle. Because the cable chosen is so long, and the waves that return reflected from the distant end are so minute, the outgoing current in the steady state has the same strength as the initially outgoing current. At a distance of $L_1 = 30$ miles say (48·28 km.), the hyp. line angle $L_1 a_1$ will be $48{\cdot}28 \times 0{\cdot}08079 = 3{\cdot}901$. The real attenuation-coefficient will be $\varepsilon^{-3{\cdot}901} = \frac{1}{49{\cdot}43} = 0{\cdot}02023$. The voltage will have fallen to $2 \times 0{\cdot}02023 = 0{\cdot}04046$ volt. The current strength will have fallen to $8{\cdot}425 \underline{/43° 13'} \times 0{\cdot}02023 = 0{\cdot}1704 \underline{/43° 13'}$ milliamperes, the current still leading the local voltage by this phase. Both the current and pressure will, however, have been retarded in the transmission by $48{\cdot}28 \times 0{\cdot}08602 = 4{\cdot}153$ radians, or 238°; so that the full expression of voltage and current for the point considered, with reference to the phase of the e.m.f. impressed at the sending end is—

$$E_P = 0{\cdot}04046 \overline{\backslash 238°} \quad \text{volt}$$
$$I_P = 0{\cdot}1704 \overline{\backslash 194° 47'} \quad \text{milliampere}$$

the ratio of which is $237{\cdot}4 \overline{\backslash 43° 13'}$ ohms, or $z_0$. Or we may express the same result by saying that the vector attenuation-coefficient is $\varepsilon^{-L_1 a} = \varepsilon^{-3{\cdot}901} . \varepsilon^{-j4{\cdot}153} = 0{\cdot}0203 \overline{\backslash 238°}$; so that the voltage at P is $E_P = 0{\cdot}0406 \overline{\backslash 238°}$ and the current at P is $= 0{\cdot}1704 \overline{\backslash 194° 47'}$ milliampere.

Table VII gives the voltage and current in each wire of the telephone circuit considered, for varying distances $L_1$ miles from the sending end.

*Table VII*

Hyperbolic Line Angles and Attenuation-Factors for Cable Circuit at the frequency of 796 ∼ or $\omega = 5000$.

| $L_1$ | | $L_1\alpha$ Kilometric Units. | | $\epsilon^{-L_1\alpha} = \epsilon^{-L_1\alpha_1} \cdot \epsilon^{-L_1\alpha_2}$ | | $E_P$ | $I_P$ |
|---|---|---|---|---|---|---|---|
| Miles. | Kilometers. | $L_1\alpha_1$. | $L_1\alpha_2$. | Attenuation-Factor. | Lag. | Volts. | Milliamperes. |
| 0 | 0 | 0 | 0 | 1·0 | 0° | 2·0 | 8·425 $\underline{/43^\circ\ 13'}$ |
| 1 | 1·609 | 0·13 | 0·1384 | 0·8781 | 7° 55′ | 1·7562 $\overline{\backslash 7^\circ\ 55'}$ | 7·398 $\underline{/35^\circ\ 18'}$ |
| 2 | 3·219 | 0·26 | 0·2769 | 0·7711 | 15° 50′ | 1·542 $\overline{\backslash 15^\circ\ 50'}$ | 6·498 $\underline{/27^\circ\ 23'}$ |
| 5 | 8·046 | 0·65 | 0·6922 | 0·5221 | 39° 40° | 1·044 $\overline{\backslash 39^\circ\ 40'}$ | 4·399 $\underline{/3^\circ\ 33'}$ |
| 0 | 16·09 | 1·30 | 1·384 | 0·2725 | 79° 20′ | 0·545 $\overline{\backslash 79^\circ\ 20'}$ | 2·296 $\overline{\backslash 36^\circ\ 7'}$ |
| 20 | 32·19 | 2·60 | 2·769 | 0·07431 | 158° 40′ | 0·149 $\overline{\backslash 158^\circ\ 40'}$ | 0·626 $\overline{\backslash 115^\circ\ 27'}$ |
| 30 | 48·28 | 3·90 | 4·153 | 0·02023 | 238° 0′ | 0·0405 $\overline{\backslash 238^\circ}$ | 0·170 $\overline{\backslash 194^\circ\ 47'}$ |
| 50 | 80·46 | 6·50 | 6·922 | 0·001503 | 396° 36′ | 0·003 $\overline{\backslash 396^\circ\ 36'}$ | 0·013 $\overline{\backslash 353^\circ\ 23'}$ |

It is evident from what has already been considered in connection with normal attenuation, that if instead of finding the current and voltage at different distances along an indefinitely extending line, we cut the line at successive distances, and connect each wire to ground potential through an impedance $z_0 \angle$ ohms, we obtain the same results. Thus, if we cut the line at 30 miles (48·28 km.) from the sending end, and bridged the ends through 474·8 $\overline{\backslash 43^\circ\ 13'}$; or 237·4 $\overline{\backslash 43^\circ\ 13'}$ between each wire and ground, the current through this terminal impedance would be 0·1704 $\overline{\backslash 194^\circ\ 47'}$ r.m.s. milliampere, while the voltage to ground at each wire end would be 0·0405 $\overline{\backslash 238^\circ}$ volt (0·081 $\overline{\backslash 238^\circ}$ volt between the wires), corresponding to normal attenuation at this distance from the sending end.

If, however, instead of grounding each wire through $z_0$ ohms., at the point where we cut the line, we ground through some other impedance; then the steady state will be disturbed by reason of the reflections set up from the terminal impedance during the unsteady state, and formulas (15) and (16) must be used, in order to determine the terminal voltage and current. In general, if the terminal impedance $z_r$ is greater than $z_0$, the current to ground will be reduced, and if $z_r$ is less than $z_0$, the current to ground will be increased, with respect to the current of normal attenuation; but the conditions evidently depend on the argument as well as on the modulus of the terminal impedance used.

In the particular case, however, when the distance $L_1$ from the sending end is so great that the current wave reflected from the grounded end makes no appreciable reappearance at the sending end, we have by (169).

$$I_A = \frac{E_A}{z_0} \quad . \quad . \quad . \quad . \quad . \quad \text{r.m.s. amperes } \angle \quad (216)$$

$$I_B = \frac{E_A}{z_0 \sinh L_1 a} \quad . \quad . \quad " \quad " \quad " \quad (217)$$

$$= \frac{E_A}{\frac{z_0 (\varepsilon^{L_1 a} - \varepsilon^{-L_1 a})}{2}} \quad . \quad " \quad " \quad " \quad (218)$$

$$= \frac{2E_A}{z_0 (\varepsilon^{L_1 a} - \varepsilon^{-L_1 a})} \quad . \quad " \quad " \quad " \quad (219)$$

But when $L_1$ is large, $\varepsilon^{-L_1 a}$ becomes very small by comparison with $\varepsilon^{L_1 a}$, and may be ignored. Consequently—

$$I_B = \frac{2E_A}{z_0 \varepsilon^{L_1 a}} = \frac{2E_A}{z_0} \varepsilon^{-L_1 a} = \frac{2E_A}{z_0} \varepsilon^{-L_1 a_1} \angle \overline{L_1 a_2} \quad \text{r.m.s. amperes} \quad (220)$$

which received current is just double that which flows to ground through a terminal impedance equal to the surge-impedance, or the normally attenuated current at the distance $L_1$. This is for the reason that the effect of grounding a line is to reflect the arriving voltage wave with 180° change of phase, or to annul that wave locally; whereas the arriving current wave is reflected with no change of phase, or is doubled in amplitude. The same proposition applies to a composite telephone line, or line of several different sections in series, provided that the length of the last section is so great that waves reflected from the grounded end do not appreciably disturb the waves as they enter that section on the sending side. That is, if the normally attenuated current at the distant point would be $I_B \angle$ r.m.s. amperes, the current flowing direct to ground at that point will be $2I_B \angle$ r.m.s. amperes.

Fig. 61 is a reproduction of a curve-sheet (Fig. 1) accompanying the paper on "Loaded Telephone Lines in Practice,"

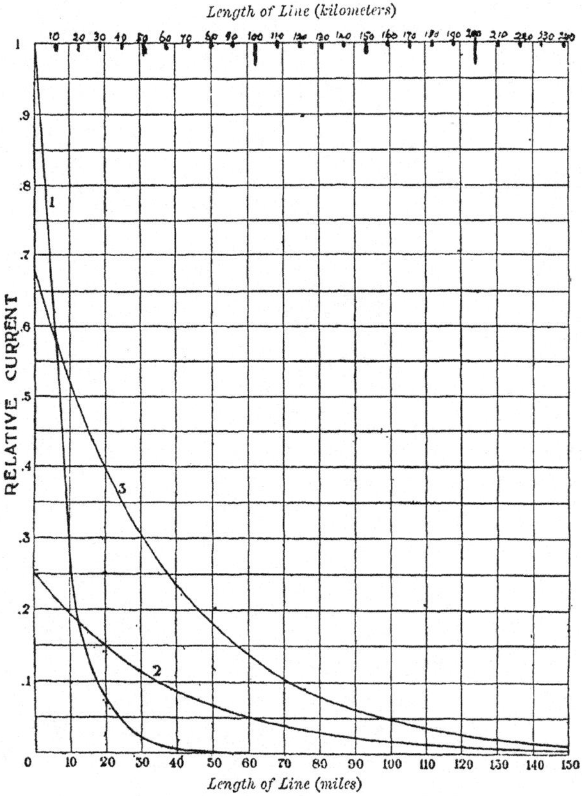


Fig. 61.—Observed Attenuation Factors and Relative Telephone Currents on unloaded and loaded cable circuits.

read by Dr. Hammond V. Hayes before the International Electrical Congress of St. Louis.* Curve 1 gives the observed attenuation-factor on an unloaded telephone cable circuit of

* *Trans. Am. Int. Elect. Congress of St. Louis* (1904), vol. iii. pp. 643, 649.

88 ohms per loop-mile (27·34 ohms per wire-km.) and 0·068 microfarad per loop-mile ($0·0845 \times 10^{-6}$ farad per wire-km.) in actual tests with standard terminal apparatus. It corresponds, therefore, to the resultant attenuation-factor of all the range of frequencies entering into telephonic transmission. If we replot Curve 1 on semi-logarithm paper, *i.e.* paper ruled with ordinary equidistant abscissas, but with logarithmic ordinates, like the distances along a slide-rule, we obtain the wavy line *b b b* (Fig. 62). This corresponds substantially with the straight line 1·0, *b'*. If the attenuation were normal for a single frequency, it would follow such a straight line. Thus the attenuation-factors in Table VII, plotted in Fig. 62, give the broken line 1·0, B. The straight line 1·0, *b'* falls to 0·01 in 36 miles, or 58 km. Consequently the real attenuation-constant $\alpha_1$ on this actual circuit was substantially—

$$\varepsilon^{-58\alpha_1} = 0·01 = \varepsilon^{-4·605}$$

or $\alpha_1 = 0·0795$ hyp. per km. (0·128 hyp. per mile). But there is one and only one simple frequency which would develop this attenuation-constant on such a cable, and it is determined by the semi-imaginary relation—

$$\alpha = \sqrt{(27·34 + j0)(0 + j0·0845 \times 10^{-6}\omega)} = 0·0795 + j0·0795 \quad . \quad \frac{\text{hyp.}}{\text{km.}} \quad (221)$$

$$\sqrt{(27·34 \angle 0^\circ)(0·0845\,\omega \times 10^{-6} \angle 90^\circ)} = 0·1124 \angle 45^\circ$$

$$\sqrt{2·31\,\omega \times 10^{-6} \angle 90^\circ} = \sqrt{0·012634 \angle 90^\circ}$$

or—

$$\omega = \frac{12634}{2·31} = 5469 \quad . \quad . \quad . \quad \text{rad/sec.}$$

corresponding to the single frequency $f = 5469/6·283 = 870·6$ ~. The line therefore behaved substantially as though a single frequency existed in the voice; or as though the pitch of the acoustic vibrations were between ♯*g''* and *a''*, at the top of the treble clef.

Assuming that such a single frequency were impressed on the circuit at the sending end, we have seen that the attenuation would not be normal as the circuit increased in length; because the impedance of the receiving apparatus is not the same as the surge-impedance. The deviations of the curve $b\,b\,b$ (Fig. 62) from the straight line $0b'$ might possibly be explained in this way.

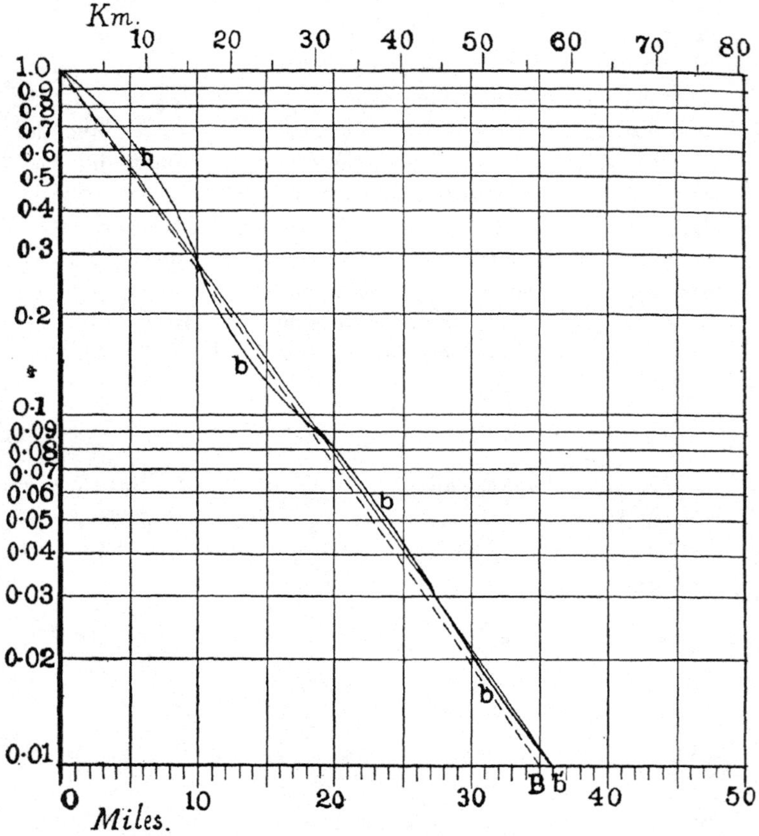


FIG. 62.—Curves showing the observed Telephonic Attenuation-Factor on an Actual Cable as compared with Normal Attenuation for a single frequency. Semilogarithm Paper, Ordinates, Logarithms of Attenuation-Factors.

*Abscissas, Distances from Sending end, Miles and Km.*

From many such measurements, telephone engineers have generally adopted the angular velocity $\omega = 5000$ as the mean angular velocity of telephony, corresponding to the frequency $f = 796 \sim$, which is taken as the standard telephonic frequency. The reason for this remarkable apparent singularity of frequency dominant in a telephone circuit is perhaps connected with the fundamental tone of the transmitter and receiver diaphragms.

Fifty kilometers (31 miles) of such cable is approximately a moderate commercial limiting telephonic range, while Figs. 61 and 62 show that the attenuation-factor is about 0·018 at this range, or $\varepsilon^{-4}$. Consequently, a first approximation to a moderate commercial limiting range on any circuit is given by the formula—

$$\mathrm{L}\alpha_1 = 4 \quad \text{. . . . . . . . hyp.} \quad (222)$$

or—

$$\mathrm{L} = \frac{4}{\alpha_1} \quad \text{. . . . . . . km.} \quad (223)$$

where $\alpha_1$ is the real attenuation-constant at $\omega = 5000$.

Applying this rough rule to the aerial circuit taken in connection with Table I, we find that $\mathrm{L} = 4/0{\cdot}004684 = 786{\cdot}8$ km. (489 miles). In the case of the cable circuit taken in connection with Table II, $\mathrm{L} = 4/0{\cdot}08079 = 49{\cdot}52$ km. (30·77 miles); and in the case of the standard cable circuit of Table III, L= 60·17 km. (37·4 m.). At this range, the normal distortion-ratio for the octave above the standard frequency would be $\varepsilon^{-1{\cdot}06}$ for the aerial circuit, $\varepsilon^{-1{\cdot}476}$ for the cable circuit of Table II, and $\varepsilon^{-1{\cdot}474}$ for the standard cable circuit of Table III. This explains why the articulation is ordinarily better over long aerial lines than over cable lines of these types, when the sound in the telephone is reduced to the commercial limit of volume.

Although the value of $\alpha_1$ for the standard telephonic frequency is, in (222) a criterion of the telephonic limiting range over a given uniform line; yet this can only be a first approximation, because it makes no allowance for the reflection of waves from the terminal receiving apparatus in the unsteady state, and the reflections are different with different types of line. Consequently neither the real component of the hyperbolic angle subtended alone by a line, nor the value of the real attenuation-constant, is an accurate criterion of the telephonic range of the line, even with standard terminal apparatus, and when the line is uniform throughout. Still less are the hyperbolic angles of the sections of a composite telephone line, or their simple vector sum, a proper criterion of the limiting range; because the angle of a line section depends, as we shall see, upon the constants of the sections with which it is connected. Strictly speaking, the criterion of telephonic range is determined by formula (59) or—

$$I_B = \frac{E_A}{z_o \sinh \theta + z_r \cosh \theta} \quad . \quad \text{r.m.s. amperes (224)}$$

where $E_A$ is the r.m.s. voltage of standard frequency impressed on the uniform line at the sending end, $z_o$ is the vector surge-impedance of the line, $z_r$ is the vector impedance of the receiving apparatus, and $\theta$ is the vector hyp. angle of the line, all at standard frequency. If the distortion-ratio of the circuit in the upper necessary frequencies is not too low, the circuit will fail to transmit satisfactory speech only when $I_B$ falls below a certain limit. Taking $E_A$ at an average standard value, this means that the circuit will fail when the receiving-end impedance $z_o \sinh \theta + z_r \cosh \theta$ exceeds a certain value. In practice, it appears that when this impedance exceeds 100,000 ohms per wire, or 200,000 per loop, even expert telephonists are unable to communicate, but when it does not exceed 12,500 ohms per wire, or 25,000 ohms per loop, commercial telephony is readily possible.

The receiving-end impedance of a simple non-composite telephone circuit may, then, be written—

$$Z_l = z_o \sinh \theta + z_r \cosh \theta = z_o \sinh \theta \left(1 + \frac{z_r}{z_o} \coth \theta\right) \quad \text{ohms} \angle \ (225)$$

The proportional increase in the impedance of a circuit due to the receiving apparatus depends thus upon the ratio $z_r/z$ and upon the cotangent of the line angle $\theta$.

As an example, let us consider a "Standard" cable of the type discussed in connection with Tables III and VI, 60·174 km. (37·4 miles) in loop-length, with a terminal receiving apparatus of 100 ohms effective resistance, and 500 ohms effective reactance, at the angular velocity of $\omega = 5000$. The angle subtended by the line alone is $\theta = 5{\cdot}7682\ \underline{/46^\circ\ 05'\ 44''} = 4{\cdot}000 + j4{\cdot}15597$ hyps. (see Table III). The wire surge-impedance at the standard frequency is, by Table VI, $285{\cdot}65\ \overline{\backslash 42^\circ\ 50'\ 36''}$. If the receiving instrument is short-circuited, the receiving-end impedance per wire of the circuit is by (26) and (225)—

$$
\begin{aligned}
z_0 \sinh\theta &= 285{\cdot}65\ \overline{\backslash 42^\circ\ 50'\ 36''}\ \sinh(5{\cdot}7682\ \underline{/46^\circ\ 05'\ 44''}) \text{ ohms.} \\
&= 285{\cdot}65\ \overline{\backslash 42^\circ\ 50'\ 36''} \times 27{\cdot}3214\ \underline{/238^\circ\ 08'\ 12''} \quad \text{,,} \\
&= 7804{\cdot}43\ \underline{/195^\circ\ 17'\ 36''} \text{ ohms,}
\end{aligned}
$$

and the receiving-end impedance of the loop would be double this quantity. Now, removing the short-circuit, and inserting the instrument impedance of $100 + j500 = 509{\cdot}90\ \underline{/78^\circ\ 41'\ 24''}$ ohms into the loop at the receiving end, or $z_r = 254{\cdot}95\ \underline{/78^\circ\ 41'\ 24''}$ ohms to ground potential in each wire, we have, by (225), the total receiving-end impedance per wire—

$$
\begin{aligned}
Z_r &= 7804{\cdot}43\ \underline{/195^\circ\ 17'\ 36''}\left(1 + \frac{254{\cdot}95\ \underline{/78^\circ\ 41'\ 24''}}{285{\cdot}65\ \overline{\backslash 42^\circ\ 50'\ 36''}} \coth\theta\right) \text{ ohms.} \\
&= 7804{\cdot}43\ \underline{/195^\circ\ 17'\ 36''}\ (1 + 0{\cdot}892517\ \overline{\backslash 121^\circ\ 32'\ 00''} \\
&\qquad\qquad \times 1{\cdot}00097\ \underline{/0^\circ\ 02'\ 04''}) \\
&= \quad \text{,,} \qquad \text{,,} \qquad (1 + 0{\cdot}89338\ \overline{\backslash 121^\circ\ 34'\ 04''}) \\
&= \quad \text{,,} \qquad \text{,,} \qquad (1 - 0{\cdot}46769 + j0{\cdot}76118) \\
&= \quad \text{,,} \qquad \text{,,} \qquad (0{\cdot}53231 + j0{\cdot}76118) \\
&= 7804{\cdot}43\ \underline{/195^\circ\ 17'\ 36''} \times 0{\cdot}92884\ \underline{/55^\circ\ 02'\ 03''} \\
&= 7249{\cdot}09\ \underline{/250^\circ\ 19'\ 39''} \quad \text{ohms,}
\end{aligned}
$$

which shows that, in this particular case, the insertion of the receiving instrument diminishes the receiving-end impedance per wire from 7804·4 to 7249·1 ohms; so that inserting this particular receiving apparatus would *increase* the strength of the current received at B by 7·65 per cent. But if the receiving apparatus, keeping an impedance of 254·95 ohms per wire, happened to possess an argument of $\overline{\backslash 42^\circ\ 48'\ 32''}$ instead of $\underline{/78^\circ\ 41'\ 24''}$, the insertion of the instrument would add 89·25 per cent. to the receiving-end impedance, or increase it to 14770

/195° 17′ 36″ ohms per wire (29,540 ohms per loop). Moreover, it is easy to see that the effect of inserting similar receiving apparatus into a cable circuit on the one hand, and into an

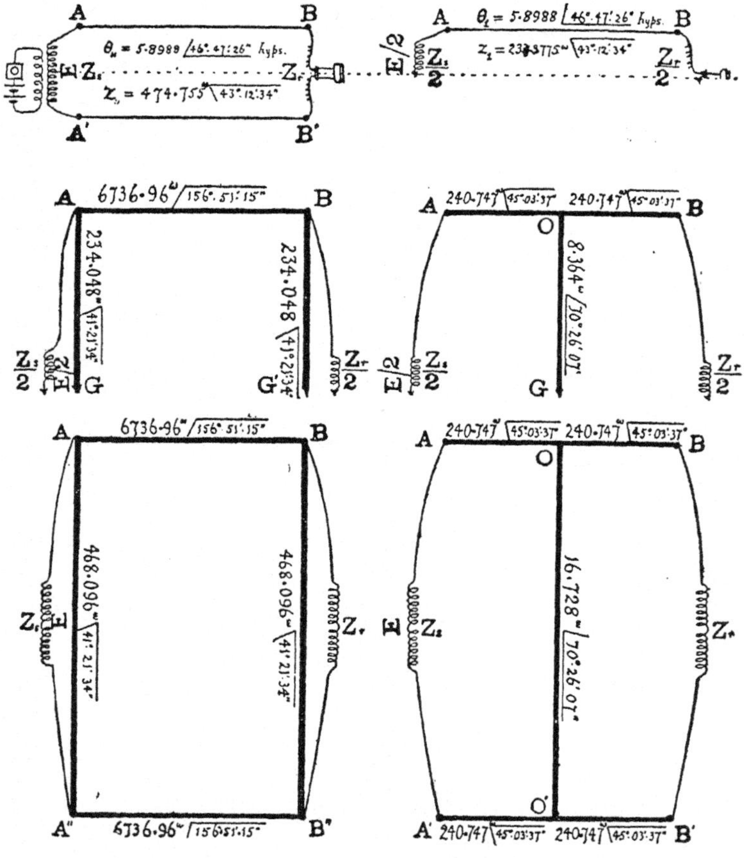

FIG. 63.—Equivalent Circuits of Lines with ground return and metallic return.
$r = 27{\cdot}96$ ohms/*w.km.*; $l = 0{\cdot}35 \times 10^{-3}$ *h/w.km.*; $g = 0$; $c = 0{\cdot}0994 \times 10^{-7}$ *f/w.km.*

aerial circuit on the other, will, in general, make a considerable relative difference in the receiving-end impedance, owing to the difference in argument of $z_0$ for the two circuits. Thus, if we insert the instrument above considered into the receiving end of an aerial line of 786·8 km. of loop-length, and of the constants discussed in connection with Tables I and IV, the effect would be to increase the receiving-end impedance nearly 20 per cent. In both circuits the real component of the line hyperbolic angle would be 4·0 hyps., and their normal attenuation-factors would each be 0·0183; but the insertion of this particular receiving apparatus would change their relative actual attenuation-factors considerably. It should be noticed, moreover, that although the impedance of an ordinary receiving sub-station set has a modulus of about 250 ohms per wire; yet its argument is ordinarily more nearly 30° than 78° 41′ 24″, as above assumed.

*Equivalent Circuits of Telephone Lines.*—The simplest types of fixed-impedance conductors capable of replacing, in all external relations, a given telephone line, at a given single frequency, are, as already observed in relation to (70) and (75*a*), the equivalent $T$ and $\Pi$. As an example, we may take the case of a cabled line 50 km. (31·068 m.) in length, with the following linear constants—

| | Per Loop Mile. | Per Wire Mile. | Per Loop km. | Per Wire km. |
|---|---|---|---|---|
| Ohms . . . | $r'' = 90$ | $r' = 45$ | $r_{''} = 55·92$ | $r = 27·96$ |
| Henrys . . | $l'' = 1·126 \times 10^{-3}$ | $l' = 0·563 \times 10^{-6}$ | $l_{''} = 0·70 \times 10^{-3}$ | $l = 0·35 \times 10^{-3}$ |
| Mhos . . . | $g'' = 0$ | $g' = 0$ | $g_{''} = 0$ | $g = 0$ |
| Farads . . . | $c'' = 0·08 \times 10^{-6}$ | $c' = 0·16 \times 10^{-6}$ | $c_{''} = 0·0497 \times 10^{-6}$ | $c = 0·0994 \times 10^{-6}$ |

We have already seen (35) that we obtain the same value of attenuation-constant, whether we use loop- or wire-constants. With the above values, and $\omega = 5000$, we find—

$$\alpha_{''} = \alpha = 0·1179766 \; \underline{/46^\circ \; 47' \; 26''} \text{ hyp. per loop-km., or wire-km.}$$

$$\theta_{''} = \theta = 5·89883 \; \underline{/46^\circ \; 47' \; 26''} \text{ hyps. for the loop or either wire.}$$

$$z_{0''} = 474·755 \; \overline{\backslash 43^\circ \; 12' \; 34''} \text{ ohms for the loop circuit.}$$

$$z_0 = 237·3775 \; \overline{\backslash 43^\circ \; 12' \; 34''} \text{ ohms for either wire.}$$

The simplest elements of a telephone circuit are indicated at the top of Fig. 63, both for the loop, and for one wire to neutral potential surface.

The equivalent $\Pi$ and $T$ of one line wire are indicated at ABGG′ and AOBG in Fig. 63. The architrave impedance is $6736{\cdot}96/\overline{156^\circ\ 51'\ 15''}$ ohms, which is also the receiving-end impedance of each line, excluding the receiving instrument $z_r$; because, if we ground the line at B, the current which will flow to ground at B will be the impressed potential at A, divided by this architrave impedance.

The equivalent circuits of the loop line are indicated at ABB″A″ and AOBB′O′A′ in Fig. 63. The former is a rectangle of impedances, the latter an $I$ of impedances. It will be seen that the rectangle ABB″A″ is merely a doublet of the single line $\Pi$, ABG′G; while the $I$, AOBB′OA′ is merely a doublet of the single-line $T$, AOBG. The receiving-end impedance of the loop circuit is evidently $2 \times 6736{\cdot}96/\overline{156^\circ\ 51'\ 15''} = 13473{\cdot}92/\overline{156^\circ\ 51'\ 15''}$ ohms, excluding the receiving instrument $z_r$.

Since, then, the equivalent circuits of metallic-circuit, or loop lines, are mere doublets of those for their component single wires, and the latter are easier to think about and discuss, we will confine our attention to the latter.

*Artificial Lines for Telephony.*—It would appear from Fig. 63 that, at the frequency considered, either the single rectangle ABB″A″, or the single $I$ AOB, A′O′B′, is the complete equivalent externally of the actual line. This being the case, it is theoretically unnecessary to employ an artificial line divided into numerous sections to represent the behaviour of the actual line. Either one of these equivalent circuits is sufficient at and for this single frequency. It is to be noted, moreover, that the rectangle is not capable of being constructed in impedances without the aid of transformers, because the argument of the architrave exceeds 90°. The $I$ is capable, however, of being constructed of resistance, inductance, and capacitance, without transformers. As a general rule, either the $T$ or the $\Pi$ of a line is capable of being constructed, sometimes the one, and sometimes the other, according to the length of the line, the linear constants, and the frequency.

Since the $T$ and $\Pi$ of a line vary with the frequency, it is evident that the $T$ or $\Pi$ which represents a line at $\omega = 5000$, fails to represent it correctly for other telephonic frequencies, and that the discrepancy depends for its amount, among other things, on the length of the line. Consequently, it is unsafe to assume that a single-section artificial line—its $T$ or $\Pi$—which happens to represent it correctly at the standard telephonic frequency, represents it adequately for all telephonic purposes. Experiments have been reported to show that 10-mile (16·1 km.) sections of artificial standard cable adjusted to the equivalent $\Pi$ at $\omega = 5000$, satisfactorily imitate such a cable, so far as acoustic behaviour is concerned; but the differences between the $\Pi$ of a 10-mile section at $\omega = 5000$ and $\omega = 10{,}000$ are not very great. Further published research is needed, in critical company with the theory, to show how few sections of artificial line can be used successfully to represent a full-range actual line for all acoustic purposes. The answer to the question seems to depend upon the relative importance and prominence of the upper frequencies.

*Influence of Increasing the Distributed Linear Inductance of a Line.*—As was first pointed out by Heaviside, the effect of distributed inductance in a telephone line is to diminish the attenuation of the telephonic current. The influence of increasing linear inductance on the attenuation-constant is clearly shown geometrically in Fig. 64, where O I is the vector linear impedance of the conductor, $z = r + jx$ ohms per km. The linear admittance of the dielectric, $y = g + jb$ mhos per km. is shown similarly at O A. The argument of $z$ is $\beta_1$, and of $y$, $\beta_2$, radians. The product $zy$, or $\alpha^2$, is then indicated at O B, whose modulus is the product of the moduli $z$ and $y$, and whose argument is $\beta_1 + \beta_2$, the sum of the arguments of $z$ and $y$. At O C is indicated the square root of O B, or the vector attenuation-constant $\alpha$ in hyps. per km., where the argument is half the argument of O B. The real part O$c$ is the hyperbolic component, and the imaginary or $j$-part $c$C is the circular component.

As the linear inductance $l$ of the line is increased, the linear reactance $x = l\omega$ increases proportionately. This has the immediate effect of increasing the modulus or length O I, and also

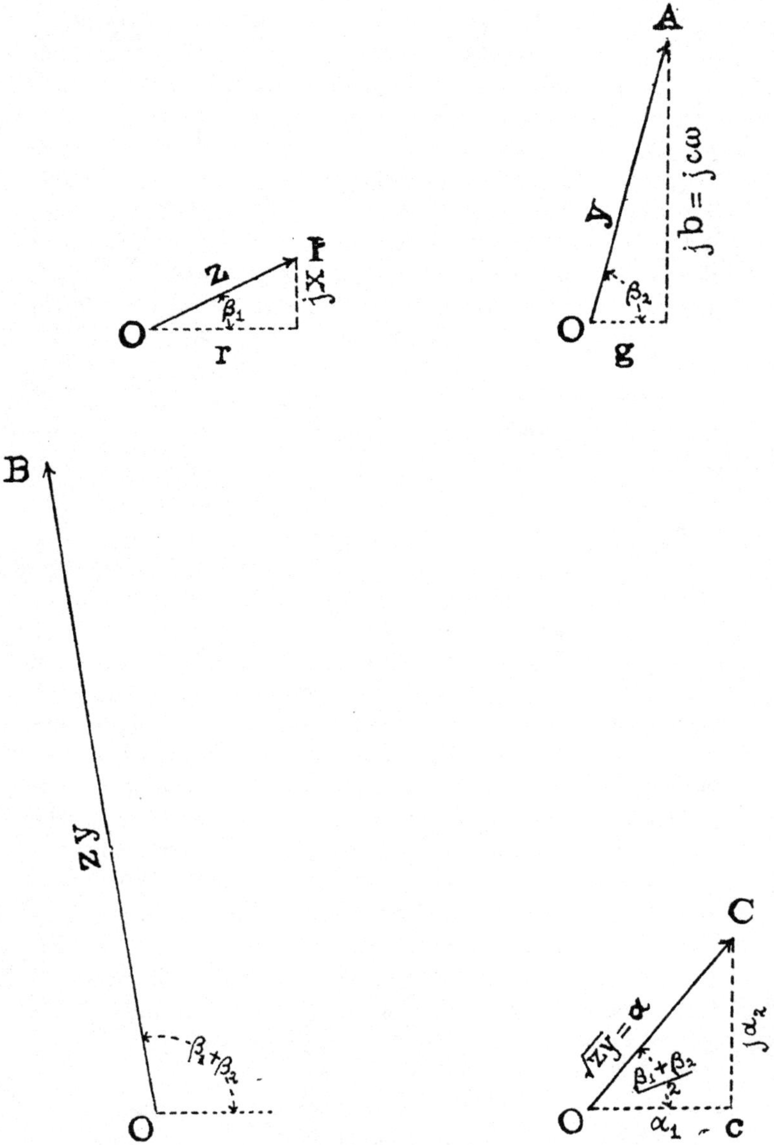


Fig. 64.—Vector Development of an Attenuation-Constant.

of increasing the argument $\beta_1$. The secondary effect of the change is to increase the modulus O C in the square-root ratio, and to increase the argument of O C by half the increase of $\beta_1$. The vector attenuation-constant $\alpha$ is thus somewhat increased; but, owing to the increase in argument, the real component $\alpha_1$ is markedly diminished; while the imaginary component $\alpha_2$ is increased. The increase in $\alpha_2$ merely diminishes the velocity of propagation, and shortens the wave-length; while the reduction in $\alpha_1$ reduces the attenuation.

*Limit to the Reduction in Attenuation with Increasing Distributed Inductance.*—If, as generally happens, there is dissipation of power in the dielectric; so that $g$ is not zero, and the argument $\beta_2$ of O A, Fig. 64, is less than 90°; then there is a limit to the benefit that can be obtained in lessening attenuation by increasing the distributed linear inductance. It can be demonstrated, either algebraically, or geometrically, that increasing $jx$ in Fig. 64 diminishes $\alpha_1$ until $\beta_1$ becomes equal to $\beta_2$. Beyond this critical point, further increase in $jx$ increases $\alpha$ instead of diminishing it.

This result is indicated geometrically in Fig. 65, which contains a direct geometrical construction for $\alpha$, having given the vector linear conductor impedance $oz'$, and the linear dielectric admittance $oy'$. Construct the triangle O$gy$ to the proper scale of linear admittance, with the base O$g$ on the axis OX. Then the angle $y$O$g = \beta_2$. On O$y$ as base, construct the triangle O$rz$, to the proper scale of linear impedance. Then the angle $z$O$y = \beta_1$ and the angle $z$OX $= \beta_1 + \beta_2$. Take O$x$ = O$z$. With centres $x$ and $z$, and equal radii, draw intersecting arcs at A, to bisect the angle $z$OX. Join OA, which is the bisecting line. Through O draw the dotted line $aa$ perpendicular to OA. With centre O, carry circular arcs from $z$ and $y$, to intersect $aa$ at the points Z and Y respectively. On ZY as diameter, construct the semicircle Z$p$Y, intersecting OA in $p$. Then O$p$ is, to scale, the vector attenuation-constant $\alpha$. Its projection O$q$ on the axis OX is $\alpha_1$, the real or hyperbolic component; while $pq$ is $\alpha_2$, the imaginary, or circular component.

As $r'z'$, the linear conductor reactance is increased, by

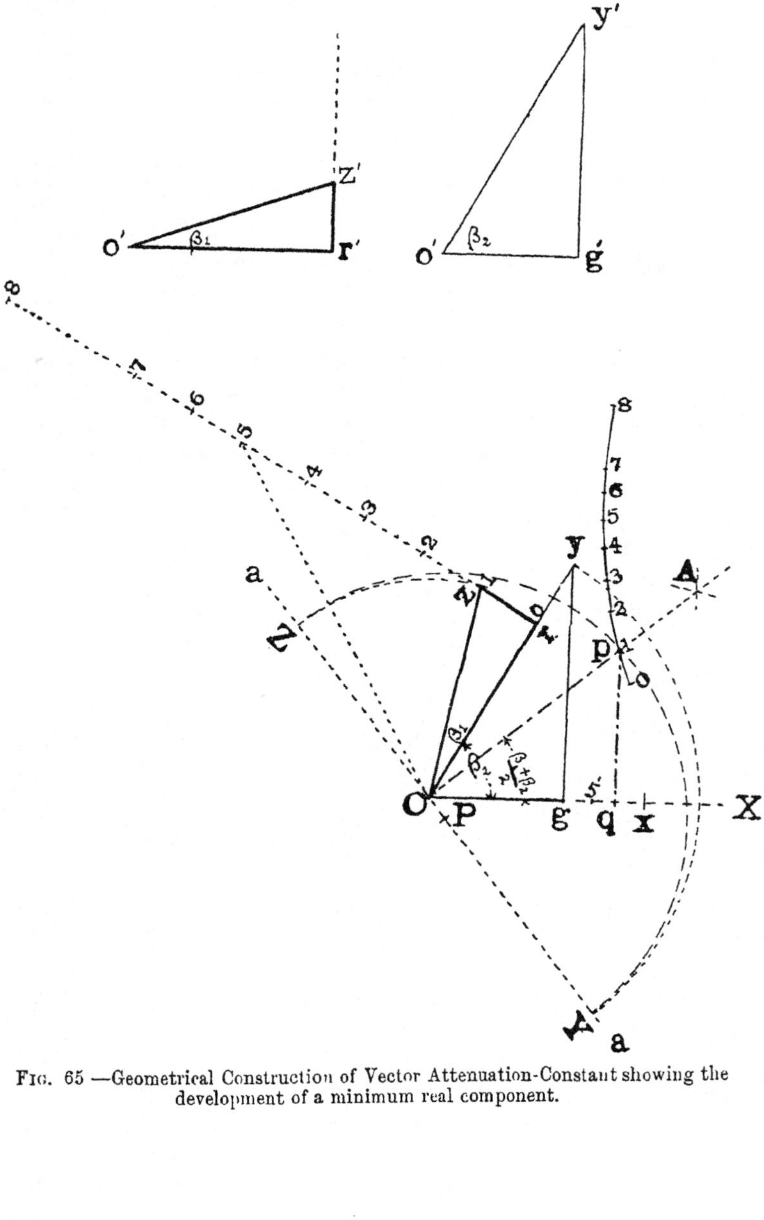


FIG. 65 —Geometrical Construction of Vector Attenuation-Constant showing the development of a minimum real component.

increasing distributed inductance on the line, the triangle $Orz$ extends, in succession, to the points 123 . . . 8, along the dotted line $r8$. By repeating the above-described construction, it will be found that the locus of the vector attenuation-constant pursues the corresponding curved path 012 . . . 8. The minimum real attenuation-constant, $O5'$, occurs at the point 5, where the argument $5Or$ of $\beta_1$ is equal to the constant argument $\beta_2$ of the dielectric admittance.

If, then, there be no dissipation of power in the dielectric, or $g = o$, there will be theoretically no limit to the reduction in $\alpha_1$ obtained by increasing $x$, assuming $r$ constant; although the proportional reduction becomes less and less as $x$ is increased. But when dielectric dissipation exists, no benefit of reduced attenuation is obtained when $\beta_1$ overtakes $\beta_2$; *i.e.* when $\frac{l}{r} = \frac{c}{g}$; or when the time-constant of the conductor is equal to the time-constant of the dielectric. In such a case, the argument of $z_0$ is zero; or the line behaves like a non-inductive resistance to all frequencies, and the real attenuation-constant $\alpha_1$ is the same at all frequencies, so that the normal distortion-ratio is 1 for all ranges and the line is distortionless.

The same proposition would hold, by symmetry, if $\beta_1$ were greater than $\beta_2$. In that case, the increase of linear capacitance $c$ would reduce the real attenuation-constant until the argument $\beta_2$ overtook the argument $\beta_1$. In either case, the maximum benefit is obtainable when $\beta_1 = \beta_2$, both as to minimum attenuation for the frequency considered and as to the same attenuation for all frequencies or absence of distortion. It is also evident that no circuit can theoretically be distortionless which has no dielectric dissipation; because such a perfect dielectric could only make $\beta_1 = \beta_2$ at infinite linear reactance $x$, since $r$ cannot be made zero.

In practice, with actual telephone circuits, $\beta_2$ is always greater than $\beta_1$, which means that inductance has to be added to the line to diminish attenuation and distortion. This is true even for aerial lines, and it is markedly true for cabled lines. Separating the wires of an aerial loop, as suggested by Heaviside, helps to reduce attenuation and distortion, by increasing $l$ while diminishing $c$; but since $l$ only increases as the logarithm of the distance separating the wires, the practical benefit obtainable in this way is comparatively small, and no marked benefit was obtained until the Pupin system was adopted of artificial inductance coils, inserted in the line at suitable regular intervals; *i. e.* until lines became regularly loaded with inductance.

*Loaded Lines.*—A regularly loaded line differs from the same line, with the same total inductance uniformly distributed, owing to the effects of lumping or imperfect distribution. The principal formulas dealing with the effects of regular inductance loading are (107) to (115). From a theoretical standpoint, we may either analyse the behaviour of the loads, as Campbell* has done, by analogy with the propagation of waves along a periodically loaded string,† studied by Godfrey; or by considering line sections replaced by their equivalent $T$'s as in Fig. 20, and (107) (Bibliography, 48); or by grounding the line at the middle of each successive load and comparing the received current in each section with that received over an equal length of smooth line. (Appendix G.) (Bibliography, 29.)

The following particular case may be taken from actual practice, as an example.

A cabled line has the following linear constants—

| | Per Loop-mile. | Per Wire-mile. | Per Loop-km. | Per Wire-km. |
|---|---|---|---|---|
| Ohms . . . | $r'' = 88$ | $r' = 44$ | $r_{,,} = 54{\cdot}68$ | $r = 27{\cdot}34$ |
| Henrys . . | $l'' = 0{\cdot}65 \times 10^{-3}$ | $l' = 0{\cdot}325 \times 10^{-3}$ | $l_{,,} = 0{\cdot}404 \times 10^{-3}$ | $l = 0{\cdot}202 \times 10^{-3}$ |
| Mhos . . . | $g'' = 1{\cdot}73 \times 10^{-6}$ | $g' = 3{\cdot}46 \times 10^{-6}$ | $g_{,,} = 1{\cdot}075 \times 10^{-6}$ | $g = 2{\cdot}15 \times 10^{-6}$ |
| Farads . . . | $c'' = 0{\cdot}072 \times 10^{-6}$ | $c' = 0{\cdot}144 + 10^{-6}$ | $c_{,,} = 0{\cdot}04724 \times 10^{-6}$ | $c = 0{\cdot}08948 \times 10^{-6}$ |

At successive distances of 2·607 km. (1·62 m.), inductance coils of 9·07 ohms effective loop resistance, and 0·1766 henrys loop inductance, are inserted in the cable.

The attenuation-constant of this cable, unloaded, at the standard telephonic frequency ($\omega = 5000$), is—

$$\begin{aligned} a &= \sqrt{(27{\cdot}34 + j1{\cdot}010)(2{\cdot}15 + j447{\cdot}4)10^{-6}} \\ &= \sqrt{(27{\cdot}359 \underline{/2^\circ\ 06'\ 57''})(447{\cdot}4 \underline{/89^\circ\ 43'\ 30''}) \times 10^{-6}} \\ &= 0{\cdot}11064 \underline{/45^\circ\ 55'\ 14''} = 0{\cdot}07697 + j0{\cdot}07948 \text{ hyp. per km.} \end{aligned}$$

* *See* Bibliography, 27. † *See* Bibliography, 19.

The surge-impedance, unloaded, is—

$$z_0 = \sqrt{(27{\cdot}359\ \underline{/2^\circ\ 06'\ 57''})/(447{\cdot}4 \times 10^{-6}\ \underline{/89^\circ\ 43'\ 30''})}$$
$$= 247{\cdot}284\ \overline{\backslash 43^\circ\ 48'\ 16''}\ \text{ohms.}$$

The hyp. angle subtended by a section of 2·607 km., unloaded, is—

$$\theta = 0{\cdot}28843\ \underline{/45^\circ\ 55'\ 14''} = 0{\cdot}20065 + j0{\cdot}20720\ \text{hyps.}$$

From which $\qquad \sinh\theta = 0{\cdot}28831\ \underline{/46^\circ\ 42'\ 52''}$,

and $\qquad \sinh\theta/\theta = 0{\cdot}99957\ \underline{/0^\circ\ 47'\ 38''}$.

The load $\Sigma$ per wire, at $\omega = 5000$, is $4{\cdot}535 + j441{\cdot}5$ ohms.

The semi-load $\sigma$ per wire, at $\omega = 5000$, is $2{\cdot}268 + j220{\cdot}75$

$$= 220{\cdot}76\ \underline{/89^\circ\ 24'\ 41''}$$

Fig. 66 shows, at AB, a section of this line wire before loading, with its length, angle, and surge-impedance. The nominal $T$ of this section is shown at $aob$; where $ao = ob = \frac{\theta}{2}z_0$ ohms, and $o$G, the staff of the $T$, or nominal total admittance, is $\frac{\theta}{z_0}$. The equivalent $T$ of the section is then determined by (74) and (75). It is shown at $a'o$G′. A semi-load $\sigma$ is now added to each arm of the equivalent $T$, producing the extended $T$, A′$o$B′. The arm of this extended $T$ is indicated, at $o$B′, as $37{\cdot}90 + j222{\cdot}56$ ohms. The next step is to revert from the extended $T$, which includes the loads, to the smooth line A″B″, its equivalent by (84) and (86). The angle of one section of loaded line is thus $0{\cdot}0621 + j0{\cdot}7384$ hyps., and the loading has reduced the real, or hyperbolic, component of the line angle, from 0·20065 to 0·0621 hyp.

It is to be observed that on the short section of 2·6 km. here considered, there is very little difference between the nominal $T$, $aob$G, and the equivalent $T$, $a'ob'$G′. The great change is brought about by the addition of $\sigma$ to each arm of the $T$. An examination into the effect of this extension shows that there is a certain amount of reactance in $\sigma$ which the $T$ will stand without unduly increasing the real part of $\theta'$, the equivalent angle of the extended $T$. Beyond this critical value, the real part of $\theta'$ runs up, at first slowly, and then very rapidly, until the real part becomes enormous.

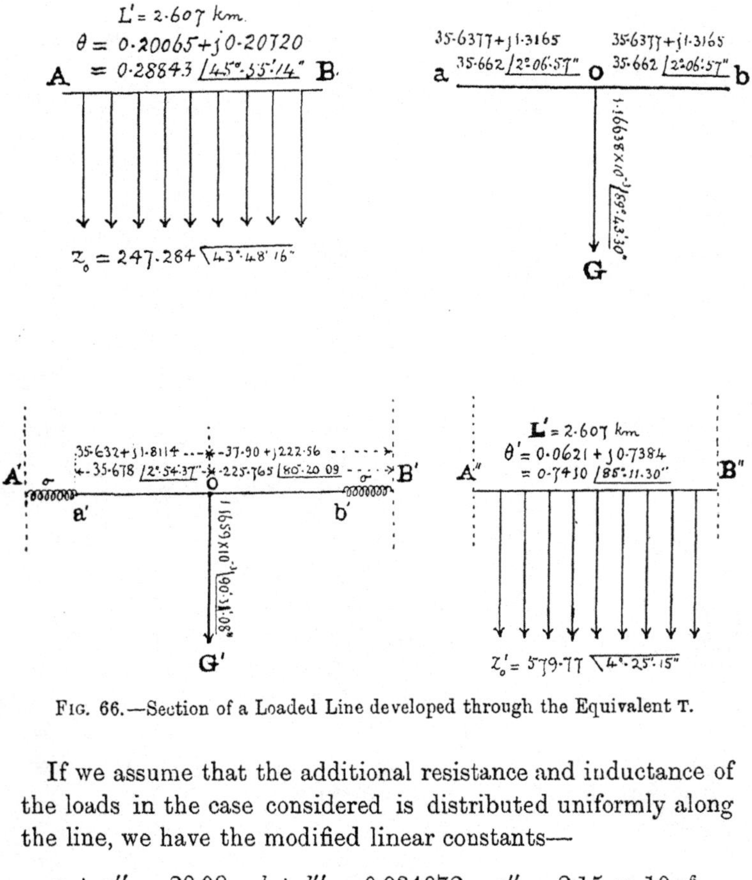


FIG. 66.—Section of a Loaded Line developed through the Equivalent T.

If we assume that the additional resistance and inductance of the loads in the case considered is distributed uniformly along the line, we have the modified linear constants—

$$r + r'' = 29{\cdot}08, \quad l + l'' = 0{\cdot}034072, \quad g'' = 2{\cdot}15 \times 10^{-6},$$
$$c'' = 0{\cdot}08948 \times 10^{-6},$$

and we obtain, for $\omega = 5000$, by (33) or (150) the corresponding smoothed attenuation-constant—

$$a'' = 0{\cdot}02414 + j0{\cdot}27702 = 0{\cdot}27807 \,\underline{/85^\circ\ 01'\ 08''} \quad \text{hyp. per km.}$$

and the smoothed angle subtended by a loaded section from mid-coil to mid-coil—

$$\theta'' = 0{\cdot}06294 + j0{\cdot}72213 = 0{\cdot}72493 \;\underline{/85^\circ\ 01'\ 08''} \text{ hyp.}$$

with a smoothed surge-impedance—

$$z_0'' = 621{\cdot}518 \;\overline{\backslash 4^\circ\ 42'\ 20''} \text{ ohms,}$$

whereas the actual attenuation-constant of the loaded line is—

$$\alpha' = 0{\cdot}02382 + j0{\cdot}28323 = 0{\cdot}28423 \;\underline{/85^\circ\ 11'\ 30''} \text{ hyp. per km.}$$

This means that the real attenuation-constant at $\omega = 5000$, of the loaded line, with its inductance added in lumps every 2·607 km., is less by 1·4 per cent. than it would have been if the same extra resistance and inductance were distributed uniformly. The possibility of this unexpected result was first demonstrated mathematically by Campbell. In general, however, if the lumps of inductance are relatively large, and are too far apart, the reverse condition sets in; namely, that the real attenuation-constant $\alpha'_1$ of the lumpy loaded line is greater than the real attenuation-constant $\alpha_1''$ of the equally loaded smoothed line, and in some cases enormously greater.

We obtain the same results as those above stated, and pointed out in Fig. 65, if we employ the Campbell formula (112) for deriving the actual loaded section-angle $\theta'$, or attenuation-constant $\alpha'$. But whichever formula we select from (107) to (112), for this purpose, we find in it a high degree of sensitiveness. That is, a relatively very small error in the steps of the computation, or in the values of the hyperbolic functions, may involve a considerable error in the result. Consequently, more than ordinary care is necessary in working with these formulas, and a higher degree of precision is needed in the tabular values of sinh $(\theta \angle)$ cosh $(\theta \angle)$ and tanh $(\theta \angle)$ than is ordinarily required. Formula (112) has, however, been carefully investigated by Campbell, who has shown* that for the case of an extra effective resistance in the load coils, amounting to half the unloaded line resistance, and negligible inductance in the

* *See* Bibliography, 27.

unloaded line, there is the following percentage of excess in $a_1'$ over $a_1''$—

| | | | | | |
|---|---|---|---|---|---|
| For $n''$ 3 coils per smoothed wave-length, $a'_1$ is | over 500% greater than $a''_1$. |
| " 4 " " " | 16% greater than $a''_1$. |
| " 5 " " " | 7 " " |
| " 6 " " " | 3 " " |
| " 7 " " " | 2 " " |
| " 8 " " " | 1 " " |

These results are affected to some extent by the ratios $r''/r$ and $l''/l$. With more than 8 coils per smoothed wave-length, the difference between $a'_1$ and $a''_1$ becomes insignificant.

The "smoothed wave-length" is defined by the formula—

$$\lambda'' = \frac{2\pi}{\sqrt{(l+l'')\omega . c\omega}} = \frac{2\pi}{\omega\sqrt{(l+l'')c}} = \frac{1}{f\sqrt{(l+l'')c}} \quad \text{km.} \quad (226)$$

where $l''$ is the extra linear inductance of the loading assumed distributed. Thus, the smooth wave-length in the case considered is—

$$\lambda'' = \frac{1}{795{\cdot}8\sqrt{0{\cdot}034072 \times 0{\cdot}08948 \times 10^{-6}}} = 22{\cdot}758 \text{ km.};$$

so that, since there is one coil per 2·607 km., there are 8·73 coils per smoothed wave-length. In the case considered, the extra linear resistance of the loads instead of being 50 per cent. of the unloaded linear resistance, is only 6·4 per cent. thereof, which changes $a'_1$ to slightly less than, instead of slightly greater than, $a''_1$.

Strictly speaking, the smooth wave-length $\lambda''$ should be obtained by the full formulas (151) and (162); but in the case of a loaded line, $r$ and $g$ become so much smaller than $l\omega$ and $c\omega$ respectively, that the shorter formula (226) suffices. For the same reason, the smoothed velocity of propagation becomes by (159)—

$$v'' = \frac{\omega}{\omega\sqrt{(l+l'')c}} = \frac{1}{\sqrt{(l+l'')c}} \quad . \quad \text{km. per sec.} \quad (227)$$

which is, therefore, to the degree of precision under discussion, constant for all frequencies impressed on the line. The number of coils which advancing (smoothed) waves of any frequency will encounter per second must therefore be—

$$N'' = \frac{v''}{L'} = \frac{1}{L'\sqrt{(l + l'')c}} \quad . \quad . \quad \text{coils per sec. (228)}$$

In the case considered, the actual velocity at $\omega = 5000$, over the unloaded line, is $v = 62{,}910$ km/sec. The actual velocity over the loaded line is $v' = 17{,}654$ km/sec. The smoothed velocity, by (227), is $v'' = 18{,}111$ km/sec. The smoothed number of coils struck per second by advancing waves is then, by (228), $N'' = 6947$.

The number of coils per smoothed wave at any frequency $f$ is—

$$n'' = \frac{\lambda''}{L'} = \frac{N''}{f} \quad . \quad \text{coils per smooth wave-length (229)}$$

Thus, in the case considered—

at $\omega = 5{,}000$ or $f = 795{\cdot}8$, $n'' = 8{\cdot}73$
„ $\omega = 10{,}000$ „ $f = 1591{\cdot}6$, $n'' = 4{\cdot}365$
„ $\omega = 12{,}566$ „ $f = 2000$, $n'' = 3{\cdot}473$
„ $\omega = 15{,}000$ „ $f = 2387{\cdot}4$, $n'' = 2{\cdot}91$

We have already seen, however, that at $n'' = 3$, $\alpha'_1$ is more than six times $\alpha_1''$, which means that there would be heavy attenuation at $\omega = 15{,}000$, and the attenuation commences to rise rapidly at $\omega = 10{,}000$. But it is found in practice that $N'' = 6947$, or roughly 7000 coils struck per second, is a satisfactory condition for commercial telephony. Hence we may consider it demonstrated that $f = 2000$ ($\omega = 12{,}566$) is approximately the highest frequency that has to be preserved for intelligible speech. A loaded cable line is, in fact, a wave sieve, that rapidly damps out and extinguishes currents of frequency higher than $N''/\pi$ cycles per second. (See Appendix G.)

It should be remembered that, in the preceding discussion, we have referred for convenience all our results to $v''$, $N''$, and $\lambda''$, the smoothed-line conditions. Near the limit of $\pi$ coils per smoothed wave-length, the actual wave-length $\lambda'$ shortens considerably, and the actual velocity $v'$, diminishes in like manner; so that there are actually $n' = 2$ coils per actual wave-length, when there are $n'' = \pi$ coils per smoothed wave-length.

In designing loads for a line, it is sufficient, according to the above principles, to provide sufficient extra inductance, assumed uniformly distributed, for the required reduction of the real attenuation-constant $\alpha_1$ to $\alpha_1''$. Then the spacing of the load coils must be such that 7000 are struck by advancing waves per second; or, in other words, that there shall be more than $\pi$ coils per smoothed wave-length $\lambda''$, at the highest frequency (about 2000 ~) which has to be preserved. There will then be $2\pi$ coils per wave at $f = 1000$ ~, $4\pi$ coils per wave at $f = 500$ ~, and so on. For accurate results, however, recourse must be had to some one of the full formulas (107) to (115).

*Effect of Leakage on Loaded Lines.*—It is observed in practice, and is noticeable in the arithmetical theory here under discussion, that when a line is heavily loaded, it is more subject to disturbance from casual extra distributed leakage than when it is in the unloaded state. Accidental leaks along the line, and defective insulation during storms, influence a loaded line more prejudicially than a similar line unloaded. An explanation of this behaviour is found in the relation of a change in the dielectric conductance argument on the attenuation-constant argument in the two cases. Thus, referring to Fig. 64, if, by reason of linear leakance $g$, the argument of $y$ falls from 90° to 88°, the result will be that the argument of $\alpha$ will fall 1°. In an unloaded cable the argument $\beta_1$ of the linear impedance is small, and the argument $\frac{\beta_1 + \beta_2}{2}$ of the attenuation-constant is in the neighbourhood of 45°; whereas in a heavily loaded cable, the argument of the linear impedance is nearly 90°, and that of the attenuation-constant in the neighbourhood of 85°. If now the argument of $\alpha$ for the unloaded line drops from 45° to 44°, the change in $\alpha_1$ is an increase of about 1·7 per cent.; but if the argument of $\alpha$ for the loaded line drops from 85° to 84°, the change in $\alpha_1$ is an increase of nearly 20 per cent. Leakance in a loaded line has, therefore, to be restricted and avoided more carefully than in an unloaded line. (For a more formal demonstration, see Appendix H.)

*Influence of Loading on the Normal Attenuation.*—Since the

immediate effect of loading a line is the same as that of adding distributed inductance, except in so far as the spacing of the loads—*i. e.* the lumpiness—may affect the result, the loading of a line reduces the real part of its hyperbolic angle, and so reduces in the same proportion the exponent $L\alpha_1''$ of its normal attenuation-factor $\varepsilon^{-L\alpha_1''}$. As has already been shown, there is no theoretical advantage obtainable, however, in carrying the loading so far that the argument of the linear conductor impedance $\beta_1''$ overtakes the average argument of the dielectric admittance $\beta_2''$. Moreover, it usually happens that the expense of loading, both in cost of coils and in the cost of space to accommodate the coils, makes it inexpedient to carry the loading up to this limit, especially because the advantage obtainable from any increase in $l''$ is necessarily partly offset by the corresponding accompanying increase in $r''$. That is, economy usually demands that the loading falls considerably short of the theoretical limit $\beta_1'' = \beta_2''$. In the case above considered, for example, the loaded linear conductor impedance is $= 172{\cdot}826 \underline{/80^\circ\,18'\,48''}$; while the loaded linear dielectric admittance is $= 447{\cdot}405 \times 10^{-6} \underline{/89^\circ\,43'\,28''}$.

*Influence of Loading on the Surge-Impedance.*—An important and inevitable effect of loading a line with regular inductances is to increase its surge-impedance of $z_0$. Thus, in the case considered, the surge-impedance of the unloaded circuit was $247{\cdot}284 \overline{\backslash 43^\circ\,48'\,16''}$ ohms per wire; while in the loaded circuit it was $579{\cdot}77 \overline{\backslash 4^\circ\,25'\,15''}$ ohms per wire. But since the receiving-end impedance is by (225), $z_0 \sinh\theta\left(1 + \frac{z_r}{z_0}\coth\theta\right)$, it is evident that, neglecting the impedance $z_r$ of the receiving apparatus, the receiving-end impedance is increased in direct proportion to $z_0$. This means that for very short lines, in which the reduction in $\sinh\theta$ due to the loading has had no opportunity to develop, loading a line makes the received current less instead of greater. This result is indicated in curve 2 of Fig. 61, which shows the received current on a loaded cable, as compared with curve 1 for the same cable unloaded. It is evident that the current received on a loaded cable of very short length is only 25 per cent. of that received on the same cable unloaded, and it is not until the cable is about 19 km. (12 m.) long that the current received over the loaded cable overtakes that received over the unloaded cable. Beyond this distance, the diminished real component of the loaded line-angle more than compensates for the increase in surge-impedance. Nevertheless, it is evident that the increase in $z_0$ is a heavy drawback on the advantage secured by loading the line.

The physical reason for the increase in $z_0$ with loading may be found in the terminal reflections of the waves running over the line in the unsteady or building-up state. It would be a tedious and complicated task to compute these reflections, sum them, and find their total effect; but in the hyperbolic theory, they are automatically and accurately integrated by the change in $z_0$ of the steady state.

Another way of looking at the matter is borrowed from the concepts associated with a power-transmission circuit. The loading of the line with reactance increases the impedance of the line, and diminishes the current flow. This reduces the $I^2R$ loss along the line, and enables the energy to be carried further without being absorbed; but the line has become raised in voltage, and calls for a change in the winding of the generator- and motor-apparatus at the terminals. These should be wound with finer wire, in more numerous turns, so as to generate and absorb less current, but at a higher voltage, than on ordinary unloaded lines. Theoretically, then, if the apparatus remained unchanged in efficiency after rewinding, such a modification would be capable of avoiding the excessive terminal reflections, and of preventing the change in $z_0$ from reducing the received current, by operating on the term $\frac{z_r}{z_0} \coth \theta$ in (220). But even if this plan could be satisfactorily carried out in practice, it would involve the use of two sets of terminal telephone apparatus, one for unloaded and the other for loaded lines, a very objectionable differentiation. A better partial solution of the difficulty has been obtained by the use of "terminal tapers"; *i.e.* graded reductions in the loading near the ends of a loaded line, whereby the terminal surge-impedance is reduced, at some sacrifice in the line-angle. The effect of such terminal papers is shown in curve 3 of Fig. 61 from Hayes's paper. Here the initial current loaded is raised to 67 per cent. of the current unloaded, without appreciable detriment to the gain in attenuation.

Another expedient in extended use at the present time for reducing the effect of the rise in surge-impedance with the loading of a line is the use of terminal transformers. The theory of such transformer reduction in $z_0$ is presented in Fig. 67. At AA′ is a loaded telephone circuit in its simplest elements. At the frequency considered, the loaded line subtends an angle of $\theta$ hyps., and has an excessive surge-impedance of $z_0 \angle$ ohms per wire, or $2z_0 \angle$ ohms per loop. The transmitter operates under an e.m.f. indicated as double, and the receiver has a loop-impedance of $2z_r \angle$ ohms. At each end of the line is a transformer with the higher-tension side to line. Let $\nu$ be the ratio of transformation; *i. e.* the ratio of the e.m.f. generated in the higher-tension winding to that generated in the lower-tension winding. Neglecting magnetic leakage, we know that $\nu$ is the ratio of the turns in the higher- to those in the lower-tension winding. In the presence of actual magnetic leakage, $\nu$ will vary slightly with the frequency.

At BB′ one wire only of the circuit is shown, worked to neutral-potential plane, and the loaded line-wire is shown replaced by its equivalent $T$, with $\rho' \angle$ ohms in each arm, and $y' \angle$ mhos in the staff, in the manner represented in Fig. 66.

At CC′ the two transformers are supposed to have been changed to level transformers, by the imaginary process of removing the higher-tension winding, and replacing it with a winding of the same number of turns as the lower-tension winding, keeping the same volume of copper and of insulation. The e.m.f. induced in these windings will now be reduced in the ratio of $1/\nu$ for the same magnetic flux in the core as before. Nevertheless, the power in the line system will be the same as before, provided that all impedances, both in- and out-side the changed winding, are reduced in the ratio $1/\nu^2$, which means that admittances must be increased, in the ratio $\nu^2$. Let, then, each of the arms of the $T$ be reduced in impedance to $\rho^1/\nu^2 \angle$

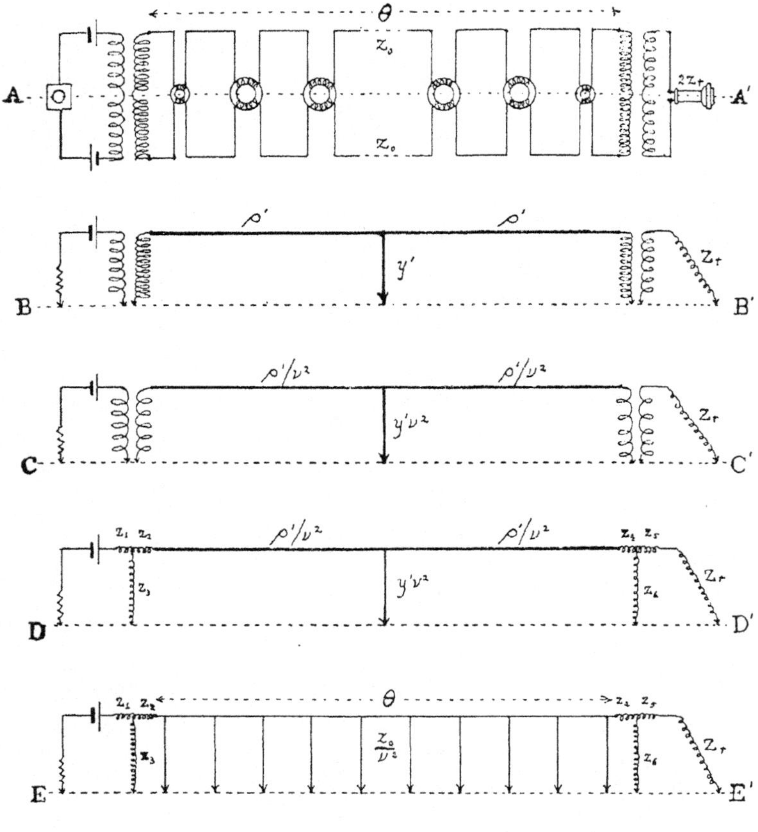


FIG. 67.—Diagrams illustrating Effects of Terminal Transformers in loaded lines.

ohms, and the admittance in the staff be increased to $y^1\nu^2 \angle$ mhos. The impedance inside the coil will be automatically changed in the proper ratio by performing the substitution above described.

A level transformer in any circuit is known to be, in all

respects, equivalent to a conductive connection $z_1 z_2$ plus a leak $z_3$, as shown at DD′. The magnitudes of the impedances $z_1$ and $z_2$ will be the same if the two level coils are symmetrical in all respects, and the magnitude of the leak $z_3$ is determined by the excitation losses of the transformer with its secondary circuit open. Consequently, at DD′, the two transformers have been virtually replaced by an impedance-$T$ at each end of the line, the line itself being modified in the manner indicated.*

Finally, replace the modified $T$ of CC′ and DD′ by its equivalent smoothed line, by (84) and (89). Then, it will be evident that the line-angle $\theta$ will be the same as at AA′ before the conversion, and the new surge-impedance will be $z_0/\nu^2 \angle$ ohms.

The circuit AA′ is therefore equivalent to a circuit without terminal transformers, but with the surge-impedance of the line reduced in the ratio $1/\nu^2$, and with a certain impedance-$T$ injected at each end of the line. The power losses in these $T$'s will offset, to a certain extent, the benefit of the change in $z_0$; but, theoretically, if the terminal transformers had no losses and perfect efficiency, they would secure the required reduction in surge-impedance to the original unloaded value, without any detrimental effects.

Precisely similar reasoning would apply if instead of employing the equivalent $T$ of the line, at BB′, we employed the equivalent $\Pi$.

*Composite Lines.*—A conducting line formed of two or more successive sections, each section having its own length, and uniformly distributed constants, may be called a *composite line*, as distinguished from a simple uniform line, which may be called a *single line*. A composite line is, therefore, made up of a series of successive single lines. In practical telephony, most long lines are composite, since in every large city the wires must go underground, and interurban lines are ordinarily aerial lines. Consequently, a very simple circuit, connecting two subscribers in different cities, would be a three-section composite line, consisting of a central aerial section, and two terminal underground sections. In practice, a composite line may

* *See* Bibliography, 16, 17, and 21.

include many sections. If the hyperbolic theory of telephony is to have general and useful application, it must embrace composite lines with a satisfactory degree of simplicity. As the theory is extensive, has already been worked out to a considerable extent,* and is likely to be worked out much further in future, only an abstract can be given here.

An obvious method of dealing arithmetically with a composite line, in order to arrive at a quantitative knowledge of its properties, is to find either the equivalent $T$ or the equivalent $\Pi$, for each successive section separately, connect these equivalent sectional conductors together, in the proper series order, and compute an equivalent $T$ or $\Pi$ of the combination, by repeated use of the star-delta theorem (Appendix E). Such a final equivalent $T$ or $\Pi$ may be called a *merger* $T$ or $\Pi$; because it is arrived at by merging successive $T$'s or $\Pi$'s. It is always possible, theoretically, to arrive at the merger $T$ or $\Pi$ of a composite line in this manner; although, in the case of a line of many sections, the process is long, tedious, and liable to arithmetical mistakes. In general, the final $T$ or $\Pi$ of a composite line is *dissymmetrical*; whereas the equivalent $T$ or $\Pi$ of a single line is *symmetrical*. That is to say, the equivalent $T$ of a composite line has its two arms unequal, and the equivalent $\Pi$ of a composite line has its two pillars unequal. Any composite line, no matter how numerous may be its component sections, and no matter how many casual loads may be applied to it, at its junctions, either as impedances inserted in the line, or as leaks to ground, must be capable of representation by one and only one $T$, or by one and only one $\Pi$; † so that one

* *See* Bibliography, 52 and 61.

† If we consider a composite telephone line as having a dissymmetrical $\Pi$, then referring to Fig. 63, it is evident that, in general, such a line should present some telephonic dissymmetry; because the pillars of the $\Pi$, acting as shunts to the terminal apparatus, are unequal at the two ends of the line. Consequently, a markedly dissymmetrical line, having, say, a long cabled section at one end and a long aerial section at the other might be expected to show some dissymmetry in telephonic operation. Although such dissymmetry has long been known in telegraphy, over dissymmetrical composite lines worked at high speeds with Wheatstone apparatus, yet the condition of dissymmetrical telephony does not seem to have been reported in any publication.

volt applied at each end, in turn, will send the same strength of current to ground at the other end. This proposition assumes that the composite line is connected directly to ground at the receiving end; also that all leaks are of constant resistance and are devoid of any variable e.m.f. such as might be caused by polarization.

It is easy to demonstrate that, by the use of hyperbolic position-angles, assigned by definite law to the successive junctions of a composite line, the resultant distribution of potential, current, impedance and power over the line, as well as the final equivalent $T$ or $\Pi$ of the composite line, may be determined by relatively simple single formulas, that involve much less time and labour to work out than do the successive steps of the merging process. The $T$ or $\Pi$ computed by hyperbolic trigonometry may be called the *hyperbolic T or Π*, in contradistinction to the *merger T or Π*. Nevertheless, the hyperbolic method requires frequent references to tables of functions of complex hyperbolic angles, and is only a swift method by reason of the existence of such tables. When such tables are not available, the labour of the hyperbolic method becomes increased by the labour of computing the needed sines, cosines and tangents, or their inverse functions; so that in the existing absence of proper tables, the merger method is usually less onerous than the hyperbolic method. Suitable tables of complex hyperbolic functions are not yet available, but are in process of formation; so that the hyperbolic theory may advantageously be studied, even though, for the present, its application may have to be deferred.

### FIRST CASE.

*Sections of the Same Attenuation-Constant and of the Same Surge-Impedance.*—If a line AB (Fig. 68) of $L_1$ km. is connected to a line CD of $L_2$ km., and each has the same attenuation-constant $\alpha$, and the same surge-resistance $z$ ohms * (conditions which imply the same linear constants), the line-angles will be

* We abbreviate $z_0$ to $z$, and $y_0$ to $y$, for convenience, when discussing that branch of the theory which relates to composite lines.

$\theta_1 = L_1\alpha$ and $\theta_2 = L_2\alpha$ hyps. respectively. Then, if we free the composite line at D, the resistance at A is—

$$R_f = z \coth (\theta_1 + \theta_2) \quad . \quad . \text{ ohms (230)}$$

while, if the composite line be grounded at D, the resistance at A is—

$$R_g = z \tanh (\theta_1 + \theta_2) \quad . \quad . \text{ ohms (231)}$$

Reciprocally, freeing and grounding the composite line at A, we get resistances $R_f$ and $R_g$ at D, respectively, of the same values as in (230) and (231).

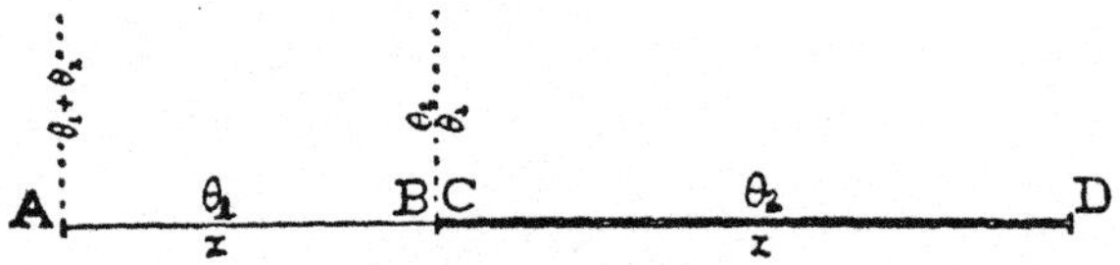


FIG. 68.—Composite Line with sections of the same Attenuation-Constant and Surge-Resistance

It is evident, then, that the composite line differs in no way, except in length, from either of the component sections. The angle subtended by the whole line AD is the sum of the component section line-angles.

### SECOND CASE.

*Sections of Different Attenuation-Constant but of the Same Surge-Impedance.*—If a section CD (Fig. 68) of $L_2$ km. be connected to a section AB of $L_1$ km., and their respective linear constants $r_2$, $g_2$ and $r_1$, $g_1$ are such that their attenuation-constants $\alpha_1$, $\alpha_2$ differ; while their surge-resistances $z$ are the same, we assign the angles subtended by the sections $\theta_1 = L_1\alpha_1$ and $\theta_2 = L_2\alpha_2$ hyps. The angle subtended by the whole line will then be $\theta_1 + \theta_2$, as in the preceding case. That is, except for a disproportionality between the section-angles and their line-lengths, two sections of different attenuation-constant, but of the same surge-resistance, connect together like two sections of one and the same type of line. This is for the reason that in the unsteady state, or period of current building prior to the formation of the steady state here discussed, there is neither wave reflection nor discontinuity of wave propagation at the junction BC, when the surge resistance or impedance $z$ is the same on each side thereof.

In order, however, to simplify the transition to more complex

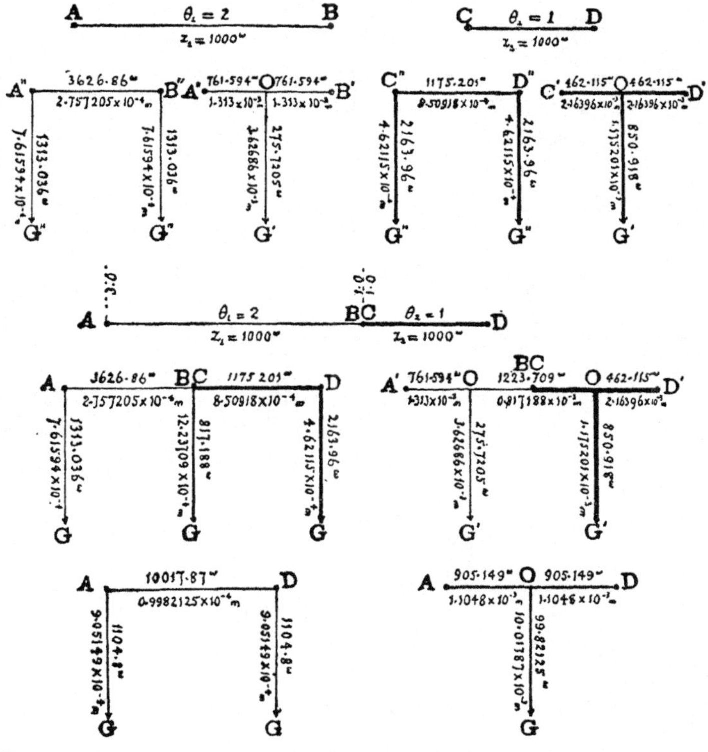


FIG. 69.—Composition of two sections with the same Surge-Resistance, but with different Attenuation-Constants.

cases later on, we may pause to consider the following case of two sections, with different $\alpha$ but the same $z$.

$L_1 = 100$ km., $r_1 = 20$ ohms/km., $g_1 = 2 \times 10^{-5}$ mho/km.

$L_2 = 100$ km., $r_2 = 10$ ohms/km., $g_2 = 10^{-5}$ mho/km.

Whence— $\alpha_1 = 0{\cdot}02$ hyp/km., $z_1 = 1000$ ohms;

$\alpha_2 = 0{\cdot}01$ hyp/km., $z_2 = 1000$ ohms.

Fig. 69 shows the two lines at AB and CD respectively. It shows the $\Pi$ and $T$ equivalent circuits of AB, at A″B″G″G″ and A′OB′G′, likewise of CD, at C″D″G″G″ and C′OD′G′. If we connect the sections together at BC, into a composite line AD, we virtually connect together some one pair of the combinations of equivalent circuits $\Pi_{AB}\Pi_{CD}$, $T_{AB}T_{CD}$, $\Pi_{AB}T_{CD}$, $T_{AB}\Pi_{CD}$. The first two combinations are shown at ABCDGGG and A′OBCOD′G′G′. If we merge together the two elements of any such pair by the formulas of Appendix E, we arrive either at the equivalent $\Pi$, ADGG; or the equivalent $T$, AODG, of the composite line.

In all of the examples to be considered, the equivalent $\Pi$ and $T$ of the various composite lines have been derived hyperbolically; but have also been checked by the merging process.

*Equivalent $\Pi$.*—In order to compute hyperbolically the equivalent $\Pi$ of the composite line AD (Fig. 69), we proceed as follows—

Ground either end of the composite line AD, say the end D. Assign the junction-angle $\theta_2$ at BC. Then the angle subtended by the composite line at A will be $\delta_A = \theta_1 + \theta_2$ hyps. The sending-end resistance of the composite line at A is, by (23)—

$$R_{gA} = z_1 \tanh \delta_A \quad . \quad . \quad . \quad . \quad \text{ohms} \quad (232)$$
$$= 1000 \tanh 3 = 995{\cdot}055 \text{ ohms.}$$
$$G_{gA} = 1/R_{gA} = y_1 \coth \delta_A \quad . \quad . \quad \text{mhos} \quad (233)$$
$$= 0{\cdot}001 \times \coth 3 = 10{\cdot}049{,}7 \times 10^{-4} \text{ mho.}$$

Then the architrave resistance AD of the composite $\Pi$ will be—

$$\rho'' = z_1 \sinh \delta_A \quad . \quad . \quad . \quad . \quad \text{ohms} \quad (234)$$
$$= 1000 \sinh 3 = 10017{\cdot}87 \text{ ohms.}$$
$$y'' = 1/\rho'' = 0{\cdot}9982125 \times 10^{-4} \text{ mho.}$$

The conductance $g''_A$ of the leak at A is by (23)—

$$g''_A = y_1 \coth \delta_A - y'' \quad . \quad . \quad . \quad \text{mho} \quad (235)$$
$$= 9{\cdot}05149 \times 10^{-4} \text{ mho.}$$

If we ground the composite line at A instead of at D, the angle subtended by the whole line at D will be $\delta_D = \theta_1 + \theta_2 = \delta_A$. The architrave resistance DA will be the same as that given in

(234). The sending-end resistance $R_{gD}$ and conductance $G_{gD}$ will be identical with $R_{gA}$ and $G_{gA}$ respectively, by (232) and (233); so that the leak-conductance $g''_D$ at D will be identical with $g''_A$ by (235). This completes the hyperbolic $\Pi$, ADGG of the composite line.

*Equivalent T*.—To find the hyperbolic equivalent $T$ of the composite line AD, Fig. 69, free the line at one end, say D. Then the angle subtended by the line at A will be, as before, $\delta_A = \theta_1 + \theta_2$ hyps.

The sending-end resistance of the line at A will be, by (20)—

$$R_{fA} = z \coth \delta_A \quad . \quad . \quad . \quad . \quad \text{ohms} \quad (236)$$
$$= 1000 \coth 3 = 1004{\cdot}97 \text{ ohms.}$$

The conductance of the leak OG is, by (71)—

$$g' = y \sinh \delta_A \quad . \quad . \quad . \quad . \quad \text{mhos} \quad (237)$$
$$= 0{\cdot}001 \sinh 3 = 10{\cdot}01787 \times 10^{-3} \text{ mhos.}$$

And its resistance is—

$$R' = 1/g' = 99{\cdot}82125 \text{ ohms.}$$

The resistance of the AO branch is, then—

$$\rho' = R_{fA} - R' \quad . \quad . \quad . \quad . \quad . \quad \text{ohms} \quad (238)$$
$$= 1004{\cdot}97 - 99{\cdot}821 = 905{\cdot}149 \text{ ohms.}$$

Similarly, if we free the composite line at A, instead of at D, the angle subtended by the line at D will be $\delta_D$. As before, $\delta_D = \theta_2 + \theta_1 = \delta_A$ hyps. The sending-end resistance offered by the line at D will then be identical with that found previously at A. The conductance of the leak will, by (71) and (237), be the same as that found from A. Finally, the resistance of the DO line-branch will, by (70) and (238), be identical with that of the AO branch ($905{\cdot}149^{\omega}$). This completes the $T$ of the composite line.

We may infer from the above reasoning, and it may be readily demonstrated formally, that when a composite line is composed of sections differing in linear constants, but having the same surge-impedance, the angle subtended by the whole line is the same at either end, and whether the distant end be freed or grounded. Consequently the equivalent $\Pi$ and $T$ of the composite line will be symmetrical. That is, the two leaks of the $\Pi$ are equal, and the two line branches of the $T$ are equal.

Conversely, it follows, from equations (84) and (95), that any composite line made up of sections differing in attenuation-constant, but with the same surge-impedance, may be replaced by an equivalent single line of uniform attenuation- and linear-constants.

THIRD AND GENERAL CASE.

*Sections with Different Surge-Impedances.*—Let a section AB of 100 km. (Fig. 70) be connected to a section CD of 300 km., and let their respective linear-constants be as follow—

$$r_1 = 20 \text{ ohms/km.}; \; g_1 = 20 \times 10^{-6} \text{ mho/km.}$$
$$r_2 = 10 \text{ ohms/km.}; \; g_2 = 2{\cdot}5 \times 10^{-6} \text{ mho/km.}$$

from which—

$$\alpha_1 = 0{\cdot}02 \text{ hyp/km.}; \; \theta_1 = 2 \text{ hyps}; \; z_1 = 1000 \text{ ohms};$$
$$\alpha_2 = 0{\cdot}005 \text{ hyp/km.}; \; \theta_2 = 1{\cdot}5 \text{ hyps}; \; z_2 = 2000 \text{ ohms},$$

so that the surge-resistances of the two sections are unequal. It follows that the angle subtended by the composite line will differ at the two ends, and will also differ according to whether the distant end is freed or grounded.

*Equivalent $\Pi$.*—Let us ground the end $A_2$ of the composite line $A_2D_2$ (Fig. 70). Then by formula (23), the sending-end resistance at B of the section BA grounded, will be—

$$R_{gB} = z_1 \tanh \theta_1 \quad . \quad . \quad . \quad . \quad \text{ohms} \quad (239)$$
$$= 1000 \tanh 2{\cdot}0 = 964{\cdot}0265 \text{ ohms.}$$

The angle of the section AB, at its end B, is $\delta_B = 2$ hyps. At the junction BC, however, the line-angle changes abruptly, owing to the change in surge-resistance, and at C, just across the junction, it is—

$$\delta_C = \tanh^{-1}\left(\frac{z_1}{z_2} \tanh \theta_1\right) = \tanh^{-1}\left(\frac{R_{gB}}{z_2}\right) \quad . \quad \text{hyps.} \quad (240)$$

That is, the hyp-tangent of the new angle is the ratio of the sending-end resistance at B to the surge-resistance of the new section CD. In this case—

$$\delta_C = \tanh^{-1}\left(\frac{964{\cdot}0265}{1000}\right) = \tanh^{-1} 0{\cdot}9640265\,;$$

or, by tables of hyperbolic tangents, $\delta_c = 0{\cdot}525608$ hyp. We mark this angle opposite to C on the line $A_2D_2$ (Fig. 70).

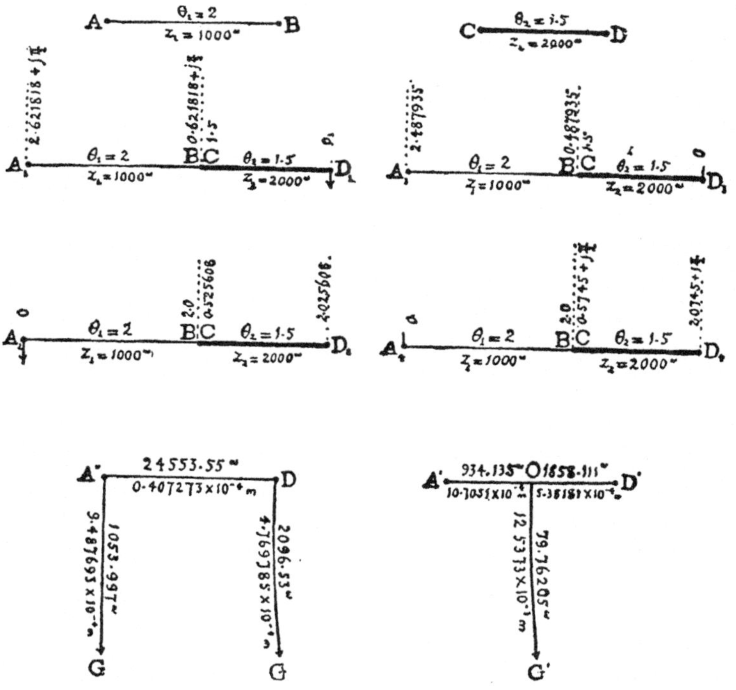


FIG. 70.—Composition of two sections of different Surge-Resistances and different Attenuation-Constants.

The angle subtended at $D_2$ by the composite line is, therefore—

$$\delta_D = \theta_2 + \delta_C = 2{\cdot}025608 \text{ hyps.}$$

The sending-end resistance of the grounded composite line is then, at $D_2$, by (23), (232), and (239)—

$$R_{gD} = z_2 \tanh \delta_D \quad . \quad . \quad . \quad . \quad . \quad . \quad . \quad . \quad . \quad \text{ohms} \quad (241)$$

$$= 2000 \tanh 2{\cdot}025608 = 1931{\cdot}58 \text{ ohms,}$$

and the sending-end conductance—

$$G_{gD} = y_2 \coth \delta_D = 1/R_{gD} \quad . \quad . \quad \text{mhos} \quad (242)$$
$$= 0{\cdot}00051771 \text{ mho.}$$

The formula for finding the architrave resistance of the equivalent $\Pi$ of the line AD is—

$$\rho'' = z_2 \sinh \delta_D . \frac{\cosh \delta_B}{\cosh \delta_C} \quad . \quad . \quad . \quad . \quad . \quad . \quad \text{ohms} \quad (243)$$
$$= 2000 \sinh 2{\cdot}025608 \times \frac{\cosh 2{\cdot}0}{\cosh 0{\cdot}525608}$$
$$= 24553{\cdot}55 \text{ ohms}$$

and— $\quad y'' = 1/\rho'' = 0{\cdot}407273 \times 10^{-4}$ mho.

Formula (243) differs from the corresponding formula (234) of the preceding case by the application of the ratio $\frac{\cosh \delta_B}{\cosh \delta_C}$ or the ratio of the cosines of the line-angles across the junction BC.

The formula for finding the conductance of the leak at D is, as before (235)—

$$g''_D = G_{gD} - y'' = 1/R_{gD} - y'' \quad . \quad . \quad \text{mhos} \quad (244)$$
$$= 4{\cdot}769785 \times 10^{-4} \text{ mho.}$$

In order to complete the equivalent $\Pi$ of the line AD hyperbolically, we must repeat the above process from the opposite end $D_1$, as shown at $A_1D_1$ (Fig. 70). The line-angle at C is $\delta_C = 1{\cdot}5$ hyps. Across the junction BC this angle changes suddenly to—

$$\delta_B = \tanh^{-1}\left(\frac{z_2 \tanh \theta_2}{z_1}\right) \quad . \quad . \quad . \quad \text{hyps.} \quad (245)$$
$$= \tanh^{-1} 1{\cdot}810296.$$

This involves at first sight an impossible result; but in all cases of a hyperbolic tangent greater than unity, we may resort to the following formulas—

$$\left.\begin{aligned} \sinh\left(x \pm j\frac{\pi}{2}\right) &= \pm j \cosh x \\ \cosh\left(x \pm j\frac{\pi}{2}\right) &= \pm j \sinh x \\ \tanh\left(x \pm j\frac{\pi}{2}\right) &= \coth x \\ \coth\left(x \pm j\frac{\pi}{2}\right) &= \tanh x \end{aligned}\right\} \quad \text{numeric} \quad (246)$$

We thus obtain—

$$\delta_B - j\frac{\pi}{2} = \coth^{-1}\left(\frac{z_2 \tanh 1{\cdot}5}{z_1}\right) \quad . \quad \text{hyps (247)}$$

$$= \coth^{-1} 1{\cdot}810296$$

$$= 0{\cdot}621818 \text{ hyp}$$

and—

$$\delta_B = 0{\cdot}621818 + j\frac{\pi}{2} \text{ hyp.}$$

This difficulty with seemingly impossible antitangents or anti-cotangents is not encountered in the a.-c. case.

We inscribe this value of $\delta_B$ opposite B on the line AD. The angle subtended by the whole line at A will then be—

$$\theta_1 + \delta_B = \delta_A = 2{\cdot}621818 + j\frac{\pi}{2} \text{ hyps.}$$

The sending-end resistance of the grounded composite line is then at $A_1$, by (232) and (241)—

$$R_{gA} = z_1 \tanh \delta_A \quad . \quad . \quad . \quad . \quad . \quad . \quad . \quad . \quad . \quad . \quad \text{ohms (248)}$$

$$= 1000 \tanh \left(2{\cdot}621818 + j\frac{\pi}{2}\right)$$

$$= 1000 \coth 2{\cdot}621618 = 1010{\cdot}64 \text{ ohms,}$$

and the sending-end conductance, as in (242)—

$$G_{gA} = y_1 \coth \delta_A$$

$$= y_1 \coth \left(2{\cdot}621818 + j\frac{\pi}{2}\right)$$

$$= 0{\cdot}001 \tanh 2{\cdot}621818 = 9{\cdot}894966 \times 10^{-4} \text{ mho.}$$

The architrave resistance, as in (243), is—

$$\rho'' = z_1 \sinh \delta_A . \frac{\cosh \delta_C}{\cosh \delta_B} \quad . \quad . \quad . \quad . \quad . \quad \text{ohms (249)}$$

$$= 1000 \cosh 2{\cdot}621818 . \frac{\cosh 1{\cdot}5}{\sinh 0{\cdot}621818}$$

$$= 24553{\cdot}55 \text{ ohms}$$

and— $$y'' = 1/\rho'' = 0{\cdot}407273 \times 10^{-4} \text{ mho.}$$

The conductance of $\Pi$ leak at A is, as in (244)—

$$g''_A = G_{gA} - y''$$

$$= 9{\cdot}487693 \times 10^{-4} \text{ mho.}$$

*Equivalent T.*—To compute the equivalent $T$ of the composite line AD (Fig. 70), free the line at one end, say $D_3$, and find the sending-end resistance at C in this condition. It is by (20), (230), and (236)—

$$R_{fC} = z_2 \coth \theta_2$$
$$= 2000 \coth 1{\cdot}5 = 2209{\cdot}59 \text{ ohms.}$$

The line-angle changes abruptly at the junction BC from $\delta_C = 1{\cdot}5$ to $\delta_B = 0{\cdot}487935$ hyp, by the condition—

$$\delta_B = \coth^{-1}\left(\frac{z_2 \coth \theta_2}{z_1}\right) = \coth^{-1}\left(\frac{R_{fC}}{z_1}\right) \quad . \quad \text{hyps. (250)}$$
$$= \coth^{-1} 2{\cdot}20959 = 0{\cdot}487935 \text{ hyp.}$$

The line-angle at the end $A_2$ is thus $\theta_1 + \delta_B = 2{\cdot}487935$ hyps. The sending-end resistance at $A_2$ is finally, by (20)—

$$R_{fA} = z_1 \coth \delta_A \quad . \quad . \quad . \quad . \quad . \quad . \quad . \quad . \quad . \quad \text{ohms (251)}$$
$$= 1000 \coth 2{\cdot}487935 = 1013{\cdot}897 \text{ ohms.}$$

The conductance of the leak OG′ is, by (237)—

$$g' = y_1 \sinh \delta_A . \frac{\cosh \delta_C}{\cosh \delta_B} \quad . \quad . \quad . \quad . \quad . \quad . \quad . \quad . \quad . \quad \text{mhos (252)}$$

$$= 0{\cdot}001 \times \sinh 2{\cdot}487935 \times \frac{\cosh 1{\cdot}5}{\cosh 0{\cdot}487935}$$

$$= 12{\cdot}5373 \times 10^{-3} \text{ mho.}$$

The resistance of the leak OG′ is, therefore, $R' = 1/g' = 79{\cdot}762$ ohms.

The resistance of the AO branch is then, by (238)—

$$\rho' = R_{fA} - R' \quad . \quad . \quad . \quad . \quad . \quad . \quad . \quad . \quad . \quad . \quad \text{ohms (253)}$$
$$= 1013{\cdot}897 - 79{\cdot}762 = 934{\cdot}135 \text{ ohms.}$$

In order to complete the equivalent $T$ of the line AD, we must repeat the above process from the opposite end, by freeing the end A, as shown at $A_4D_4$ (Fig. 70). The line-angle at B is $\delta_B = 2{\cdot}0$. Across the junction BC this angle changes suddenly to—

$$\delta_C = \coth^{-1}\left(\frac{z_1 \coth \theta_1}{z_2}\right) \quad . \quad . \quad . \quad . \quad . \quad . \quad . \quad \text{hyps. (254)}$$

$$= \coth^{-1}\left(\frac{1037{\cdot}315}{2000}\right) = \coth^{-1} 0{\cdot}5186575.$$

In order to avoid an impossible operation, apply formula (246)—

$$\delta_C - j\frac{\pi}{2} = \tanh^{-1} 0{\cdot}5186575 = 0{\cdot}57450 \text{ hyp}$$

$$\delta_C = 0{\cdot}5745 + j\frac{\pi}{2} \text{ hyps.}$$

The line-angle at the end $D_4$ is thus $\theta_2 + \delta_C = 2{\cdot}0745 + j\frac{\pi}{2}$ hyps.

The sending-end resistance at $D_4$ is finally, by (20) and (251)—

$$R_{fD} = z_2 \coth \delta_D \quad . \quad . \quad . \quad . \quad . \quad . \quad . \quad . \quad \text{ohms (255)}$$

$$= 2000 \coth\left(2{\cdot}0745 + j\frac{\pi}{2}\right)$$

$$= 2000 \tanh 2{\cdot}0745 = 1937{\cdot}873 \text{ ohms.}$$

The conductance of the leak OG′ is, therefore, by (237) and (252)—

$$g' = y_2 \sinh \delta_D \cdot \frac{\cosh \delta_B}{\cosh \delta_C} \quad . \quad . \quad . \quad . \quad . \quad . \quad . \quad . \quad \text{mhos (256)}$$

$$= 0{\cdot}001 \sinh\left(2{\cdot}0745 + j\frac{\pi}{2}\right) \cdot \frac{\cosh 2{\cdot}0}{\cosh\left(0{\cdot}5745 + j\frac{\pi}{2}\right)}$$

$$= 0{\cdot}001 \cosh 2{\cdot}0745 \cdot \frac{\cosh 2{\cdot}0}{\sinh 0{\cdot}5745} = 12{\cdot}5373 \times 10^{-3} \text{ mho.}$$

The resistance of the leak OG′ is, therefore, $R' = 1/g' = 79{\cdot}762$ ohms. The resistance of the DO branch is then, by (70) and (253)—

$$\rho' = R_{fD} - R' \quad . \quad . \quad . \quad . \quad . \quad . \quad . \quad . \quad . \quad \text{ohms (257)}$$

$$= 1937{\cdot}873 - 79{\cdot}762 = 1858{\cdot}111 \text{ ohms.}$$

This completes the *T* of the composite line.

It may be inferred from the preceding reasoning that for the case of a composite line of two sections with different surge-impedances, the receiving-end impedance of the line in the absence of receiving instruments, which is the architrave of the line-*Π*, has the same value from each end of the line. The leak of the composite line-*T* has also one and the same value, computed from either end. Both the *Π* and the *T* are, however, dissymmetrical. Each requires two separate computations and line-angle distributions, one from each end.

*Summary of Two-Section Formulas.*—If we expand formulas (234) and (243), we obtain for the architrave of the composite line $\Pi$—

$$\rho'' = z_1 \sinh \theta_1 \cosh \theta_2 + z_2 \cosh \theta_1 \sinh \theta_2 \quad . \quad . \quad \text{ohms} \quad (258)$$

$$= \frac{z_1 + z_2}{2} \sinh (\theta_1 + \theta_2) + \frac{z_1 - z_2}{2} \sinh (\theta_1 - \theta_2) \text{ ohms*} \quad (259)$$

$$= z_1 \sinh \theta_1 \frac{\sinh \delta_D}{\sinh \delta_C} \quad . \quad \text{ohms (line grounded at A)} \quad (260)$$

$$= z_2 \sinh \delta_D \frac{\cosh \delta_B}{\cosh \delta_C} \quad . \quad \text{ohms (line grounded at A)} \quad (261)$$

$$= z_2 \sinh \theta_2 \frac{\sinh \delta_A}{\sinh \delta_B} \quad . \quad \text{ohms (line grounded at D)} \quad (262)$$

$$= z_1 \sinh \delta_A \frac{\cosh \delta_C}{\cosh \delta_B} \quad . \quad \text{ohms (line grounded at D)} \quad (263)$$

Similarly, if we expand formulas (237) and (252), we obtain—

$$g' = y_1 \sinh \theta_1 \cosh \theta_2 + y_2 \cosh \theta_1 \sinh \theta_2 \quad . \quad . \quad \text{mhos} \quad (264)$$

$$= \frac{y_1 + y_2}{2} \sinh (\theta_1 + \theta_2) + \frac{y_1 - y_2}{2} \sinh (\theta_1 - \theta_2) \text{ mhos} \quad (265)$$

$$= y_1 \sinh \theta_1 \frac{\sinh \delta_D}{\sinh \delta_C} \quad . \quad . \quad . \quad \text{mhos (line freed at A)} \quad (266)$$

$$= y_2 \sinh \delta_D \frac{\cosh \delta_B}{\cosh \delta_C} \quad . \quad . \quad . \quad \text{mhos (line freed at A)} \quad (267)$$

$$= y_2 \sinh \theta_2 \frac{\sinh \delta_A}{\sinh \delta_B} \quad . \quad . \quad . \quad \text{mhos (line freed at D)} \quad (268)$$

$$= y_1 \sinh \delta_A \frac{\cosh \delta_C}{\cosh \delta_B} \quad . \quad . \quad . \quad \text{mhos (line freed at D)} \quad (269)$$

*Single Lines Equivalent to a Dissymmetical $\Pi$ or $T$.*—It is evident that formulas (79), (80), (94), and (95) apply only to a symmetrical $\Pi$ or $T$. Moreover, it may be seen that no single smooth and uniform line can correspond to a dissymmetrical $\Pi$

* Formulas (258) and (259) were first published as receiving-end impedances of a two-section composite line by Dr. G. di Pirro. *See* Bibliography, 52.

or $T$. This means that, in general, no single smooth and uniform line can be the counterpart of a composite line having sections of different surge-resistance. But if we reduce a dissymmetrical $\Pi$ to a symmetrical $\Pi$ and a terminal leak, we may apply equations (95) and (96) to transform the symmetrical $\Pi$ into an equivalent single line. It follows that any composite line may be resolved into one, and only one, uniform smooth line of the same length with a leak permanently applied to one end; or to an infinitude of such single uniform smooth lines having a leak at each end.

Similarly, the $T$ of a composite line may be reduced to a symmetrical $T$ plus a line-impedance at one end. By the use of equations (84) and (88), we may substitute a single smooth uniform line for the symmetrical $T$. Consequently, any composite line may be resolved into one, and only one, uniform smooth line of the same length with a line-impedance at one end; or, to an infinitude of such single uniform smooth lines having a line-impedance at each end.

*Composite Line of "n" Sections.*—To compute the equivalent $\Pi$ of a composite line of $n$ successive sections, ground the line at the A end and develop the line-angles towards the opposite end, following the process of (240). Find the architrave impedance according to formula (260) or (261). This may be regarded as formula (26) modified by the application of $(n-1)$ ratios of cosines in (261), or of $(n-1)$ ratios of sines in (260). The opposite end leak admittance will then be the sending-end admittance minus the architrave admittance. The process must be repeated after grounding the line at the distant end and developing line-angles towards A.

To compute the equivalent $T$, free the line at the A end and develop the line-angles towards the opposite end, following the process of (250). Find the $T$-leak admittance by following formula (266) or (267). This may be regarded as formula (71) modified by the application of $n-1$ ratios of cosines in (267), or of $n-1$ ratios of sines in (266); that is, one such ratio for each junction. The opposite-end line-branch impedance will then be the sending-end impedance minus the leak impedance.

The process must be repeated after freeing the line at the distant end and developing line-angles towards A.

One complete equivalent circuit, say the $\Pi$, of a composite line of $n$ sections calls then for the determination $n-1$ line-angles first in one direction and then in the other. The formulas are well adapted to logarithmic computation. If, however, only the receiving-end impedance of the composite line is required, then we need only develop the line-angles in one direction over the line, so as to apply one of the architrave formulas, and neglect the pillars of the $\Pi$, as at $AF_0$, Fig. 71.

## LOADED COMPOSITE LINES.

*Definitions.*—Loads in a line may be either *regular* or *casual.* Regular loads are such as are applied at regular intervals, in order to improve the current delivery on telephone lines. Casual loads are of an irregular or incidental character, such as might occur at section-junctions or at the ends of a composite line. In the former case they would be *intermediate* casual loads, and in the latter case, *terminal* casual loads. Only casual loads will be here discussed; because it is easy, with the aid of formulas already discussed on page 45, to substitute an equivalent smooth unloaded line for any uniformly loaded line.

Loads may also be divided into two classes: namely, (1) those applied in series with the line, or *impedance* loads, such as coils of impedance or resistance; and (2) those applied in derivation to the line, or *leak* loads.

## INFLUENCE OF LOCATION OF AN IMPEDANCE LOAD ON THE RECEIVING-END IMPEDANCE OF A COMPOSITE LINE.

It can be shown that if a single smooth uniform line is terminally loaded with a given impedance, the change in the receiving-end impedance due to the load is the same, whichever end of the line the load may be applied to; *i.e.* whether the load is applied at the sending or at the receiving end. In the case of a composite line, however, this proposition generally fails. The effect of a resistance coil of 100 ohms on the receiving-end resistance of the three-section composite line comprising the two sections AB and CD above discussed, in series with a third section EF of angle $\theta_3 = 0{\cdot}5$ hyp. and surge-resistance $z_3 = 2500$ ohms (Fig. 72), is shown in Fig. 71. Without the load, the receiving-end resistance of the line, or the architrave of its equivalent $\Pi$, is 44,247 ohms. If the load is added at the A end of the line, the receiving-end resistance becomes 48,619·7 ohms; but if added at the F end, it is only 46,192. When the same coil is inserted as an intermediate load, its influence on the receiving end resistance is not so great. In alternating-current composite

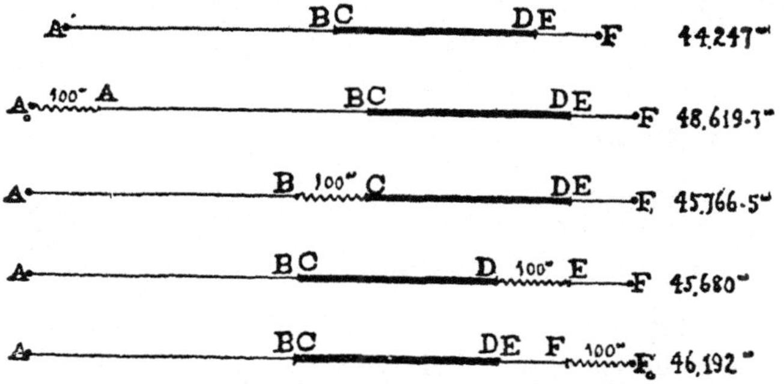


FIG. 71.—Diagram showing the Influence of the Location of an Impedance Load on the Receiving-end Resistance of a Three-section Composite Line.

lines, the opportunities for such variations are more marked. In all cases, however, the application of a terminal impedance $\sigma$ to a line (single or composite), increases the receiving-end or architrave impedance of that line in the ratio $\frac{R_g + \sigma}{R_g}$; where $R_g$ is the sending-end impedance of the line at the loaded end before the load is applied. This is true whether the loaded end is made the sending or receiving end of the circuit. For single lines, $R_g$ has the same value at either end, and therefore the ratio of increase in receiving-end impedance is the same at whichever end of a single line the load $\sigma$ is applied; whereas, for composite lines, we have seen that $R_g$ is different, in general, at the two ends.

Rules have been worked out for dealing with any succession of casual loads in a composite line, including loads applied through the medium of transformers. For detailed information, reference may be made to the original papers on the subject.* In general, one additional ratio, or fractional coefficient, must be added to the expression of the architrave impedance of the equivalent $\Pi$, for each additional load, whether in series or in derivation. Thus, in the case of the three-section line of Fig. 72 with its double terminal load (which is a combined leak and impedance) and one intermediate load, the architrave impedance of the equivalent $\Pi$ has the following expression, working from the grounded end at A—

$$\rho'' = z_0 \sinh \delta_H \cdot \frac{\cosh \delta_D}{\cosh \delta_E} \cdot \frac{\cosh \delta_B}{\cosh \delta_C} \cdot \frac{R_{gF}}{R_{gG}} \quad . \quad . \quad \text{ohms} \quad (270)$$

$$= 1936{\cdot}87 \sinh 1{\cdot}535312 \cdot \frac{\cosh 2{\cdot}09295}{\cosh 1{\cdot}03531} \cdot \frac{\cosh 2{\cdot}0}{\cosh 0{\cdot}59295} \cdot \frac{1809{\cdot}74}{1565{\cdot}14}$$

$$= 51{,}615 \text{ ohms.}$$

All of the preceding formulas in relation to composite lines, although illustrated by real numbers for continuous-current cases, are applicable, through the medium of corresponding complex numbers, to any single-frequency alternating-current case.

The preceding theory of composite lines shows that the practice of computing the equivalent length of a composite circuit by "Transmission Equivalents" or their reciprocals, is fallacious. The transmission equivalent of a type of line is the ratio of the real attenuation-constant of "standard" cable, to the real attenuation-constant of the line considered, both at the standard telephonic frequency. Thus, if the real attenuation-constant of standard cable is 0·066 hyp. per km., then a cable whose real attenuation-constant is, say 0·132, has a transmission equivalent of 0·5, so that half a km. of this cable has the same normal attenuation-factor as one km. of the standard cable. It

* *See* Bibliography, 52, 61.

is evident that if the limiting commercial range of standard cable is, say 50 km., then the corresponding range of the cable considered will be $50 \times 0{\cdot}5 = 25$ km., neglecting reflections at the receiving apparatus. This merely means that two different types of line have the same normal attenuation-factor when the real components of their hyperbolic angles are equal. We may not, however, safely infer that the proposition extends to composite lines; because we have seen that the angle subtended

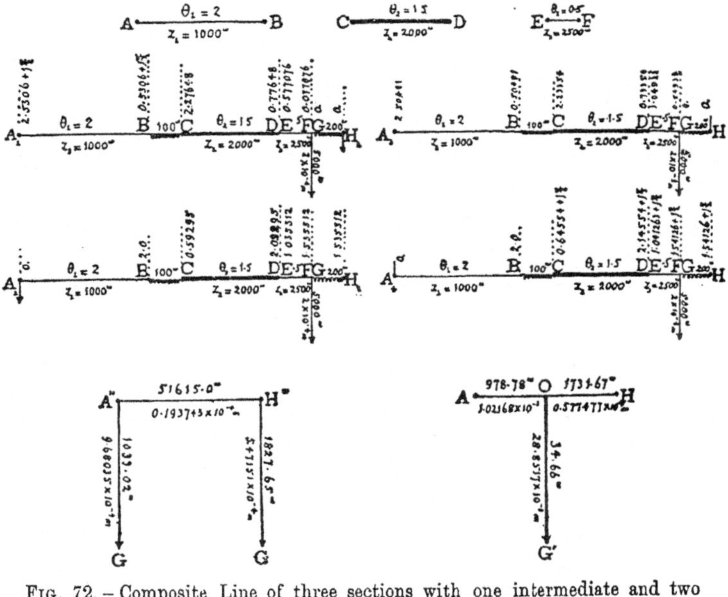


FIG. 72. – Composite Line of three sections with one intermediate and two terminal loads.

by a composite line is not the sum of the angles of its component sections; but a more complex function of those angles, depending upon the order of succession, and on the individual surge-impedances. Only in the particular case when the surge-impedances of the sections are the same, will the angle subtended by the composite line be the sum of the section angles. While, therefore, the kilometer, or mile, of standard cable is a valuable unit in the absence of anything better, and is a valid unit, neglecting terminal reflections, when single lines only are considered, it may become a misleading unit when composite lines are employed.

A correct standard, in the present state of our knowledge, would be either of the following—

(1) The total hyperbolic angle of the composite line in hyps., including the angle subtended by the receiving apparatus, corrected for discontinuities at junctions.

(2) The total receiving-end impedance of the composite line including the receiving apparatus, corrected for discontinuities at junctions. In the latter case, the architrave impedance of the proper equivalent $\Pi$ including receiving apparatus might be substituted.

In either case, a correction might, strictly speaking, be needed for differences in the impressed e.m.f. at the generating end, seeing that this impressed e.m.f. supplied by the generating apparatus depends to some extent on the sending-end impedance. Further published research is needed in this direction.

## CHAPTER IX

### THE APPLICATION OF HYPERBOLIC FUNCTIONS TO WIRE TELEGRAPHY

SINCE uniform telegraph lines possess evenly distributed conductor-resistance and dielectric leakance, it follows that when subjected to steady and continuous e.m.f., either at one end or both, in any assigned manner, the steady-state distributions over them of potential, current, and power, are subject to hyperbolic trigonometry. Thus, in all tests of a telegraph line with continuous e.m.f., hyperbolic formulas apply, on the assumption that the line is uniform, contains no faults, and that the linear dielectric leakance follows some known law of time. In underground and submarine cables, for instance, the apparent linear leakance is well known to diminish rapidly with the duration of impressed e.m.f. by the so-called "polarization" process. Even if the line is faulty, the fault or faults may be regarded as leak-loads of magnitudes subject to variation, applied at junctions between sections of otherwise uniform line. The above class of cases may be described as class (1) of steady-state tests.

When we consider electric signalling over a telegraph line, the process falls into one of three other classes, according to the particular conditions.

(2) If the line is short, and the speed of signalling employed much below the theoretically attainable limit; so that the steady state of the circuit is closely approximated to at each signal; then hyperbolic trigonometry is applicable to the circuit.

(3) If, on the contrary, the line is long, or the speed and type of signalling such that the steady state of the line is hardly ever attained; then the hyperbolic formulas developed in

preceding chapters with special reference to the steady state are inapplicable. Dr. Fleming, has, however, recently published * an application of hyperbolic functions to the process of computing "curves of arrival" over long submarine cables, as an abbreviation of the method, originally developed by Lord Kelvin, for calculating the behaviour of such lines in the unsteady state.

(4) An intermediate condition presents itself when the line is worked nearly to the limit of its practicable signalling speed, and when the method of signalling is such that a steady state is constantly approached, although never closely attained. That is, the signalling may be roughly imitated by a regular succession of alternating-current impulses, impressed on the line at the sending end, with discontinuities occurring at somewhat irregular intervals; so that the steady state is never actually reached, but is repeatedly tended to. In this case, the hyperbolic theory does not apply definitely, but it may apply tentatively; that is, experiment may show that a certain approximate relation exists between the actual limiting signalling speed, and the theoretical steady-state limiting speed of pure alternating-current impulses.

We shall only consider examples from Classes (1), (2), and (4).

*Class 1: Steady-state Tests.*—If a uniform line AB, without faults is freed at the distant end B, we know, by (20), that its apparent resistance at A is $R_f = r_0 \coth \theta$ ohms, and when it is grounded at B, its apparent resistance at A becomes, by (23), $R_g = r_0 \tanh \theta$ ohms. Consequently, the surge-resistance of the line is, by (27), $r_0 = \sqrt{R_f R_g}$, and the angle of the line is, by (29), $\theta = \tanh^{-1}\sqrt{\frac{R_g}{R_f}}$ hyps. Again, if an ammeter or other device, of resistance $R_r$, is inserted in the circuit to ground at B, the angle subtended by the instrument will be, assuming $r_0 > R_r$—

$$\theta' = \tanh^{-1}\frac{R_r}{r_0} \quad . \quad . \quad . \quad . \quad \text{hyps.} \quad (271)$$

* *The Propagation of Electric Currents in Telephone and Telegraph Conductors*, by J. A. Fleming, F.R.S., chap. v

and the angle of the line and instrument together becomes $(\theta + \theta')$ hyps. The apparent resistance of A under these conditions becomes—

$$R'_g = r_0 \tanh(\theta + \theta') \quad . \quad . \quad \text{ohms} \quad (272)$$

The receiving-end resistance, as judged from the current in the ammeter at B, and the e.m.f. impressed at A, is—

$$R_l = \frac{r_0 \sinh(\theta + \theta')}{\cosh \theta'} \quad . \quad . \quad . \quad \text{ohms} \quad (273)$$

the line being assumed devoid of all earth-currents or polarization. If $\theta'$ is not over 0·1, cosh $\theta'$ may be taken as unity, without much error, and we have approximately—

$$R_l \cong r_0 \sinh(\theta + \theta') \quad . \quad . \quad . \quad \text{ohms} \quad (274)$$

Consequently, if instead of having the line freed and grounded at B, making two successive observations at A, we have the line merely grounded at B through an ammeter, and insert an ammeter and voltmeter at A, thus making a single test, with simultaneous observations at both ends, we obtain from (56) and (58)—

$$(\theta + \theta') \cong \cosh^{-1}\left(\frac{R_l}{R'_g}\right) \quad . \quad . \quad . \quad \text{hyps.} \quad (275)$$

from which, $r_0$, $\theta'$, and $\theta$ can obviously be successively deduced.

As a simple arithmetical case, let us take a uniform single line 300 km. long, with the linear constants $r = 6$ ohms per km., and $g = 1{\cdot}5 \times 10^{-6}$ mho per km. The attenuation-constant of the line is $\sqrt{9 \times 10^{-6}} = 0{\cdot}003$ hyp. per km., and its surge-resistance $\sqrt{6000000/1{\cdot}5} = 2000$ ohms. Its line angle is 0·9 hyp.

With the line first freed, and then grounded at B, and both tests taken successively at A, we should have $R_f = 2000 \coth 0{\cdot}9 = 2792{\cdot}2$ ohms and $R_g = 2000 \tanh 0{\cdot}9 = 1432{\cdot}6$ ohms. The surge-resistance of the line is therefore computed to be $r_0 = \sqrt{2792{\cdot}2 \times 1432{\cdot}6} = 2000$ ohms, while the angle of the line is computed to be—

$$\theta = \tanh^{-1}\sqrt{\left(\frac{1432{\cdot}6}{2792{\cdot}2}\right)} = \tanh^{-1}\sqrt{0{\cdot}51307}$$

$$= \tanh^{-1} 0{\cdot}71630 = 0{\cdot}9 \text{ hyp.}$$

From these results we obtain at once, $R = r_0\theta = 1800$ ohms in conductor-resistance, or 6 ohms per km., and $g = \theta/r_0 = 0{\cdot}45 \times 10^{-3} = 450 \times 10^{-6}$ mho in dielectric leakance, or 1·5 micromho per km.

In the second case, with the line grounded at B through a milliammeter of 200 ohms, the angle subtended by the instrument is $\tanh^{-1}(0{\cdot}1) = 0{\cdot}10033$. The total angle of line and instrument subtended at A is 1·00033 hyps. If 50 volts were applied at the A-end of the line, a current of 32·82 milliamperes might be expected at A, and 21·37 milliamperes at B. The sending-end resistance would then be $R'_g = 50/0{\cdot}03282 = 1523{\cdot}5$ ohms, and the receiving-end resistance $R_l = 50/0{\cdot}02137 = 2339{\cdot}7$ ohms. From these observations, exchanged telegraphically, the total angle of the line would be, approximately, by (275): $(\theta + \theta') = \cosh^{-1}\left(\frac{2339{\cdot}7}{1523{\cdot}5}\right) = \cosh^{-1} 1{\cdot}53578 = 0{\cdot}99374$ hyp., as against the correct value of 1·00033. The inferred value of the surge-resistance is then $1523{\cdot}5/\tanh 0{\cdot}99374 = 2007{\cdot}3$ ohms, as against 2000. The inferred value of $\theta'$ is also $\tanh^{-1}(200/2007{\cdot}3) = 0{\cdot}09997$ hyp. as against 0·100 hyp. leaving the line-angle $\theta = 0{\cdot}89377$ hyp. as against the correct value of 0·9.

With alternating-current testing, the first case of measurements at A with the B end freed and grounded successively, is perfectly applicable, by extending the formulas into two dimensions. The second case is not strictly applicable, because the phase of the received current with respect to the impressed e.m.f. at A is not measured.

*Class 2: Steady-state Signalling. Best Resistance of a Receiving Instrument.*—Electromagnetic receiving instruments in wire telegraphy may be divided into two classes, namely: (*a*) those, as of the D'Arsonval movable-coil type, in which the magneto-mechanical force, or torque, is directly proportional to the ampere-turns in the coil; and (*b*) those, like simple polarized or non-polarized relays, in which the magneto-mechanical force, or torque, may be nearly proportional to the square of the ampere-turns at low magnetic saturation, but, as the saturation increases, may fall to perhaps a lower power than the first. In either case, the magneto-mechanical force may be expressed by—

$$F = a\,(I_B n_1)^p \quad . \quad . \quad . \quad . \quad \text{dynes or dyne-perp. cm.} \quad (276)$$

where F is the force in dynes, or the torque in dynes acting perpendicularly to a radius of 1 cm., $I_B$ is the received current strength in amperes, $n_1$ is the number of turns of wire in the winding, $a$ a constant of the instrument depending on its construction, and $p$ some real exponent not greater than 2. The received current $I_B$ is found by (57). The number of turns $n_1$ in a given winding-space is well known to be sensibly proportional to $\sqrt{R_r}$, the square root of the resistance in ohms of the winding, provided that the size of insulated copper wire selected lies within the fairly wide range where the ratio of covered diameter to the bare diameter of wire is sensibly constant. Consequently we have, approximately, with $a'$ a modified instrument constant—

$$F = a'\left(\frac{E_A\sqrt{R}}{r_0 \sinh\theta + R_r \cosh\theta}\right)^p \text{ dynes or dyne-perp. cm.} \quad (277)$$

In order to make this mechanical force a maximum by varying $R_r$, we differentiate F with respect to $R_r$, in the usual way, and equate to zero. We then find—

$$R_r = r_0 \tanh\theta \quad . \quad . \quad . \quad . \quad \text{ohms} \quad (278)$$

That is, the best resistance, with respect to mechanical force, for the electromagnetic winding of a receiver, that does its work during the steady state, is equal to the sending-end resistance of the line to ground at B, no matter what may be the exponent $p$, which expresses the relation between the torque and the ampere-turns.*

When the speed of signalling rises, so that the steady state fails to be approached, the resistance determined by (278) is, in general, too low for the best action.

*Class 4: Limiting Signalling Speeds on Long Lines.*—When the signalling speed is carried up to, or near to, the practicable

* *See* Ayrton and Whitehead paper, No. 12 in the Bibliography.

limit on long lines, and especially on long submarine cables, it is important to determine theoretically how the limit is affected by the terminal apparatus.

It was first demonstrated by Lord Kelvin that, on any long cable, a certain interval of time elapsed from the instant of applying a continuous e.m.f. at the sending end to the first appearance of even an infinitesimal current flow to ground at the receiving end. This silent interval he showed to be—

$$a = \frac{\tau}{\pi^2} \log_\epsilon\left(\frac{4}{3}\right) = 0{\cdot}023\,\tau \quad . \quad . \quad \text{seconds} \quad (279)$$

where $\tau$ is the time-constant of the cable, defined by the relation—

$$\tau = \mathrm{CR} = \mathrm{L}c\,.\,\mathrm{L}r = \mathrm{L}^2 cr \quad . \quad . \quad \text{seconds} \quad (280)$$

Moreover, cables of the same time-constant have the same signalling speed, with the same terminal apparatus. Under given terminal conditions, the signalling speed is inversely proportional to the time-constant. In simplex practice, with the Kelvin siphon-recorder as receiving instrument, a hand double cable key as the sending instrument, and a condenser in the receiving circuit, the following empirical formula has been found satisfactory—

$$n = \frac{10}{\tau} \quad . \quad . \quad . \quad . \quad \text{* letters per second} \quad (281)$$

The last formula, which has been much used in submarine telegraphy, is satisfactory, so far as regards a cable operated under stereotyped conditions, and embodies the results of Kelvin's famous investigation. It is, however, incomplete in so far as it throws no light upon the influence of variations in the terminal apparatus. We proceed to consider how the terminal apparatus would effect the signalling speed, if the process of signalling consisted in using a simple alternating-current generator at the sending end, of assigned e.m.f. amplitude, the speed of the generator, and the frequency of its e.m.f. being increased, until the instrument at the receiving end was unable to give legible indications.

* *Encyclopædia Britannica:* "Submarine Telegraphy." (10th ed.)

We must, then, assign some relation between the frequency, or number of impulses per second of the hypothetical alternator, and the average number of impulses per second impressed rhythmically on the cable in practice by the ordinary sending apparatus. The deductions from the alternating-current theory will not be strictly applicable to dot-and-dash alphabet sending, but will be tentatively admissible, as a working hypothesis, subject to experimental check.

If we connect a long submarine cable AB, as shown in Fig. 73, to an alternating-current generator producing a maximum cyclic e.m.f. $E_m$ at the sending end, through a terminal impedance $z_s$,

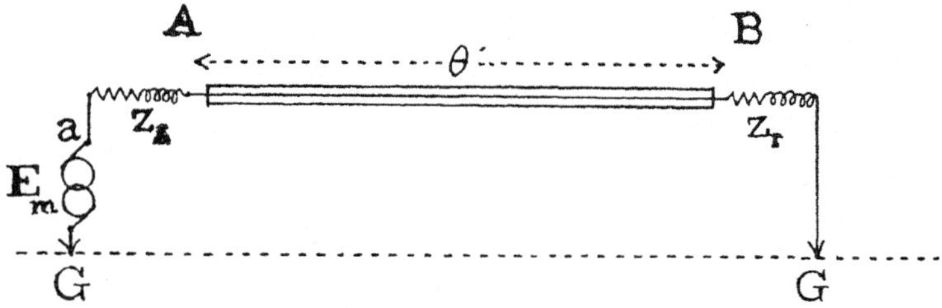


FIG. 73.—Hypothetical Alternating-Current Generator impressing maximum cyclic potential, at signalling frequency, on sending end of a cable which is grounded through receiving apparatus at distant end.

the receiving end may be considered as connected to ground through an impedance $z_r$ of the receiving instrument. Then if the alternator, assumed bipolar, is driven at angular velocity $\omega$ radians per second, the angle subtended by the cable taken separately is, by (19) and (150)—

$$\theta = \sqrt{j\,C\,R\,\omega} = \sqrt{j\,\tau\,\omega} = \sqrt{\tau\,\omega}\,\underline{/45^\circ} \text{ hyps.} \angle \quad (282)$$

and its surge-impedance is, by (152)—

$$z_0 = \sqrt{\frac{r}{j\,c\,\omega}} = \sqrt{\frac{r}{c}} \cdot \frac{1}{\sqrt{\omega}}\overline{\backslash 45^\circ} \quad \text{ohms} \angle \quad (283)$$

assuming, as we safely may, that the ordinary submarine cable has relatively negligible linear inductance and leakance at low frequencies.

The receiving-end impedance of the cable and receiving apparatus is, by (59)—

$$Z_l = z_0 \sinh\theta + z_r \cosh\theta \quad . \quad . \quad \text{ohms} \angle \quad (284)$$

But on any cable worked nearly to its speed-limit, $\theta$ is so large that $\sinh\theta = \cosh\theta$ very nearly; so that—

$$Z_l = (z_0 + z_r)\sinh\theta \quad . \quad . \quad \text{ohms} \angle \quad (285)$$

If now the receiving instrument is of such a nature and degree of sensitiveness that pure alternating-current signals can be read satisfactorily when the received current strength has a maximum cyclic strength of $I_{mB}$ amperes; then—

$$I_{mB} = \frac{E_{mA}}{Z_l} = \frac{E_{mA}}{(z_0 + z_r)\sinh\theta} \quad \text{amperes} \angle \quad (286)$$

where $E_{mA}$ is the maximum cyclic e.m.f. impressed on the sending end of the cable at A. This will differ from the maximum cyclic e.m.f. $E_m$ at alternator terminals, owing to the IZ drop of pressure in the inserted impedance $z_s$, which is determined as follows—

The final sending-end impedance of the cable is by (202b)—

$$z_A = z_0 \tanh(\theta + \theta') \quad . \quad . \quad \text{ohms} \angle \quad (287)$$

where $\theta'$ is the angle subtended by the receiving apparatus. But $\theta$ is so large that $\tanh\theta = 1$, and therefore, a fortiori, $\tanh(\theta + \theta') = 1$. Consequently, on long cables—

$$z_A = z_0 \quad . \quad . \quad . \quad . \quad \text{ohms} \angle \quad (288)$$

The maximum cyclic e.m.f. impressed on the cable is—

$$E_{mA} = E_m \frac{z_0}{z_s + z_0} \quad \text{max. cy. volts} \angle \quad (289)$$

Or the maximum cyclic current at B is—

$$I_{mB} = \frac{E_m}{\left(1 + \frac{z_s}{z_0}\right)(z_0 + z_r)\sinh\theta} \quad \text{max. cy. amperes} \angle \quad (290)$$

If, then, we assign a limiting value to $E_m$, such as $\pm$ 50 volts, it is evident that the speed and frequency of the alternator at A can be increased until the current $I_{mB}$ just falls to the lowest satisfactory strength agreed upon. Increasing the frequency will alter $z_s$ to some extent, and will also alter $z_0$ inversely as the square root, by (283); but sinh $\theta$ will increase much more rapidly. When $\theta$ is large—

$$\sinh\theta = \frac{\varepsilon^{\frac{\theta}{\sqrt{2}}}}{2} \angle \frac{\theta}{\sqrt{2}} \quad . \quad . \quad . \quad . \quad \text{numeric } \angle \quad (291)$$

so that sinh $\theta$ increases with $\theta$ exponentially.

For stereotyped terminal values of $E_m$, $z_s$, and $z_r$, it is evident that on any cable the limiting alternating-current frequency and speed are fixed, to a first approximation, by sinh $\theta$, or $\varepsilon^{\frac{\theta}{\sqrt{2}}}$; so that, under such conditions, all cables must have approximately the same value of $\varepsilon^{\frac{\theta}{\sqrt{2}}}$, and, therefore, the same value of $\theta$. But, by (282), $\theta = \sqrt{\tau\omega}\ \underline{/45^\circ}$ hyps. Consequently, if we call $\theta_0$ this limiting value of $\theta$, we have, with $\omega_0$ the associated limiting angular velocity—

$$\theta_0 = \sqrt{\tau\omega_0}\ \underline{/45^\circ} \quad . \quad . \quad . \quad . \quad \text{hyps. } \underline{/45^\circ} \quad (292)$$

and—

$$\omega_0 = \frac{\theta_0^2}{j\tau} \quad . \quad . \quad . \quad . \quad . \quad \text{radians per sec.} \quad (293)$$

or—

$$f_0 = \frac{\theta_0^2}{2\pi} \cdot \frac{1}{j\tau} = \frac{\theta_0^2}{2\pi} \cdot \frac{1}{jCR} \text{ cycles per sec.} \quad (294)$$

where $f_0$ is the corresponding limiting impressed frequency. That is, the pure alternating-current frequency on any long cable of negligible linear inductance and leakance, under stereotyped conditions of terminal apparatus, and neglecting relatively small variations in $z_0$, is inversely proportional to the time-constant $\tau = CR$ of the cable, a deduction in accordance with Kelvin's law. Our tentative alternating-current signalling theory agrees, therefore, with the fundamental theory, and with practical observations of alphabet-signalling, at least to this extent.

The next question is how properly to connect the frequency

of the hypothetical alternator with the frequency of alphabetical signalling. In Fig. 74, the two particular words *Submarine Telegraph* are analysed in terms of their international cable-alphabet signals. Each black rectangle, above or below the zero line, represents the impression of continuous e.m.f. on the sending end of the line, assuming the simple case of a sending-key directly connected to the cable. Ordinates then represent impressed voltage, and abscissas elapsed time. Dot- and dash-elements occupy equal intervals, so that each signal occupies one dot-element. The canonical interval between signals in a letter is one dot-element, and that between successive letters three dot-elements. Between successive words there may be six dot-elements. The word *submarine* thus covers 59 dot-elements, and *telegraph* 61. The two together $59 + 61 + 6 = 126$. The signals as received over a cable of 1·47 seconds time-constant are shown beneath, at CC.

*Dot-Frequency and Reversal-Frequency.*—There are two hypothetical uniform signalling frequencies. One is indicated at A A or A′ A′ (Fig. 74), the other at B B. The A A type may be called pure dot-signalling. In this type of signalling, the impulses have all the same sign, and the complete period is two dot-elements. The A A type would be equivalent to a certain complex alternating-current e.m.f., superposed on a positive continuous e.m.f. of half the A amplitude. The A′ A′ type would be equivalent to the same alternating current e.m.f., superposed on a negative continuous e.m.f. of half the A′ amplitude. The B B type may be called pure reversal-signalling. It corresponds to a complex alternating e.m.f. of period equal to four dot-elements. Reversal-signalling has, therefore, just half the frequency of dot-signalling, and requires no associated continuous e.m.f.

There has been some debate as to which of these hypothetical types of rhythmic signalling more nearly corresponds to practical alphabet-signalling. On behalf of dot-signalling, it may be urged that it employs the actual frequency of the sending keys, disregarding the direction of current. On behalf of reversal-signalling, it may be urged that on long submarine cables the dots do not appear. They are smoothed out by retardation, and can only be inserted in their proper places by the trained intelligence of the receiving operator; whereas the reversals actually show on the record, and form the sign-posts, so to speak, by which the operator is guided to interpretation.

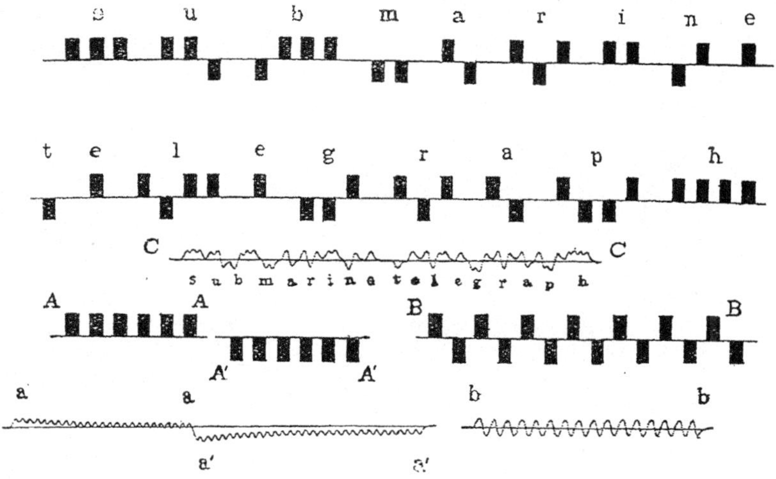


FIG. 74.—Signals of Impressed E.M.F. at Sending and Receiving Ends in the two particular words *Submarine Telegraph*, as compared with either "dot-signals" or "reversals."

As for the actual comparison of the sending record with either dot-signals or reversals, there is no marked preponderance of evidence. Thus, in the two particular words of Fig. 74, which have been taken at random, compared with dot-signals, there are—

8 cases of 2 successive dots or dashes, in *s*, *u*, *b*, *m*, *i*, *l*, *g*, *p*, *h*.
3 „ 3 „ „ „ *s*, *b*, *h*.
1 „ 4 „ „ „ *h*.

Compared with reversals, there are—

11 cases of a cycle or complete reversal, in *u*, *b*, *a*, *r*, *n*, *l*, *g*, *r*, *a*, *p*, *p*.
3 „ cycle and a half *r*, *l*, *r*.
0 „ 2 cycles.

We shall see that so far as concerns the cable alone, it is a matter of indifference whether we refer alphabetical signalling to reversal-frequency, or to dot-frequency; but, in regard to the influence of the terminal apparatus, there is a considerable difference. We shall, therefore, discuss both standards, but lay emphasis on the reversal standard, since cable engineers are believed to prefer the latter.*

In the two words of Fig. 74 there happens to be 18 letters, and 126 dot-elements, in all, or, allowing a space before the next word, 132 dot-elements. This is at the rate of 7·35 dot-elements per average letter. In sentences taken from an English newspaper, an average letter occupies about 7·3 dot-elements including the average necessary spacing. In unintelligible 10-letter code words, with a diminished number of vowels, the average rises to about 8·0 dot-elements. We may take as a working mean 7·7 dot-elements per letter. This corresponds to 3·85 cycles in pure dot-signalling, or 1·925 cycles in pure reversals. If, therefore, we have as a speed of signalling $n$ letters per second, the corresponding hypothetical frequencies will be—

for dot-signalling, $f' = 3{\cdot}85\,n$ cycles per sec. (295)

for reversal-signalling, $f'' = 1{\cdot}925\,n$ „ „ „ (296)

We have already seen in (281) that a well-known empirical formula for cable-speeds under stereotyped terminal conditions is that $n = 10/\tau$. Consequently, the limiting frequency $f_\circ$ is—

for dot-signalling, $f_\circ' = \dfrac{38{\cdot}5}{\tau}$ . cycles per sec. (297)

for reversal-signalling, $f_\circ'' = \dfrac{19{\cdot}25}{\tau}$ „ „ (298)

Substituting the value of frequency in terms of a limiting cable angle in (292), we find—

for dot-signalling,

$$\theta_\circ' = \sqrt{2\pi \times 38{\cdot}5}\ \underline{/45^\circ} = 15{\cdot}55\ \underline{/45^\circ} = 11 + j11 \text{ . hyps.}\angle \quad (299)$$

for reversal signalling,

$$\theta_\circ'' = \sqrt{2\pi \times 19{\cdot}25}\ \underline{/45^\circ} = 11{\cdot}0\ \underline{/45^\circ} = 7{\cdot}777 + j7{\cdot}777 \text{ hyps.}\angle \quad (300)$$

* *See* Bibliography, 46.

That is, at dot-signalling frequency, the normal attenuation-factor is $\varepsilon^{-11} = 0{\cdot}000{,}0167$, or to nearly 1/600 of 1 per cent.; while at reversals-frequency, it is $\varepsilon^{-7{\cdot}777} = 0{\cdot}000{,}42$, or to less than 1/20 of 1 per cent. These values of $\theta_0$ may be regarded as the equivalents of formula (281) in alternating-current cable theory.

*Influence of Impedance at the Sending End.*—We may now consider the effect of modifications in the terminal apparatus upon the speed of signalling according to our tentative alternating-current theory.

In Fig. 73 let the cable have a linear conductor-resistance $r = 6{\cdot}7$ ohms per nautical mile (naut) and a linear capacitance $c = 0{\cdot}42 \times 10^{-6}$ farad per naut. Then, by (283), the surge-impedance of the cable is $3994/\sqrt{\omega}$. With a length of 1229 nauts, the time-constant of the cable is $\tau = 4{\cdot}25$ seconds. When working at the rate of 150 letters per minute = 2·5 letters per second, the corresponding reversals-frequency would be 4·812 cycles per second by (296), and the angular velocity $\omega'' = 30{\cdot}23$ radians per second. The angle subtended by the cable would then be, by (282), $\theta'' = \sqrt{128{\cdot}5}\ \underline{/45^\circ} = 11{\cdot}35\ \underline{/45^\circ}$ hyps. The surge-impedance is then $726{\cdot}4\ \overline{\backslash 45^\circ}$ ohms. If the terminal impedance $z_s$ happened to be a condenser of, say, 40 microfarads capacitance, its impedance would be $-j/c\omega'' = -j827{\cdot}1$ ohms. The total impedance offered to the alternator at the sending end is, therefore, $z_s + z_0 = 513{\cdot}7 - j(513{\cdot}7 + 827{\cdot}1) = 1436\ \overline{\backslash 69^\circ\ 02'}$ ohms. The ratio of the potential at the end of the cable at A to the potential at the terminal of the alternator would be, by (289)—

$$\frac{E_{mA}}{E_m} = \frac{726{\cdot}4\ \overline{\backslash 45^\circ}}{1436\ \overline{\backslash 69^\circ\ 02'}} = 0{\cdot}506\ \underline{/24^\circ\ 02'}.$$

That is, the potential impressed on the cable would be nearly 50 per cent. less than that which would be impressed if the condenser $z_s$ were short-circuited. The lowered potential would also lead the potential at alternator-terminal $a$ by 24° 02′. The reduction in impressed potential would either require a compensating increase in the generated e.m.f. of the alternator, or a reduction in the frequency and equivalent signalling-speed in order to restore the received current to its proper strength.

If, however, in place of a condenser, we insert a reactance-coil at the sending end, with an effective resistance of 15 ohms, and an inductance of 12·5 henrys, the impedance of the coil at this frequency will be $z_s = 15 + j378$ ohms. The total impedance at the sending end will then be $528{\cdot}7 - j135{\cdot}7 = 545{\cdot}8 \diagdown 14° 24'$. The ratio of the impressed voltage at A to the generated voltage at $a$, is then $\frac{E_{mA}}{E_m} = \frac{726{\cdot}4 \diagdown 45°}{545{\cdot}8 \diagdown 14° 24'}$ $= 1{\cdot}33 \diagdown 30° 36'$ or the effect of inserting a magnetic reactance in $z_s$, instead of a condensive reactance, is to raise the impressed potential 33 per cent., instead of lowering it 50 per cent. This would mean increasing the received current at B, and the speed of signalling could be slightly but distinctly increased, to bring up sinh $\theta$, and restore the original current strength agreed upon.

*Best Resistance of the Receiving Instrument.*—It has already been shown, in connection with (278), that the best resistance for an electromagnetic receiver winding to possess, when operated by continuous currents in the steady state is $z_0$ tanh $\theta$. It is shown in Appendix J, that whether the mechanical force, or mechanical torque, exerted in the receiver varies directly with the ampere-turns, or the square of the ampere-turns, or any intermediate power thereof, the largest maximum cyclic force, or torque, exerted by the receiving instrument will be obtained when the reactance of the receiving apparatus balances and annuls the surge-reactance or reactance component of the surge-impedance, and when, moreover, the resistance of the winding in the receiver is equal to the surge-resistance or resistance-component of the surge-impedance, increased by any other receiving-apparatus resistance present. That is, the most powerful reversal-signals or dot-signals will be respectively obtained when—

$$\sigma = r_0 + R'_r \quad . \quad . \quad . \quad . \quad . \quad \text{ohms (301)}$$

where $\sigma$ is the resistance of the receiving-coil or coils, as measured with continuous currents or with alternating currents of signalling frequency, $r_0$ is the real, or resistance-component of the surge-impedance $z_0$, and $R'_r$ is the resistance-component of any reactive apparatus in the receiving circuit for balancing the surge-reactance. If $R'_r$ is so small that it may be neglected, then the receiving-instrument reactance should be equal and opposite to the surge-resistance of the cable, and the receiving instrument resistance should be equal to the surge-resistance of the cable at signalling frequency. In other words, the receiving-circuit impedance should be equal to the surge-impedance in modulus, and have the equal but opposite argument.

As an illustration of these principles to dot- and to reversal-alternating-current signalling, the following Table gives, in parallel columns, the receiving-end impedance of the 4·25-second cable previously discussed, both for the pure reversal-frequency of $\omega'' = 30{\cdot}23$, and the dot-frequency of $\omega' = 60{\cdot}46$ radians per second, corresponding to the signalling frequency of 2·5 letters per second; on the supposition that the receiving instrument is connected directly between cable and ground.

| | Reversals | Dots |
|---|---|---|
| Frequency $f$, cycles/sec. | 4·81 | 9·62 |
| Angular Velocity $\omega$, radians/sec. | 30·23 | 60·46 |
| Surge-Impedance $z_0$ ohms. | $726{\cdot}4\ \overline{\diagdown 45^\circ}$ | $513{\cdot}7\ \overline{\diagdown 45^\circ}$ |
| ,, ,, $r_0 - j\,x_0$ | $513{\cdot}7 - j513{\cdot}7$ | $363{\cdot}2 - j363{\cdot}2$ |
| Receiving inst. res. $\sigma$, ohms. | 513·7 | 363·2 |
| $(z_0 + z_r)$ | $1027{\cdot}4 - j513{\cdot}7$ | $726{\cdot}4 - j363{\cdot}2$ |
| ,, | $1148{\cdot}5\ \overline{\diagdown 26^\circ\ 34'}$ | $812\ \overline{\diagdown 26^\circ\ 34'}$ |
| Angle of Cable $\theta$, hyps. | $8{\cdot}014 + j8{\cdot}014$ | $11{\cdot}334 + j11{\cdot}334$ |
| ,, ,, | $11{\cdot}334\ \underline{/45^\circ}$ | $16{\cdot}03\ \underline{/45^\circ}$ |
| Sinh $\theta$, numeric. | $1510\ \underline{/459^\circ\ 10'}$ | $41900\ \underline{/649^\circ\ 30'}$ |
| Receiving-end Impedance, ohms. | $1{,}734{,}000\ \underline{/432^\circ\ 36'}$ | $34{,}030{,}000\ \underline{/622^\circ\ 56'}$ |

The above Table indicates that doubling the frequency impressed on the cable has had the effect of increasing sinh $\theta$ 28 times, and of increasing the receiving-end impedance 20 times.

If we assume that the maximum cyclic amplitude of current

required for effective reversal-frequency signalling is 15 microamperes, and for effective dot-frequency signalling 0·75 microampere; then the maximum cyclic potential $E_{mA}$ impressed on the sending end of the cable would be 26·01 volts with reversals, and 25·52 volts with dots. It must be remembered, however, that 15 microamperes on each side of zero would be able to give a distinct record on a sensitive siphon recorder in good adjustment; whereas 0·75 microampere could not be expected to give any perceptible amplitude. A condenser would be needed in the circuit, either at the sending or receiving end, with dot-signalling, to shut off the continuous-current component mentioned on page 188. In pure reversal-signalling there is also need for such a condenser owing to the constant departure, towards dots or towards dashes in succession, from the pure reversal régime.

If now we insert a reactance into the receiving circuit, equal to the surge-reactance, without appreciably increasing the resistance, then we obtain the following conditions—

| | Reversals | Dots |
|---|---|---|
| Frequency $f$, cycles/sec. as before | 4·81 | 9·62 |
| Surge Impedance $r_o - jx_o$ ohms. | $513{\cdot}7 - j513{\cdot}7$ | $363{\cdot}2 - j363{\cdot}2$ |
| Receiving Apparatus Impedance $\sigma + jx_r$ ohms. | $513.7 + j513{\cdot}7$ | $363{\cdot}2 + j363{\cdot}2$ |
| Extra Inductance in receiving circuit, henrys. | 17·0 | 6·0 |
| $(z_o + z_r)$ ohms. | $1027{\cdot}4 \underline{/0^\circ}$ | $726{\cdot}4 \underline{/0^\circ}$ |
| Sinh $\theta$ | $1510 \underline{/459^\circ\ 10'}$ | $41900 \underline{/649^\circ\ 30'}$ |
| Receiving-end Impedance, $Z_l$ ohms. | $1,551,000 \underline{/459^\circ\ 10'}$ | $30,440,000 \underline{/649^\circ\ 30'}$ |
| Max. cyclic voltage $E_{mA}$ | 23·26 | 22·83 |

The effect, then, of cancelling the surge-reactance of the cable is to diminish the receiving-end impedance, or the required impressed voltage at A, by about 10 per cent. Or, if the impressed voltage were held constant for both cases, about 3 per cent. increase in angular velocity and theoretical speed of signalling might be derived instead.

The extra magnetic reactance at the receiving end might be introduced in shunt to the siphon-recorder instead of in series therewith, with substantially the same range of benefit.

A set of experimental connections on an artificial line of 4·25-seconds time-constant is indicated in Fig. 75. On this line, according to formula (281), we should expect a maximum practical working speed of 10/4·25 = 2·35 letters per second = 141 letters per minute under stereotyped simplex operation.

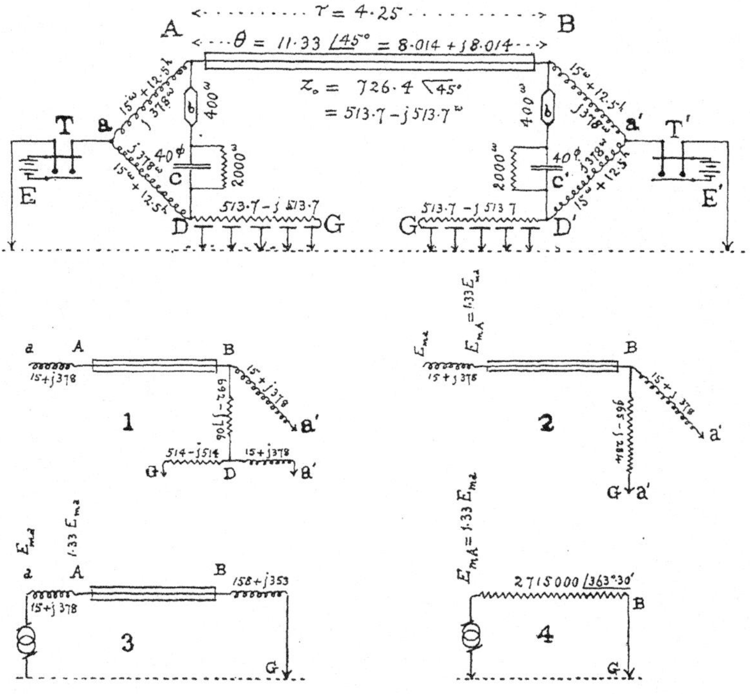


FIG. 75.—Diagrams indicating Limiting Receiving-end Impedance for satisfactory signals in a particular case, referred to equivalent reversals-frequency.

With duplex connections, owing to the shunting of the receiving instrument, the speed ordinarily falls slightly, or the sending-battery voltage has to be raised at each end in order to restore simplex speed. With the duplex connections shown, however, the limiting practical working speed was found to be about 150 letters per minute in each direction, or 2·5 letters per second. By (296) this corresponds to $f''=4·81$ cycles per second equivalent reversals-frequency, or $\omega''=30·23$ radians per second.

In key diagram 1, of Fig. 75, is represented the impedance at each end of the cable for reversals-frequency. The impedance of the shunted condenser is $292 - j706$ ohms, so that the impedance of the path BD is $692 - j706$ ohms. In key diagram 2, the branches DG and D$a'$, having a joint impedance of $273 + j422$ ohms, the total impedance in the path BG is $z_1 = 965 - j284$ ohms. This is still shunted by the impedance B$a'$ of $z_2 = 15 + j378$ ohms. In key diagram 3, the total impedance in the resultant path to ground at B is $158 \times j353 = z_r$ ohms. The current in the receiving instrument $\sigma$ is, however, less than that leaving the cable at B in the ratio $z_2/(z_1 + z_2)$. The receiving-end impedance, as judged at the receiving instrument, is therefore increased in the ratio $(z_1 + z_2)/z_2$ and becomes—

$$Z_l' = (z_o + z_r) \sinh \theta \left(\frac{z_1 + z_2}{z_2}\right) \quad . \quad . \quad . \quad . \quad . \quad . \quad \text{ohms} \angle \quad (302)$$

In this case—

$$Z_l' = (513·7 - j513·7 + 158 + j353) \times 1510 \underline{/459° \ 10'} \times \frac{984 + j94}{15 + j378}$$

$$= 2{,}715{,}000 \ \underline{/363° \ 30'}$$

The last result is indicated in key diagram 4, where the maximum cyclic voltage impressed on the sending end of the cable at A delivers current to ground at B through an impedance of 2·715 megohms, with a lag of 363° 30′. The voltage at A is, however, raised above that impressed on the bridge at $a$, in the ratio 1·33.

We may, therefore, say that according to the tentative alternating-current and hyperbolic theory of signalling speed over long submarine cables, using either simplex or duplex connections, and not more than 50 volts in the sending battery, the limiting speed is that which brings the receiving-end impedance up to 2·7 megohms with respect to the maximum cyclic alternating e.m.f. of reversals-frequency impressed on the sending end of the cable, and with the most sensitive type of modern siphon recorder, or to 2·0 megohms with respect to the transmitter e.m.f.

In resolving the impedance $z_r$ at the receiving end for the case indicated in Fig. 75, we have assumed the bridge apex $a'$ to be permanently connected to ground through the transmitting apparatus T′. If we cut off this ground connection at $a'$, by opening the key T′, we increase the impedance $z_r$ at the receiving end; but it is demonstrated in Appendix K, that when a proper duplex balance has been obtained, the quantity $(z_o + z_r)\left(\frac{z_1 + z_2}{z_2}\right)$ remains constant, which means that whether the bridge apex $a'$ is freed or grounded at B, or remains in any intermediate state, the receiving-end impedance $Z'_l$ including the effect of the shunt on the receiving instrument, is unchanged, and the received signals are also unchanged, either in strength or in phase. It is ordinarily somewhat easier to compute $Z'_l$ with the ground connection cut at T′, than with the ground connection completed as shown.

If, in the signalling arrangement of Fig. 75, the system were computed for dot-frequency, or twice the reversals-frequency, the corresponding receiving-end impedance $Z'_l$ would be nearly 60 megohms.

On the basis of dot-frequency, if we apply an impressed e.m.f. of amplitude E volts with the aid of a battery, then, as shown at AA Fig. 76, the effect is the same as though an e.m.f. of E/2 volts were sustained on the cable throughout the succession of dots, and superposed on this, would be the well-known Fourier series of alternating e.m.f.'s—

$$S = \frac{E}{2} \cdot \frac{4}{\pi}\left\{\sin \omega' t + \frac{1}{3} \sin 3\,\omega' t + \frac{1}{5} \sin 5\,\omega' t + \ldots \right. \quad \text{volts} \quad (303)$$

where $\omega'$ is the angular-velocity of the dot-frequency $f'$, $t$ is the time, and $\pi = 3{\cdot}14159$. The first term only, or fundamental sinusoid, is represented at *oaaa*, Fig. 76. Its amplitude is $2/\pi$ of the battery voltage OA. Consequently, if a battery of 10 volts is applied at regular make-and-break intervals, the effect will be the same as though 5 volts were steadily applied, and on this a series of Fourier harmonic e.m.f.'s, the first term of which would have a maximum cyclic value of 6·365 volts, of frequency equal to the dot-frequency $f'$. The next term would have a maximum cyclic e.m.f. of 2·12 volts, and a frequency of $3f'$, and so on. All these theoretical alternating e.m.f.'s would be in steady operation at the sending end, and each would send its own current into the cable, independently of the rest. That is, the various hypothetical harmonic currents of the

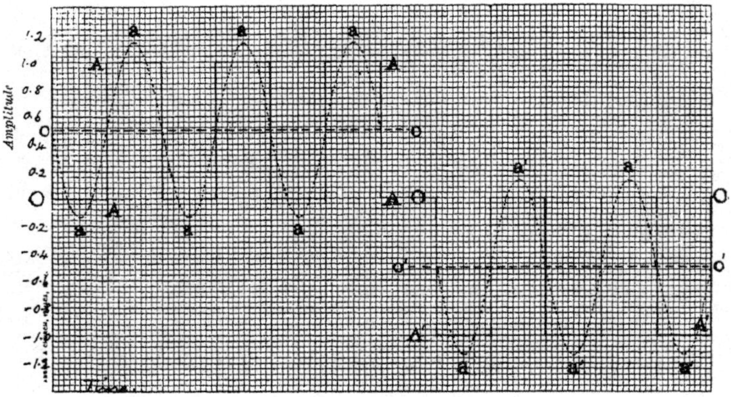


FIGS. 76 and 77.—Dot Signals and Dash Signals with their Equivalent Fundamental Alternating E.M.F.

Fourier series, having different frequencies, would not interfere with one another. The total maximum cyclic amplitude would be the rectangular summation of the respective individual amplitudes (Fig. 54). It can easily be shown, however, that when the angle subtended by the cable exceeds 4 hyps., as must happen on a long cable operated at or near full speed, the current received to ground at the distant end of the line from the triple-harmonic frequency component is insignificantly small, and quite negligible. In other words, only the fundamental frequency component arrives at the receiving end. All the currents of higher frequencies are absorbed in transmission. The higher harmonic currents have very appreciable amplitude at the sending end of the line, but are regarded, in the tentative alternating-current theory, as mere idle currents, that neither help nor hinder the fundamental working-currents.

The theoretical condition in sending a regular succession of dashes is indicated at A′A′, Fig. 77. Here the same maximum cyclic fundamental e.m.f. $o\ a'a'$ is operating at dot-frequency, in opposite phase to that at $a\ a$; but the continuous component

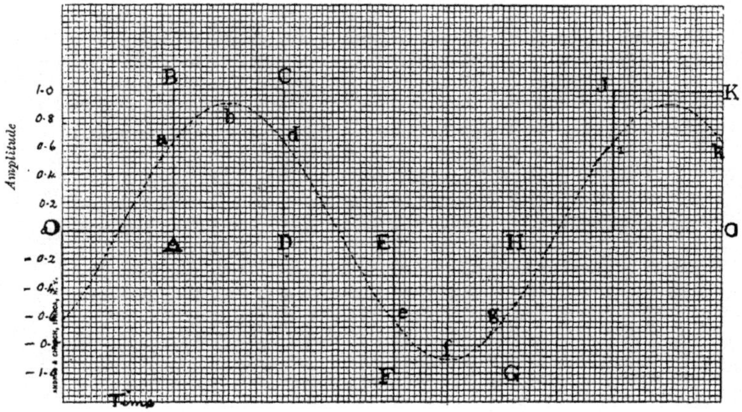


FIG. 78.—Reversals and their Equivalent Fundamental Alternating E.M.F.

$o'\ o'$ has reversed its sign from that at $o\ o$, Fig. 76. Consequently, in changing from a series of dots to a series of dashes, the equivalent alternating e.m.f. is not altered in amplitude, but both it and the continuous component E/2 are reversed.

The corresponding theoretical condition or reversal-signalling is indicated in Fig. 78. Here there is no continuous e.m.f., and the fundamental alternating component of reversals-frequency has a maximum cyclic amplitude of $2\sqrt{2}/\pi = 0{\cdot}90$, with respect to the impressed e.m.f. of the sending-battery. If, therefore, the sending-battery has, say, 50 volts e.m.f., and is applied in a uniform succession of reversals, the maximum cyclic amplitude of the fundamental reversals-frequency is 45 volts.

If, then, in the arrangement of Fig. 75, the sending-battery E had a terminal e.m.f. under working conditions of 50 volts, the equivalent reversals alternator would deliver 45 maximum cyclic volts at the bridge apex $a$, and 60 at the end A of the cable. The strength of current received through the instrument at B would be $60/2715000 = 22{\cdot}1$ microamperes maximum cyclic current.

These principles are well illustrated and checked in the records $aa$, $a'a'$, and $bb$ of Fig. 74, which were received over the Canso-Rockport cable. The frequency of the dot series was 5·34 dot-cycles per second, representing $\omega' = 33{\cdot}55$. The frequency of the reversals $bb$ was 2·67 reversal cycles per second, representing $\omega'' = 16{\cdot}75$. The maximum cyclic amplitude of the dot-signals is approximately 60 microamperes, and that of the reversal-signals 200 microamperes. Each series consists of sine-waves as nearly as the eye can detect. The sending e.m.f. was 30 volts, and in the duplex connections, double-block condensers of 50 microfarads each occupied the bridge-arms. The shunted siphon recorder had a resistance of about 220 ohms, and was in circuit with 30 microfarads. The cable has a time constant of 1·47 seconds. At the dot-frequency, its angle would be $7{\cdot}0 \underline{/45^\circ}$ hyps. and its surge-impedance $865 - j865$ ohms. At the reversals-frequency, its angle would be $4{\cdot}96 \underline{/45^\circ}$ hyps. and its surge-impedance $1225 - j1225$ ohms. The observed amplitudes of the dot and reversal signals are in fair accordance with the formulas above given.

Summing up, then, the facts concerning the tentative hyperbolic theory of long submarine cable-signalling, that theory is in accordance with the simple C.R. law, when no consideration is given to terminal apparatus. The theory undertakes to define the amplitudes of received signals when those signals consist either of uniform dots, dashes, or reversals. On the assumption that alphabetical signalling follows the conditions of reversal-signalling of the same impulse frequency, the limiting alphabetical frequency with 50 volts of battery e.m.f. is found when the receiving-end impedance rises to 2·7 megohms, reckoned from the sending end of the cable. A wide field is, however, open to experimental research for determining the actual relation of alphabetical signalling to alternating current-signalling.

Herr Béla Gáti has undertaken to show * that a comparatively small extra distributed inductance in a submarine cable is capable of producing a relatively large reduction in the real component of the angle subtended by the cable at frequencies near 1600 cycles per second, assuming no material increase in linear leakance. He has also published experiments and measurements on telegraph and telephone lines confirming the hyperbolic theory.

* *See* Bibliography, 57, 65.

## CHAPTER X

### MISCELLANEOUS APPLICATIONS OF HYPERBOLIC FUNCTIONS TO ELECTRICAL ENGINEERING PROBLEMS

WE propose in this chapter to consider two applications of hyperbolic functions to problems in electrical engineering, quite different from those we have studied, without attempting to discuss them in detail.

When a condenser is connected in series with resistance and inductance in such a manner that both the resistance and the inductance have to be taken into consideration, it is well known that the conditions of current flow when the circuit is energized or de-energized divide themselves into three classes according to the amount of resistance present—

(1) Oscillatory current flow;
(2) Aperiodic current flow;
(3) Ultraperiodic current flow.

If we denote the dissipative ohmic resistance of the circuit by $r = 2\rho$, and the surge-resistance $\sqrt{\frac{l}{c}}$ by $z$, then if $\rho$ is less than $z$ the circuit is oscillatory. If $\rho$ is greater than $z$ the circuit is non-oscillatory and ultraperiodic. If $\rho = z$, the circuit is in the intermediate condition, being aperiodic.

Considering first the periodic case, if we construct an impedance triangle, Fig. 79, on a base $\rho$ and with hypothenuse $z$ the perpendicular side will be the reactance $x = l\omega$. Then the discharge frequency in the presence of the resistance will be this reactance divided by the inductance $l$, and will be less than the frequency in the absence of resistance, in the ratio $x/z$. The triangle will have, at the base, an angle $\varphi = \tan^{-1}\left(\frac{x}{\rho}\right)$.

To find the conditions during discharge, let $OU_0$, Fig. 81, drawn to scale from the original O, along the X-axis, represent the initial voltage of the condenser that is about to send discharging current through the circuit. Draw $O\overline{U}_0$ such that $UO\overline{U}_0 = 90° - \varphi$. Draw $OE_0$ reverse to $OU_0$ and $O\overline{E}_0$ related in the same manner to $OE_0$ that $O\overline{U}_0$ is to $OU_0$. Then $O\overline{U}_0$ will be the initial vector discharging e.m.f. and $O\overline{E}_0$ the initial vector e.m.f. of self-induction in the circuit. The projections of

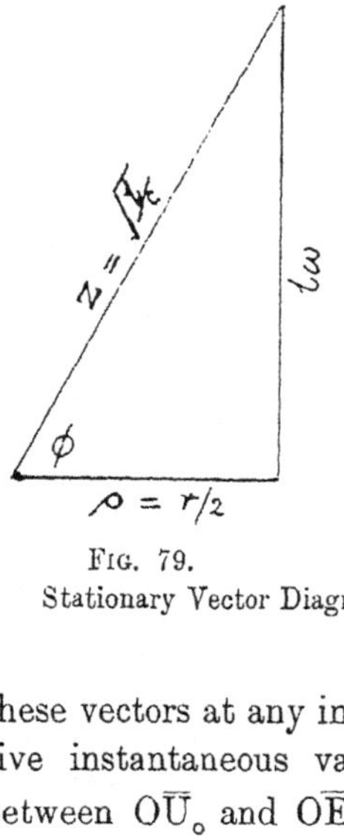


FIG. 79.

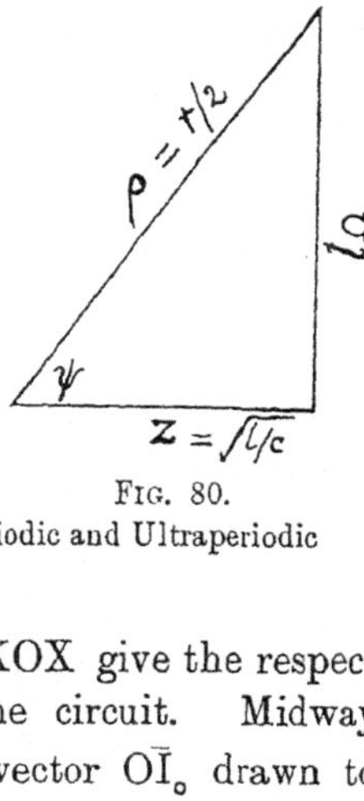


FIG. 80.

Stationary Vector Diagrams of Impedance in Periodic and Ultraperiodic Circuits respectively.

these vectors at any instant on the axis XOX give the respective instantaneous values existing in the circuit. Midway between $O\overline{U}_0$ and $O\overline{E}_0$ lies the current vector $O\overline{I}_0$ drawn to current scale, the ratio $\frac{O\overline{U}_0}{O\overline{I}_0}$ being made equal to $z$. Multiplying $\overline{I}_0$ by the resistance $2\rho = r$ in the circuit, we obtain the vector OR, which being reversed in direction to $OI_0$ represents $-I_0 r$. Now let all four vectors rotate positively, or counterclockwise, at the angular velocity $\omega$ which exists in the presence of resistance. The projections of these various vectors on XOX reveal at any moment the quantities that would exist in the circuit except for the attenuation or damping. Each projected quantity must therefore be subjected to the damping coefficient

$\varepsilon^{-\frac{t\rho}{l}}$, where $l$ is the inductance in the circuit and $t$ the time in seconds elapsed from the commencement of discharge. Or each vector, instead of rotating in a pure circle, and independently subjected to damping, may be allowed to rotate in an equiangular spiral as indicated in Fig. 81.

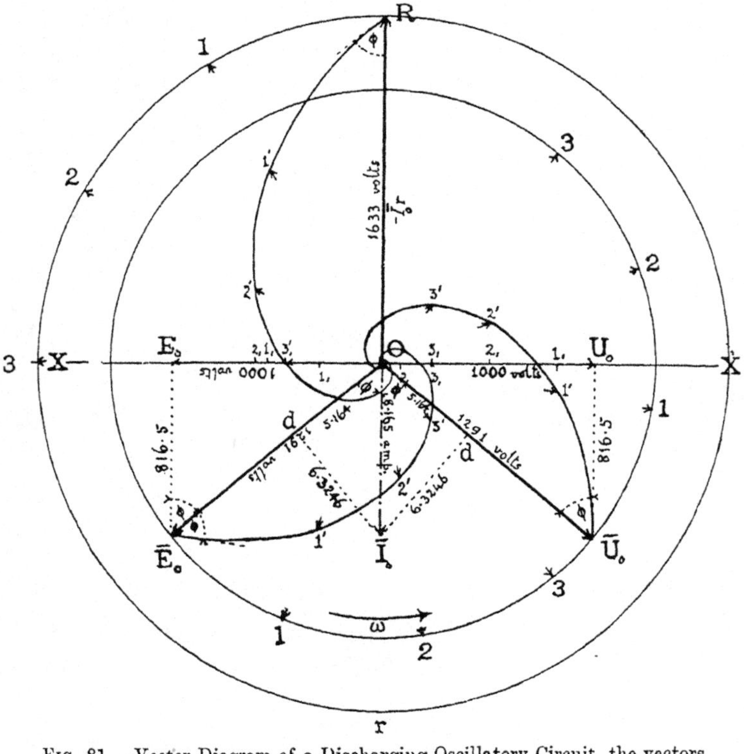


FIG. 81.—Vector Diagram of a Discharging Oscillatory Circuit, the vectors rotating on circles if subjected to independent damping.

If the circuit is ultraperiodic, so that $\rho > z$, construct an impedance triangle, Fig. 80, with $z$ as base and $\rho$ as hypothenuse. The perpendicular will be a hyperbolic reactance $l\Omega$, such that dividing this reactance by the inductance $l$ we obtain the hyperbolic angular velocity $\Omega$. The angle $\psi$ at the base of the triangle corresponds to the angle $\varphi$ in the periodic case. In Fig. 82 we have the analogue of Fig. 81 with respect to rectangular hyperbolas instead of to circles. $O\overline{U}_0$ is the initial vector discharging

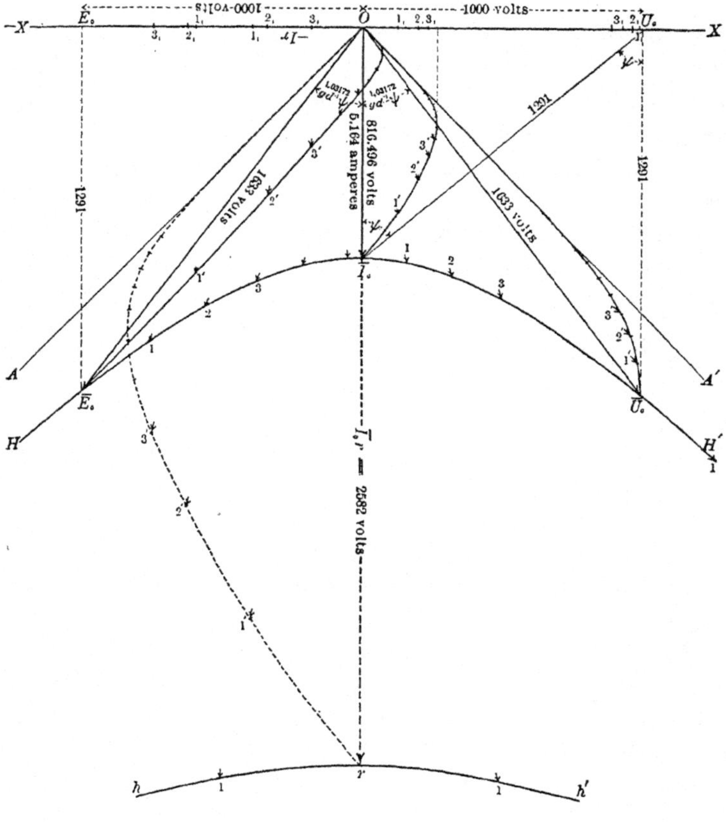


FIG. 82.—Vector Diagram of a Discharging Ultraperiodic Circuit, the vectors rotating on rectangular hyperbolas if subjected to independent damping.

e.m.f. the projection $OU_0$ of which on the XOX-axis, is the initial discharging e.m.f. actually existing in the condenser. The hyperbolic angle of the sector $O\overline{I}_0\overline{U}_0$ has its Gudermannian angle the circular angle $\psi$ of the impedance triangle. The vector current $O\overline{I}_0$ lies midway between the vector discharging e.m.f. $O\overline{U}_0$ and the vector e.m.f. of self-induction $O\overline{E}_0$. It is drawn to current scale, the ratio $\frac{OU_0}{OI_0}$ being made equal to $z$. Multiplying $\overline{I}_0$ by the resistance $r$ of the circuit, we obtain the vector $Or$, which should be reversed in direction if there were room on the diagram, in order to represent the vector $-I_0 r$ drop. As in the circular case, the three vector e.m.f.'s $O\overline{U}_0$, $O\overline{E}_0$, $-Or$ have always zero as their vector sum. In the periodic case they all rotate in circles with uniform circular angular velocity $\omega$. In the ultraperiodic case, they all rotate in rectangular hyperbolas with uniform hyperbolic angular velocity $\Omega$. The damping coefficient may be applied to each projected quantity separately, or the vectors may instead be considered to rotate in the spiral curves shown. A considerable number of trigonometric formulas which apply to the rotatory vector diagram of Fig. 81, also apply to that of Fig. 82 when hyperbolic functions are substituted for the corresponding circular functions (Bibliography, 70).

The important theorem that a hyperbola takes the place of the circle when the discharge changes from the periodic to the ultraperiodic type was first published by Dr. Alexander Macfarlane (Bibliography, 18).

*Inverse Hyperbolic Functions.*—Let an indefinitely long perfectly conducting cylinder be supported parallel to an indefinitely extending perfectly conducting plane, but separated therefrom by a uniform conducting medium of resistivity $\rho$ absohm-cm. (or C.G.S. absolute magnetic units of resistance in a cube of 1 cm. between opposed faces). A cross-section of the system is indicated in Fig. 83, where DEF is the conducting cylinder of radius $\sigma$ cm., with its centre at C, which is perpendicularly distant $d$ cm. from the conducting plane Z′OZ Let us take in imagination two such sections at a distance of 1 cm. apart along the cylinder, so as to comprise between them a slab of the system 1 cm. thick. Then the resistance of the medium in this 1 cm. slab between the cylinder and the plane will be equal to that of a 1 cm. slab of the rectilinear system shown in Fig. 84, where EF, a conducting strip $D=2\pi\sigma$ cm.

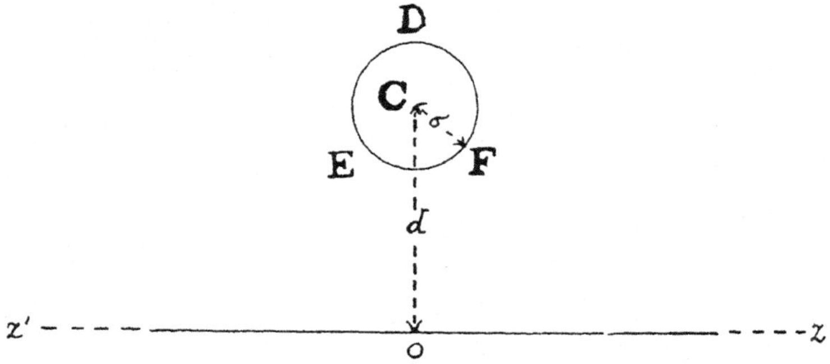


FIG. 83.—Section of a Conducting Cylinder DEF parallel to an indefinitely extending plane $z'oz$.

wide, is placed parallel to a similar strip Z'Z, but separated therefrom by the distance $\sigma \cosh^{-1}\left(\frac{d}{\sigma}\right)$ cm. between insulating walls shown shaded in the diagram. Thus the linear resistance of the system in Fig. 83, is the same as that of the system in

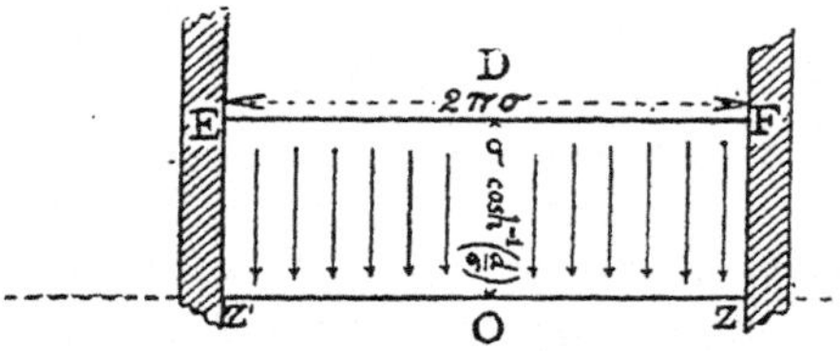


FIG. 84.—Equivalent Slab Section corresponding to infinite plane and parallel cylinder of Fig. 83.

Fig. 84. In the diagram of Fig. 83, all the stream lines of current are circles and the equipotential lines are orthogonally intersecting circles. In the diagram of Fig. 84 the stream lines are all parallel straight lines and the equipotential lines likewise.

From the linear resistance of a conducting medium we may pass by a well-known transition to the linear capacitance of the corresponding dielectric medium. The linear capacitance of two parallel cylinders, whether of the same diameter or of different diameters, then follows at once in terms of inverse hyperbolic functions or anti-hyperbolic functions. These formulas are rigid at all distances, whereas the ordinary linear capacitance formulas are only approximate formulas, in which the error is practically insignificant when the distance between them is more than say 25 times the diameter of each cylinder (Bibliography, 55).

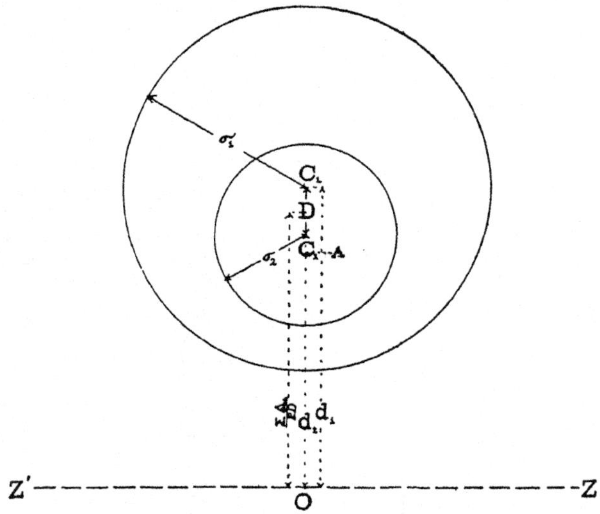


FIG. 85.—Two Parallel Eccentric Cylinders, one enclosing the other, and the inferred common zero-potential plane.

Inverse hyperbolic functions apply very conveniently also to eccentric parallel cylinders, as indicated in Fig. 85.

The geometrical transformation of a rectilinear area into a conjugate circular area is represented in Fig. 86. Here in $v\,w$ plane we have abscissas from $v=-1$ up to $v=1$, and ordinates from $w=-\frac{\pi}{2}$ to $w=+\frac{\pi}{2}$. If we apply to this plane the tanh function in the form—

$$y + jz = \tanh (v + jw)$$

we obtain the curvilinear diagram shown. Corresponding portions of these two diagrams are shaded alike. Thus if *pqrs* and *tuvx* are the sections of two parallel wires in the *yz* plane, then the flow lines of current in a conducting medium or of displacement in a dielectric medium follow the circular

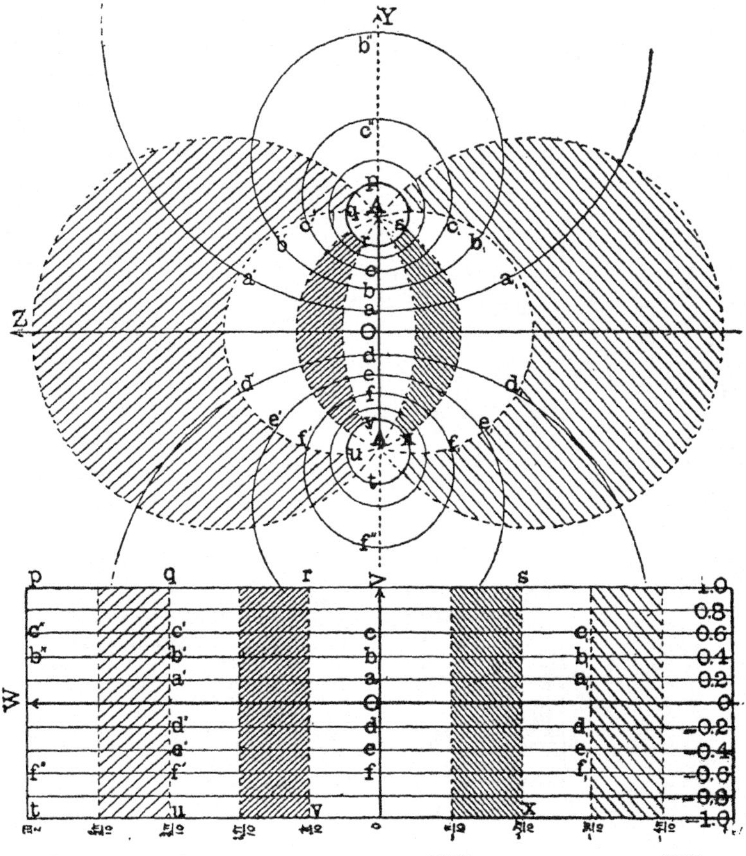

FIG. 86.—Graphical Comparison of $(v + w\sqrt{-1})$ and of $\tanh(v + w\sqrt{-1})$.

segments such as $c'b'a'd'e'f'$ and the equipotential lines are the circles intersecting these. The total flow would be the same as between the flat surfaces *pqrs* and *tuvx* in the rectilinear diagram. Corresponding amounts of the flow occur in corresponding shaded areas and corresponding potential drops can be traced in the same way.

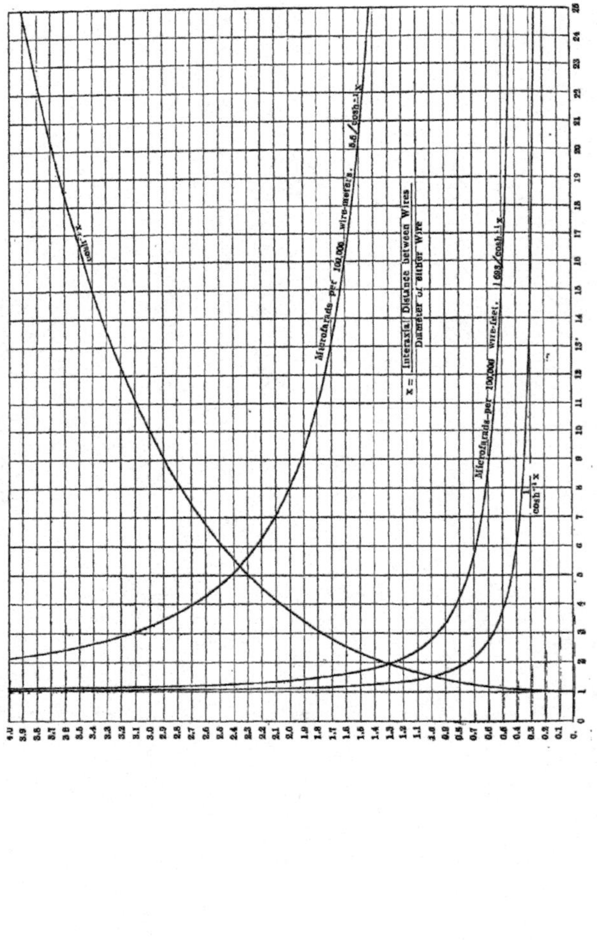


FIG. 87.—Linear Capacity per 100,000 Wire-Feet and Wire-Meters.
Graphs of $\cosh^{-1}x$, $1/\cosh^{-1}x$, and linear capacitances of bare, equal, parallel, round wires in air for interaxial distances up to 25 diameters.

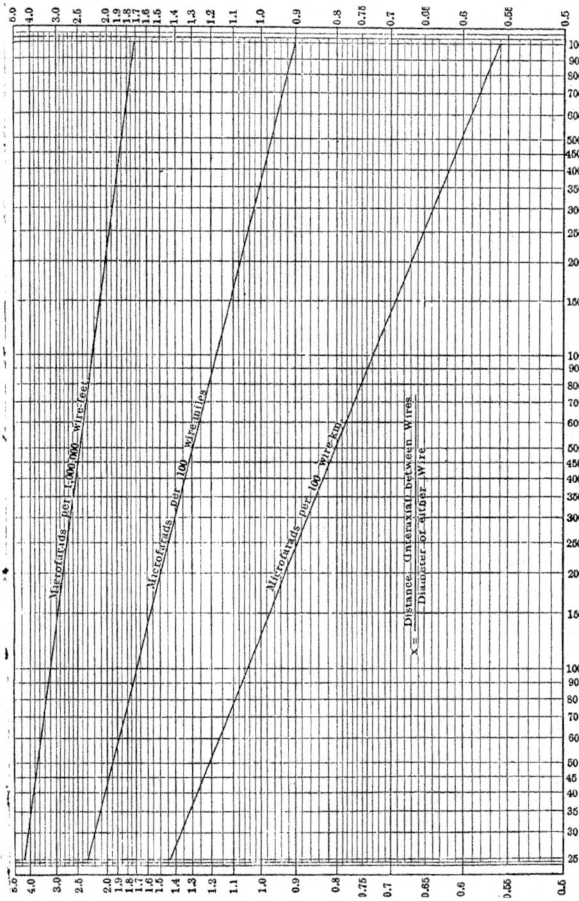


FIG. 88.—Linear Capacitance for Different Interaxial Distances.

Graphs of linear capacitance of bare, equal, parallel, round wires in air for interaxial distances from 25 to 10,000 diameters, in microfarads per 1,000,000 "wire-feet," per 100 "wire-miles" and per 100 "wire-kilometers"; $y = \frac{16 \cdot 93}{\cosh^{-1} x}$ mf. per 1,000,000 wire-feet;

$$y = \frac{8 \cdot 941}{\cosh^{-1} x} \text{ mf. per 100 wire-miles}; \; y = \frac{5 \cdot 555}{\cosh^{-1} x} \text{ mf. per 100 wire-kilometers.}$$

The application of these inverse hyperbolic formulas to linear capacities between equal parallel wires, when the wires are very close together, is illustrated in Fig. 87, which gives the linear capacitance per wire-meter and wire-foot for various interaxial distances up to 25 diameters.* The linear capacitance per loop-meter or loop-foot will be respectively half of the amounts shown.

Finally, Fig. 88 gives by inspection the linear capacitance of pairs of parallel wires for various separations greater than 25 diameters, and at which the error in the ordinary formulas becomes negligible for practical purposes.

Since the external linear inductance of any uniform linear system of insulated conductors expressed in abhenrys per cm. must be the reciprocal of the linear capacitance of the same system of conductors expressed in statfarads † per cm., it is evident that inverse hyperbolic functions may also be applied to computing the linear inductances of cylindrical conductors in close mutual proximity.

Dr. C. V. Drysdale has devised and published an interesting application of hyperbolic functions to the theory of the magnetism of linear magnets. (*See* Bibliography, 71.)

Hyperbolic functions have also been applied for a long time to the problem of eddy currents in conducting laminæ subjected to simple harmonic magnetic forces in their plane. (*See* Bibliography, 4, 62.)

Hyperbolic functions have also been applied by Professors J. J. Thomson, Alexander Russell, and others, to several electrical problems other than those discussed in the preceding pages. (*See* Bibliography, 4, 36, 56.)

* *See* Bibliography, 55, 67, 68.

† The abhenry is an unauthoritative name provisionally suggested for the absolute unit of inductance in the C.G.S. magnetic system, and the statfarad is a similarly provisional name suggested for the absolute unit of capacitance in the C.G.S. electrostatic system.

# APPENDIX A

*Transformation of Circular into Hyperbolic Trigonometrical Formulas.*

THE following identities are well known—

$$\sin x = \frac{\varepsilon^{jx} - \varepsilon^{-jx}}{j2} \quad . \; . \; . \; . \quad \text{numeric} \quad (1)$$

$$\cos x = \frac{\varepsilon^{jx} + \varepsilon^{-jx}}{2} \quad . \; . \; . \; . \quad ,, \quad (2)$$

$$\tan x = \frac{1}{j} \cdot \frac{\varepsilon^{jx} - \varepsilon^{-jx}}{\varepsilon^{jx} + \varepsilon^{-jx}} \quad . \; . \; . \quad ,, \quad (3)$$

$$\sinh x = \frac{\varepsilon^{x} - \varepsilon^{-x}}{2} \quad . \; . \; . \; . \; . \quad ,, \quad (4)$$

$$\cosh x = \frac{\varepsilon^{x} + \varepsilon^{-x}}{2} \quad . \; . \; . \; . \; . \quad ,, \quad (5)$$

$$\tanh x = \frac{\varepsilon^{x} - \varepsilon^{-x}}{\varepsilon + \varepsilon^{-x}} \quad . \; . \; . \; . \; . \quad ,, \quad (6)$$

where $x$ is any real quantity and $j = \sqrt{-1}$.

In formulas (1), (2) and (3) substitute $jy$ for $x$. Thus—

$$\sin x = \sin jy = \frac{\varepsilon^{-y} - \varepsilon^{y}}{2j} = j \cdot \frac{\varepsilon^{y} - \varepsilon^{-y}}{2} = j \sinh y \quad \text{numeric} \quad (7)$$

$$\cos x = \cos jy = \frac{\varepsilon^{-y} + \varepsilon^{y}}{2} = \frac{\varepsilon^{y} + \varepsilon^{-y}}{2} = \cosh y \quad ,, \quad (8)$$

$$\tan x = \tan jy = \frac{1}{j} \cdot \frac{\varepsilon^{-y} - \varepsilon^{y}}{\varepsilon^{-y} + \varepsilon^{y}} = j \frac{\varepsilon^{y} - \varepsilon^{-y}}{\varepsilon^{y} + \varepsilon^{-y}} = j \tanh y \quad ,, \quad (9)$$

Consequently in any transformation formula of simple circular trigonometry containing sines, cosines, or tangents of any real numerical argument $x$, we may substitute

| | | | | | |
|---|---|---|---|---|---|
| $j \sinh y$ | or also | $j \sinh x$ | for $\sin x$ . . | numeric | (10) |
| $\cosh y$ | ,, | $\cosh x$ | ,, $\cos x$ . . | ,, | (11) |
| $j \tanh y$ | ,, | $j \tanh x$ | ,, $\tan x$ . . | ,, | (12) |

since a *transformation* equation in terms of $y$ will hold true generally. We may thus derive the corresponding formula of hyperbolic trigonometry. For example, taking the well-known formula—

$$\tan x = \frac{\sin x}{\cos x} \quad . \quad . \quad . \quad . \quad \text{numeric} \quad (13)$$

we obtain
$$j \tanh x = \frac{j \sinh x}{\cosh x} \quad . \quad . \quad . \quad \text{,,} \quad (14)$$

or
$$\tanh x = \frac{\sinh x}{\cosh x} \quad . \quad . \quad . \quad \text{,,} \quad (15)$$

Again taking $\cos^2 x + \sin^2 x = 1 \quad . \quad . \quad . \quad$ ,, (16)

we obtain $\cosh^2 x + j^2 \sinh^2 x = 1 \quad . \quad . \quad . \quad$ ,, (17)

or $\cosh^2 x - \sinh^2 x = 1 \quad . \quad . \quad . \quad$ ,, (18)

Thus all the regular formulas of circular trigonometry may be transformed.

# APPENDIX B

*Short List of Important Trigonometrical Formulas showing the Hyperbolic and Circular Equivalents.*

| HYPERBOLIC | CIRCULAR |
|---|---|
| $\operatorname{cosech}\theta = \dfrac{1}{\sinh\theta}$ | $\operatorname{cosec}\beta = \dfrac{1}{\sin\beta}$ |
| $\operatorname{sech}\theta = \dfrac{1}{\cosh\theta}$ | $\sec\beta = \dfrac{1}{\cos\beta}$ |
| $\coth\theta = \dfrac{1}{\tanh\theta}$ | $\cot\beta = \dfrac{1}{\tan\beta}$ |
| $\cosh^2\theta - \sinh^2\theta = 1$ | $\cos^2\beta + \sin^2\beta = 1$ |
| $\operatorname{sech}^2\theta = 1 - \tanh^2\theta$ | $\sec^2\beta = 1 + \tan^2\beta$ |
| $\sinh 2\theta = 2\sinh\theta\cosh\theta$ | $\sin 2\beta = 2\sin\beta\cos\beta$ |
| $\cosh 2\theta = \cosh^2\theta + \sinh^2\theta$ | $\cos 2\beta = \cos^2\beta - \sin^2\beta$ |
| $\tanh 2\theta = \dfrac{2\tanh\theta}{1+\tanh^2\theta}$ | $\tan 2\beta = \dfrac{2\tan\beta}{1-\tan^2\beta}$ |
| $\coth 2\theta = \dfrac{\coth^2\theta + 1}{2\coth\theta}$ | $\cot 2\beta = \dfrac{\cot^2\beta - 1}{2\cot\beta}$ |
| $\tanh\dfrac{\theta}{2} = \dfrac{\sinh\theta}{1+\cosh\theta} = \dfrac{\cosh\theta - 1}{\sinh\theta} = \sqrt{\dfrac{\cosh\theta - 1}{\cosh\theta + 1}}$ | $\tan\dfrac{\beta}{2} = \dfrac{\sin\beta}{1+\cos\beta} = \dfrac{1-\cos\beta}{\sin\beta} = \sqrt{\dfrac{1-\cos\beta}{1+\cos\beta}}$ |
| $\sinh(\theta_1 \pm \theta_2) = \sinh\theta_1\cosh\theta_2 \pm \cosh\theta_1 \sinh\theta_2$ | $\sin(\beta_1 \pm \beta_2) = \sin\beta_1\cos\beta_2 \pm \cos\beta_1 \sin\beta_2$ |
| $\cosh(\theta_1 \pm \theta_2) = \cosh\theta_1 \cosh_2 \pm \sinh\theta_1 \sinh\theta_2$ | $\cos(\beta_1 \pm \beta_2) = \cos\beta_1\cos\beta_2 \mp \sin\beta_1 \sin\beta_2$ |
| $\tanh(\theta_1 \pm \theta_2) = \dfrac{\tanh\theta_1 \pm \tanh\theta_2}{1 \pm \tanh\theta_1 \tanh\theta_2}$ | $\tan(\beta_1 \pm \beta_2) = \dfrac{\tan\beta_1 \pm \tan\beta_2}{1 \mp \tan\beta_1\tan\beta_2}$ |
| $\coth(\theta_1 \pm \theta_2) = \dfrac{\coth\theta_1\coth\theta_2 \pm 1}{\coth\theta_2 \pm \coth\theta_1}$ | $\cot(\beta_1 \pm \beta_2) = \dfrac{\cot\beta_1\cot\beta_2 \mp 1}{\cot\beta_2 \pm \cot\beta_1}$ |
| $\sinh(\theta_1+\theta_2) + \sinh(\theta_1-\theta_2) = 2\sinh\theta_1 \cosh\theta_2$ | $\sin(\beta_1+\beta_2) + \sin(\beta_1-\beta_2) = 2\sin\beta \cos\beta_2$ |
| $\sinh(\theta_1+\theta_2) - \sinh(\theta_1-\theta_2) = 2\cosh\theta_1 \sinh\theta_2$ | $\sin(\beta_1+\beta_2) - \sin(\beta_1-\beta_2) = 2\cos\beta_1 \sin\beta_2$ |
| $\sinh\theta = \dfrac{\epsilon^{\theta} - \epsilon^{-\theta}}{2}$ | $\sin\beta = \dfrac{\epsilon^{j\beta} - \epsilon^{-j\beta}}{2j}$ |
| $\cosh\theta = \dfrac{\epsilon^{\theta} + \epsilon^{-\theta}}{2}$ | $\cos\beta = \dfrac{\epsilon^{j\beta} + \epsilon^{-j\beta}}{2}$ |
| $\tanh\theta = \dfrac{\epsilon^{\theta} - \epsilon^{-\theta}}{\epsilon^{\theta} + \epsilon^{-\theta}}$ | $\tan\beta = \dfrac{\epsilon^{j\beta} - \epsilon^{-j\beta}}{j(\epsilon \quad + \epsilon^{-j\beta})}$ |
| $\sinh\theta = \theta + \dfrac{\theta^3}{3!} + \dfrac{\theta^5}{5!} +$ | $\sin\beta = \beta - \dfrac{\beta^3}{3!} + \dfrac{\beta^5}{5!} -$ |
| $\cosh\theta = 1 + \dfrac{\theta^2}{2!} + \dfrac{\theta^4}{4!} +$ | $\cos\beta = 1 - \dfrac{\beta^2}{2!} + \dfrac{\beta^4}{4!} -$ |
| $\tanh\theta = \theta - \dfrac{\theta^3}{3} + \dfrac{2\theta^5}{15} - \dfrac{17\theta^7}{315} +$ | $\tan\beta = \beta + \dfrac{\beta^3}{3} + \dfrac{2\beta^5}{15} + \dfrac{17\beta^7}{315} +$ |
| $\dfrac{d\sinh\theta}{d\theta} = \cosh\theta$ | $\dfrac{d\sin\beta}{d\beta} = \cos\beta$ |
| $\dfrac{d\cosh\theta}{d\theta} = \sinh\theta$ | $\dfrac{d\cos\beta}{d\beta} = -\sin\beta$ |
| $\dfrac{d\tanh\theta}{d\theta} = \operatorname{sech}^2\theta$ | $\dfrac{d\tan\beta}{d\beta} = \sec^2\beta$ |

## APPENDIX C

*Fundamental Relations of Voltage and Current at any Point along a Uniform Line in the Steady State.*

CONSIDER a uniform conductor, say a lead-covered insulated wire, as indicated in Fig. C. At a distance $x$ km. from the home end, let the steady potential to ground or sheath be $e$ volts, and the steady current along the conductor be $i$ amperes.

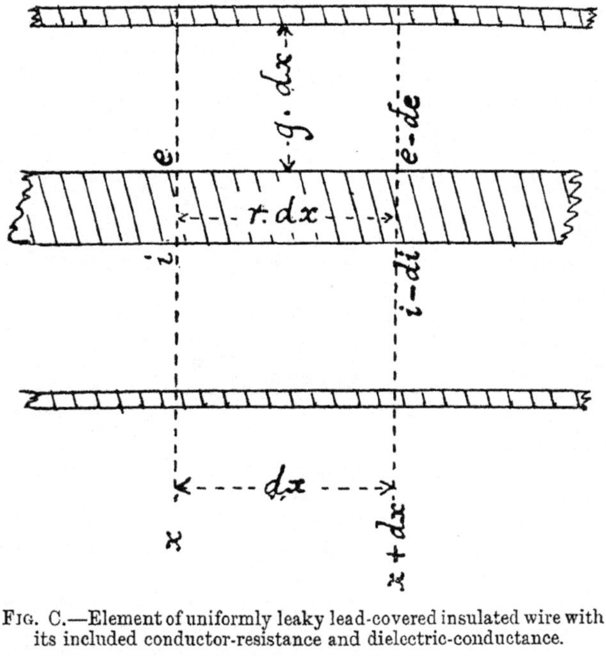


FIG. C.—Element of uniformly leaky lead-covered insulated wire with its included conductor-resistance and dielectric-conductance.

Then if we consider a short element of the conductor $dx$ km long, and if $r$ is the linear conductor-resistance, the conductor-resistance in the element will be $rdx$ ohms. Likewise, if $g$ be the linear leakance of the insulator in mhos per km., the leakance in the element will be $gdx$ mhos between conductor and sheath. The drop of potential in the element will be $i.rdx$ volts, and the drop of current in the element will be $e.gdx$ amperes.

That is—

$$-de = i\,r\,dx \quad . \; . \; . \; . \; . \; . \quad \text{volts} \qquad (1)$$

or—

$$\frac{de}{dx} = -ir \quad . \; . \; . \; . \; . \quad \text{volts/km.} \qquad (2)$$

and—

$$-di = e\,g\,dx \quad . \; . \; . \; . \; . \quad \text{amperes} \qquad (3)$$

or—

$$\frac{di}{dx} = -eg \quad . \; . \; . \quad \text{amperes/km.} \qquad (4)$$

Differentiating (2) and (4) again with respect to $x$ we obtain—

$$\frac{d^2e}{dx^2} = -r\frac{di}{dx} = gr\,.\,e \quad . \; . \; . \; . \; . \quad \frac{\text{volts}}{(\text{km})^2} \qquad (5)$$

and—

$$\frac{d^2i}{dx^2} = -g\frac{de}{dx} = gr\,.\,i \quad . \; . \; . \quad \frac{\text{amperes}}{(\text{km})^2} \qquad (6)$$

The solutions of (5) and (6) are known to be (Bibliography, 7)—

$$e = \text{A}\cosh(\sqrt{gr}\,.\,x) + \text{B}\sinh(\sqrt{gr}\,.\,x) \quad . \quad \text{volts} \qquad (7)$$

$$i = \text{C}\cosh(\sqrt{gr}\,.\,x) + \text{D}\sinh(\sqrt{gr}\,.\,x) \quad \text{amperes} \qquad (8)$$

The bracketed quantities in formulas (7) and (8) are hyperbolic angles; while A, B, C and D are constants depending on the terminal conditions of the line. If the steady conditions at the sending end are known to be an impressed e.m.f. of $E_A$ volts and an entering current of $I_A$ amperes; then—

$$e = E_A \cosh L_1\alpha - I_A r_o \sinh L_1\alpha \quad . \; . \quad \text{volts} \qquad (9)$$

and—

$$i = I_A \cosh L_1\alpha - \frac{E_A}{r_o}\sinh L_1\alpha \quad . \; . \quad \text{amperes} \qquad (10)$$

where $L_1$ is the distance $x$ from the sending end, $\alpha = \sqrt{gr}$ and $r_o = \sqrt{\frac{r}{g}}$. If, on the other hand, the conditions at the receiving end, B, are known to be a terminal e.m.f. of $E_B$ volts and an escaping current of $I_B$ amperes; then (7) and (8) become—

$$e = E_B \cosh L_2\alpha + I_B r_o \sinh L_2\alpha \quad . \; . \quad \text{volts} \qquad (11)$$

and— $$i = I_B \cosh L_2 a + \frac{E_B}{r_0} \sinh L_2 a \quad . \quad . \quad \text{amperes} \quad (12)$$

where $L_2$ is the distance of the point $x$ from the receiving end of the line.

Thus, if the distant end B of the line is freed (Fig. 10), then in (11) and (12) we know that the current to ground at that end vanishes; so that $I_B = 0$, and those equations become—

$$e = E_B \cosh L_2 a \quad . \quad . \quad . \quad . \quad \text{volts} \quad (11a)$$

$$i = \frac{E_B}{r_0} \sinh L_2 a \quad . \quad . \quad . \quad \text{amperes} \quad (12a)$$

where $e$ and $i$ are respectively the e.m.f. and current at the sending end of the line of length $L_2$ km. Consequently, the resistance offered by this line at P is—

$$R_f = \frac{e}{i} = r_0 \coth L_2 a \quad . \quad . \quad . \quad \text{ohms} \quad (13)$$

But although (11a), (12a) and (13) are arrived at over a length $L_2$ km. between the point P and the distant free end B, yet the reasoning they present is general, and applies if the point P be moved back to A (Fig. 10). That is, they apply to any line of length L and angle $\theta$; so that, substituting $E_A$ for $e$, and $I_A$ for $i$, we have—

$$E_B = \frac{E_A}{\cosh \theta} = E_A \operatorname{sech} \theta \quad . \quad . \quad . \quad \text{volts} \quad (14)$$

$$R_f = r_0 \coth \theta \quad . \quad . \quad . \quad . \quad . \quad . \quad \text{ohms} \quad (15)$$

Again, if the distant end B of the line is grounded, then in (11), the terminal e.m.f. $E_B = 0$, and this equation becomes—

$$e = I_B \, r_0 \sinh L_2 a \quad . \quad . \quad . \quad \text{volts} \quad (11b)$$

or the current $I_B$, received to ground at B, is—

$$I_B = \frac{e}{r_0 \sinh L_2 a} \quad . \quad . \quad . \quad \text{amperes} \quad (16)$$

where $e$ is the e.m.f. impressed on the section $L_2$ at P. If we move P back to A, the length becomes L, and the angle of the line becomes $La = \theta$; while $e$ becomes $E_A$. Hence—

$$I_B = \frac{E_A}{r_0 \sinh \theta} \quad . \quad . \quad . \quad . \quad \text{amperes} \quad (17)$$

To find the sending-end current at A with the line grounded at B, we may take (12), make $E_B = 0$, and substitute the value of $I_B$ from (16); or, we may take (9), and make $e = 0$ therein. In either case, we find—

$$I_A = \frac{E_A}{r_0 \tanh \theta} = \frac{E_A}{r_0} \coth \theta \quad . \quad \text{amperes} \quad (18)$$

In the same manner, if we wish to find the potential and current under any assigned terminal conditions, we have only to apply those conditions to the proper equation in (9), (10), (11) or (12), and arrive at the required result.

# APPENDIX D

*Algebraic Proof of Equivalence between a Uniform Line and its T Conductor, both at the Sending and Receiving Ends.*

LET A′O′B′G′, Fig. D, be the equivalent $T$ of a line having an angle $\theta$ hyps. and a surge-resistance $z_o$ ohms or surge-conductance $y_o = 1/z_o$ mhos. Let the two branch resistances be each made equal to $\rho' = z_o \tanh\frac{\theta}{2}$ ohms, and the conductance of the leak be made $g' = y_o \sinh\theta$ mhos. Then if the $T$ be grounded at B through a resistance $\sigma$ ohms, the apparent resistance of the line at A will be—

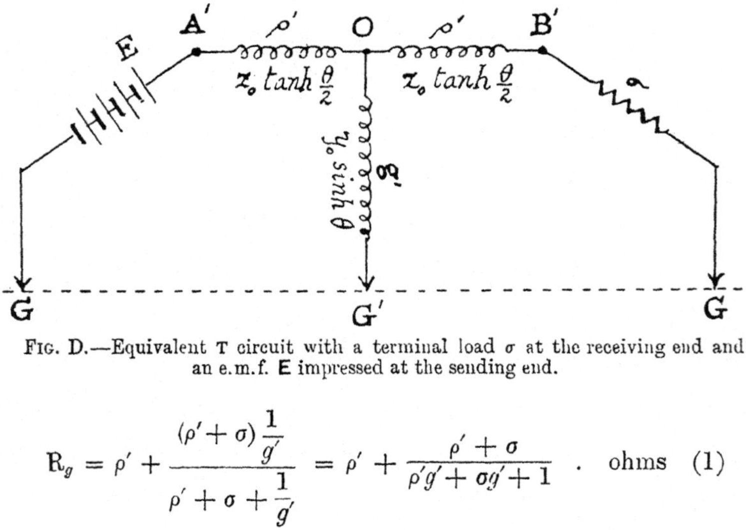


FIG. D.—Equivalent T circuit with a terminal load $\sigma$ at the receiving end and an e.m.f. E impressed at the sending end.

$$R_g = \rho' + \frac{(\rho' + \sigma)\frac{1}{g'}}{\rho' + \sigma + \frac{1}{g'}} = \rho' + \frac{\rho' + \sigma}{\rho' g' + \sigma g' + 1} \quad . \text{ ohms} \quad (1)$$

Substituting $z_o \tanh\frac{\theta}{2}$ for $\rho'$, $y_o \sinh\theta$ for $g'$, and $z_o \tanh\theta'$ for $\sigma$, by (54) we have—

$$R_g = z_o \tanh\frac{\theta}{2} + \frac{z_o \tanh\frac{\theta}{2} + z_o \tanh\theta'}{\tanh\frac{\theta}{2}\sinh\theta + \tanh\theta'\sinh\theta + 1} \text{ ohms} \quad (2)$$

$$= z_0 \left\{ \tanh \frac{\theta}{2} + \frac{\tanh \frac{\theta}{2} + \tanh \theta'}{\cosh \theta + \sinh \theta \tanh \theta'} \right\} \quad \text{ohms} \quad (3)$$

$$= z_0 \left\{ \frac{\sinh \theta}{1 + \cosh \theta} + \frac{\frac{\sinh \theta}{1 + \cosh \theta} + \tanh \theta'}{\cosh \theta + \sinh \theta \tanh \theta'} \right\} \quad \text{"} \quad (4)$$

$$= z_0 \left\{ \frac{\sinh \theta \cosh \theta + \sinh^2 \theta \tanh \theta' + \sinh \theta + \tanh \theta' (1 + \cosh \theta)}{(1 + \cosh \theta)(\cosh \theta + \sinh \theta \tanh \theta')} \right\} \quad \text{ohms} \quad (5)$$

$$= z_0 \left\{ \frac{(1 + \cosh \theta)(\sinh \theta + \cosh \theta \tanh \theta')}{(1 + \cosh \theta)(\cosh \theta + \sinh \theta \tanh \theta')} \right\} =$$

$$z_0 \frac{\tanh \theta + \tanh \theta'}{1 + \tanh \theta \tanh \theta'} = z_0 \tanh (\theta + \theta') \quad . \quad \text{ohms} \quad (6)$$

This result is in accordance with formula (56) for a terminally loaded uniform line.

The current entering the line at A will therefore be, if $E_A$ is the impressed e.m.f., $I_A = \frac{E_A}{z_0 \tanh (\theta + \theta')}$ amperes, and the current reaching ground at B will be—

$$I_B = I_A . \frac{\frac{z_0}{\sinh \theta}}{z_0 \tanh \frac{\theta}{2} + z_0 \tanh \theta' + \frac{z_0}{\sinh \theta}} \quad \text{amperes} \quad (7)$$

$$= I_A . \frac{1}{\sinh \theta \tanh \frac{\theta}{2} + \sinh \theta \tanh \theta' + 1} \quad \text{"} \quad (8)$$

$$= I_A . \frac{1}{1 + (\cosh \theta - 1) + \sinh \theta \tanh \theta'} \quad \text{"} \quad (9)$$

$$= I_A . \frac{1}{\cosh \theta + \sinh \theta \tanh \theta'} \quad . \quad \text{"} \quad (10)$$

$$= I_A . \frac{\cosh \theta'}{\cosh \theta \cosh \theta' + \sinh \theta \sinh \theta'} \quad \text{"} \quad (11)$$

$$= I_A . \frac{\cosh \theta'}{\cosh (\theta + \theta')} \quad . \quad . \quad . \quad . \quad . \quad \text{"} \quad (12)$$

$$= \frac{E_A \cosh \theta'}{z_0 \sinh (\theta + \theta')} \quad . \quad . \quad . \quad . \quad \text{"} \quad (13)$$

which is in accordance with formula (57).

## APPENDIX E

### *Equivalence of a Line Π and a Line T.*

LET $z_A$ $z_B$ $z_G$ (Fig. E) be the three impedances of the star, $y_A$ $y_B$ $y_G$ be the three corresponding admittances—

$$\frac{1}{z_A}, \frac{1}{z_B}, \frac{1}{z_G} \text{ respectively.}$$

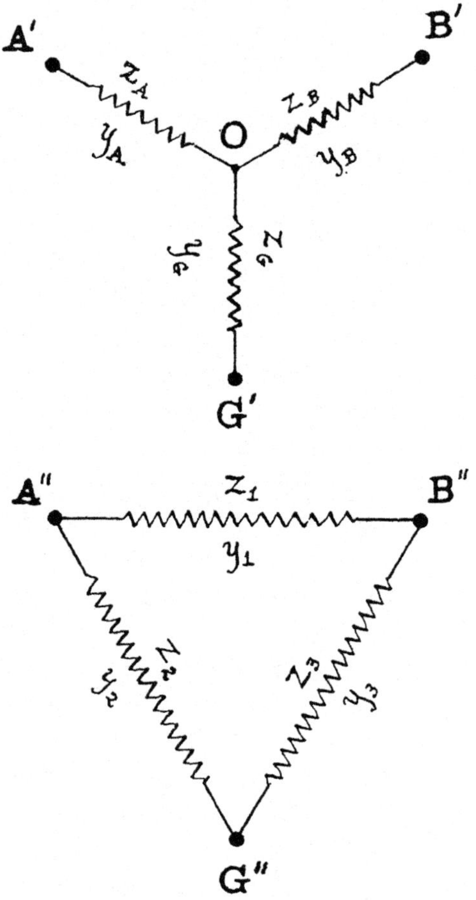


FIG. E.—Externally equivalent star and delta of resistances or impedances.

Let $z_1$ $z_2$ $z_3$ be the three impedances of the delta,
$y_1$ $y_2$ $y_3$ be the three corresponding admittances—
$\frac{1}{z_1}, \frac{1}{z_2}, \frac{1}{z_3}$ respectively.

Then the star will be the external equivalent of the delta * if—

$$z_A = \frac{z_1 z_2}{z_1 + z_2 + z_3} = \frac{z_1 z_2}{\Sigma z} \quad . \quad . \quad . \quad . \quad \text{ohms} \quad (1)$$

$$z_B = \frac{z_1 z_3}{z_1 + z_2 + z_3} = \frac{z_1 z_3}{\Sigma z} \quad . \quad . \quad . \quad . \quad \text{,,} \quad (2)$$

$$z_G = \frac{z_2 z_3}{z_1 + z_2 + z_3} = \frac{z_2 z_3}{\Sigma z} \quad . \quad . \quad . \quad . \quad \text{,,} \quad (3)$$

where $\Sigma z$ represents the sum of the three delta-impedances. The delta will likewise be the external equivalent of the star if—

$$y_1 = \frac{y_A \; y_B}{y_A + y_B + y_G} = \frac{y_A \; y_B}{\Sigma y} \text{ mhos;}$$

$$\text{or } z_1 = \frac{\Sigma y}{y_A \; y_B} \quad \text{ohms} \quad (4)$$

$$y_2 = \frac{y_A \; y_G}{y_A + y_B + y_G} = \frac{y_A \; y_G}{\Sigma y} \text{ mhos;}$$

$$\text{or } z_2 = \frac{\Sigma y}{y_A \; y_G} \quad \text{,,} \quad (5)$$

$$y_3 = \frac{y_B \; y_G}{y_A + y_B + y_G} = \frac{y_B \; y_G}{\Sigma y} \text{ mhos;}$$

$$\text{or } z_3 = \frac{\Sigma y}{y_B \; y_G} \quad \text{,,} \quad (6)$$

where $\Sigma y$ represents the sum of the three star-admittances.

If $\quad y_A = y_o/\tanh\frac{\theta}{2} \quad y_B = y_o/\tanh\frac{\theta}{2} \quad y_G = y_o \sinh\theta$:

then by (4)—

$$z_1 = \frac{y_o\left(\frac{2}{\tanh\frac{\theta}{2}} + \sinh\theta\right)}{\frac{y_o^2}{\tanh^2\frac{\theta}{2}}} = z_o \tanh\frac{\theta}{2}\left(2 + \sinh\theta\tanh\frac{\theta}{2}\right)$$

$$\text{ohms} \quad (7)$$

* *See* Bibliography, 22.

$$= z_o \tanh\frac{\theta}{2}\left(2 + \frac{\sinh^2\theta}{1+\cosh\theta}\right) = z_o \tanh\frac{\theta}{2}\left(1+\cosh\theta\right) \quad \text{ohms} \quad (8)$$

$$= z_o \sinh\theta \quad . \quad . \quad . \quad . \quad . \quad . \quad . \quad . \quad . \quad . \quad . \quad . \quad . \quad ,, \quad (9)$$

which is the impedance to be inserted in the architrave of the $\Pi$. Again by (5)—

$$y_2 = \frac{y_A\, y_G}{\Sigma y} = \frac{y_o^2 \dfrac{\sinh\theta}{\tanh\dfrac{\theta}{2}}}{y_o\left(\dfrac{2+\sinh\theta\tanh\dfrac{\theta}{2}}{\tanh\dfrac{\theta}{2}}\right)} = y_o \frac{\sinh\theta}{2 + \dfrac{\sinh^2\theta}{1+\cosh\theta}}$$

$$= y_o \frac{\sinh\theta}{1+\cosh\theta} = y_o \tanh\frac{\theta}{2} \quad . \quad . \quad . \quad . \quad (10)$$

which is the admittance to be inserted in the A pillar of the $\Pi$, and, by symmetry, also in the B pillar.

# APPENDIX F

*Analysis of Artificial Lines in Terms of Continued Fractions.*

BEFORE taking up the analysis of an artificial line, it is desirable to consider two indispensable propositions in the algebra of continued fractions.

A continued fraction of the type—

$$F_n(a, b) = \cfrac{1}{a + \cfrac{1}{b + \cfrac{1}{a + \cfrac{1}{b + \cdots}}}} \quad \text{(to } n \text{ stages)} \quad . \quad . \quad . \quad \text{numeric } \angle \quad (1)$$

is called an *alternating continued fraction*; because the two terms $a$ and $b$ appear alternately in the successive denominators. On the other hand, a continued fraction of the type—

$$F_n(c) = \cfrac{1}{c + \cfrac{1}{c + \cfrac{1}{c + \cfrac{1}{c + \cdots}}}} \quad \text{(to } n \text{ stages)} \quad . \quad . \quad . \quad \text{numeric } \angle \quad (2)$$

may be called a *constant continued fraction*. The quantities $a$, $b$, and $c$, here considered, are numerical constants, positive or negative, integral or fractional, real, imaginary or complex.*

If we multiply the four-stage continued fraction (1) by any constant factor $d$, we obtain by successive steps—

$$d \times F_n(a, b) = \cfrac{d}{a + \cfrac{1}{b + \cfrac{1}{a + \cfrac{1}{b + \cdots}}}} = \cfrac{1}{\left(\frac{a}{d}\right) + \cfrac{1}{(bd) + \cfrac{d}{a + \cfrac{1}{b + \cdots}}}} = \cfrac{1}{\left(\frac{a}{d}\right) + \cfrac{1}{(bd) + \cfrac{1}{\left(\frac{a}{d}\right) + \cfrac{1}{(bd) + \cdots}}}} \quad . \quad . \quad \text{numeric } \angle \quad (3)$$

* Certain limiting cases of the propositions, including Strehlke's theorem (Bibliography, 2), are defined in the original publication. (*See* Bibliography, 47.)

This process is perfectly general, and may be carried on to any number of stages. It leads to the following conclusions, which may be also demonstrated in other ways—

(*a*) The effect of multiplying a continued fraction by a constant $d$ is to divide all the odd denominators by $d$, and to multiply all the even denominators by $d$.

(*b*) The effect of dividing a continued fraction by a constant $d$ is to multiply all the odd denominators by $d$, and to divide all the even denominators by $d$.

(*c*) The effect of multiplying an alternate continued fraction $F_n(a, b)$, by a constant $d$ is to produce a new alternate continued fraction, in which the odd denominators are $a/d$, and the even denominators are $bd$.

(*d*) The effect of multiplying an alternate continued fraction $F_n(a, b)$, such as (1), by the particular constant $d = \sqrt{\frac{a}{b}}$, is, by the last preceding proposition, to produce a new alternate continued fraction, in which the odd denominators are $a\sqrt{\frac{b}{a}} = \sqrt{ab}$ and the even denominators are $b\sqrt{\frac{a}{b}} = \sqrt{ab}$. Thus the new alternate continued fraction reduces to a constant continued fraction $F_n(\sqrt{ab})$ for the particular case $d = \sqrt{\frac{a}{b}}$

or—

$$\sqrt{\frac{a}{b}}\, F_n(a, b) = F_n(\sqrt{ab}) \quad . \quad . \quad \text{numeric} \angle \quad (4)$$

and—

$$F_n(a, b) = \sqrt{\frac{b}{a}}\, F_n(\sqrt{ab}) \quad . \quad \text{,,} \quad \text{,,} \quad (5)$$

so that—

$$\cfrac{1}{a + \cfrac{1}{b + \cfrac{1}{a + \cfrac{1}{b + \cdots}}}} \text{ (to } n \text{ stages)} = \sqrt{\frac{b}{a}} \times \cfrac{1}{\sqrt{ab} + \cfrac{1}{\sqrt{ab} + \cfrac{1}{\sqrt{ab} + \cfrac{1}{\sqrt{ab} + \cdots}}}} \text{ (to } n \text{ stages)} \quad \text{,,} \quad \text{,,} \quad (6)$$

That is, any alternate continued fraction may be expressed as a factor multiplied by a constant continued fraction of the same number of stages, the constant term of the latter being the geometrical mean $\sqrt{ab}$ of the two alternate terms of the former, and the factor being $\sqrt{\frac{b}{a}}$ the square-root of the ratio of the alternates.

*Terminally Loaded Alternate Continued Fractions.*—If an alternate continued fraction of $n$ stages in $a$ and $b$ terminates in a denominator $m$, thus—

$$F_n(a, b)_{\frac{1}{m}} = \cfrac{1}{a + \cfrac{1}{b + \cfrac{1}{a + \cdots \cfrac{1}{b + \cfrac{1}{a + \cfrac{1}{m}}}}}} \quad \text{(to } n \text{ stages)} \qquad \text{numeric } \angle \quad (7)$$

it is called a terminally loaded alternate continued fraction, and the final fraction $1/m$ is called the terminal load of the continued fraction.

*Expression of a Constant Continued Fraction in Hyperbolic Functions of an Auxiliary Angle.*—It may be proved * that—

$$F_n(c) = \cfrac{1}{c + \cfrac{1}{c + \cfrac{1}{c + \cfrac{1}{c + \cdots}}}} \quad \text{(to } n \text{ stages)} \quad = \frac{\sinh nu}{\cosh (n+1)u} \quad \text{if } n \text{ is even} \quad (8)$$

or—

$$= \frac{\cosh nu}{\sinh (n+1)u} \quad \text{if } n \text{ is odd} \quad (9)$$

where $\sinh u = \frac{c}{2}$ or $u = \sinh^{-1}\left(\frac{c}{2}\right)$ numeric $\angle$ (10)

Consequently, any constant continued fraction $n$ stages may be expressed as a sine-cosine ratio of $n$ and $n + 1$ times a certain auxiliary hyperbolic angle $u$, defined by (10). It follows,

* *See* Bibliography, 47.

therefore, from the preceding that any alternate continued fraction of $n$ stages may be expressed as a factor, multiplied by sine-cosine ratio of $n$ and $n+1$ times a certain auxiliary hyperbolic angle $u$.

*Example.*—Consider the particular 3-stage alternate continued

$$\text{fraction } F_3(a,b) = \cfrac{1}{0{\cdot}00025 + \cfrac{1}{500 + \cfrac{1}{0{\cdot}00025}}}$$

Here $a = 0{\cdot}00025$ and $b = 500$. This can be transformed by proposition ($d$) formula (6) into the constant continued fraction—

$$\sqrt{\frac{500}{0{\cdot}00025}} \times \cfrac{1}{\sqrt{0{\cdot}125} + \cfrac{1}{\sqrt{0{\cdot}125} + \cfrac{1}{\sqrt{0{\cdot}125}}}}.$$

This, again, becomes, by (9), $\sqrt{2{,}000{,}000} \times \dfrac{\cosh 3u}{\sinh 4u}$

$$\text{where } \sinh u = \frac{\sqrt{0{\cdot}125}}{2} = \frac{0{\cdot}353554}{2} = 0{\cdot}176777$$

or, by tables, $u = 0{\cdot}17586$ . . . . hyp.

$$\text{so that } F_3(a,b) = 1414{\cdot}2138 \times \frac{\cosh 0{\cdot}52758}{\sinh 0{\cdot}70344}$$

$$= 2117{\cdot}7,$$

a result easily checked by direct arithmetical solution of $F_3(a,b)$. In this comparatively simple case, it is easier to solve the alternate by direct arithmetic; but in the case of an alternate continued fraction of unwieldy constants, and many stages, it is much easier to obtain the solution by the hyperbolic process.

*Terminally Loaded Alternate Continued Fractions Expressed in Terms of Hyperbolic Functions.*—Considering the following notation for an *ascending* series of terminally loaded alternate continued fractions—

$$F_1(a,b)_{\frac{1}{m}} = \cfrac{1}{a + \cfrac{1}{m}} \quad . \quad . \quad . \quad . \quad . \quad \text{numeric } \angle \quad (11)$$

$$F_2(b,a)_{\frac{1}{m}} = \cfrac{1}{b + \cfrac{1}{a + \cfrac{1}{m}}} \qquad \text{numeric} \angle \quad (12)$$

$$F_3(a,b)_{\frac{1}{m}} = \cfrac{1}{a + \cfrac{1}{b + \cfrac{1}{a + \cfrac{1}{m}}}} \qquad \text{,,} \quad \text{,,} \quad (13)$$

etc.; we may write consistently with the above notation—

$$F_0(b,a)_{\frac{1}{m}} = \frac{1}{m} \quad . \quad . \quad . \quad . \quad . \quad . \quad . \quad \text{numeric} \angle \quad (14)$$

It is easy to show that if $n_{,}$ is an odd value, and $n_{,,}$ an even value of $n$—

$$F_{n_,}(a, b)_{\frac{1}{m}} = \sqrt{\frac{b}{a}} \cdot \frac{\cosh(nu + u')}{\sinh\{(n+1)u + u'\}} \qquad \text{numeric} \angle \quad (15)$$

$$F_{n_{,,}}(b, a)_{\frac{1}{m}} = \sqrt{\frac{a}{b}} \cdot \frac{\sinh(nu + u')}{\cosh\{(n+1)u + u'\}} \qquad \text{,,} \quad \text{,,} \quad (16)$$

where, as before—

$$u = \sinh^{-1}\left(\frac{\sqrt{ab}}{2}\right) = \sinh^{-1}\left(\sqrt{\frac{a}{2} \cdot \frac{b}{2}}\right) \qquad \text{,,} \quad \text{,,} \quad (17)$$

and— $$u' = \tanh^{-1}\left(\frac{\cosh u}{m\sqrt{\frac{a}{b}} - \sinh u}\right) \quad . \quad . \qquad \text{,,} \quad \text{,,} \quad (18)$$

When $a$, $b$, and $m$ are positive real quantities, (18) becomes uninterpretable if the denominator becomes smaller than the numerator. That is, $m$ must not be less than $\varepsilon^u\sqrt{\frac{b}{a}}$.

If $m$ is less than $\varepsilon^u\sqrt{\frac{b}{a}}$, we may write—

$$u' = j\frac{\pi}{2} + u'' \quad . \quad . \quad . \quad . \quad . \quad . \qquad \text{numeric} \quad (19)$$

whence $$u'' = \tanh^{-1}\left(\frac{m\sqrt{\frac{a}{b}} - \sinh u}{\cosh u}\right) \quad . \quad . \quad . \qquad \text{,,} \quad (20)$$

where $u$ is positive and real. Formulas (15) and (16) then become—

$$F_{n_{,}}(a,b)_{\frac{1}{m}} = \sqrt{\frac{b}{a}} \cdot \frac{\sinh(nu + u'')}{\cosh\{(n+1)u + u''\}} \qquad \text{numeric (21)}$$

$$F_{n_{,,}}(b,a)_{\frac{1}{m}} = \sqrt{\frac{a}{b}} \cdot \frac{\cosh(nu + u'')}{\sinh\{(n+1)u + u''\}} \qquad \text{,, (22)}$$

In the particular case where $m = b/2$, $u'' = 0$; so that—

$$F_{n_{,}}(a,b)_{1/\frac{b}{2}} = \sqrt{\frac{b}{a}} \cdot \frac{\sinh nu}{\cosh(n+1)u} \quad . \quad . \quad . \quad \text{numeric (23)}$$

$$F_{n_{,,}}(b,a)_{1/\frac{b}{2}} = \sqrt{\frac{a}{b}} \cdot \frac{\cosh nu}{\sinh(n+1)u} \quad . \quad . \quad . \quad \text{,, (24)}$$

Moreover, if we add $\frac{b}{2}$ to $F_{n_{,}}(a, b)$ as in the particular case—

$$\frac{b}{2} + F_3(a,b) = \frac{b}{2} + \cfrac{1}{a + \cfrac{1}{b + \cfrac{1}{a}}} \quad . \quad . \quad . \quad . \quad . \quad . \quad . \quad \text{numeric (25)}$$

we have by (6) and (9)—

$$\frac{b}{2} + F_{n_{,}}(a,b) = \sqrt{\frac{b}{a}}\left\{\frac{\sqrt{ab}}{2} + \frac{\cosh nu}{\sinh(n_{,}+1)u}\right\} \qquad \text{numeric (26)}$$

$$= \sqrt{\frac{b}{a}}\left\{\sinh u + \frac{\cosh n_{,}u}{\sinh(n_{,}+1)u}\right\} \qquad \text{,, (27)}$$

$$= \sqrt{\frac{b}{a}}\{\cosh u \cdot \coth(n_{,}+1)u\} \qquad \text{,, (28)}$$

Again, if we add $\frac{b}{2}$ to $F_{n_{,}}(a, b)_{1/\frac{b}{2}}$ as in the particular case—

$$\frac{b}{2} + F_3(a,b)_{1/\frac{b}{2}} = \frac{b}{2} + \cfrac{1}{a + \cfrac{1}{b + \cfrac{1}{a + \cfrac{1}{b/2}}}} \quad . \quad . \quad \text{numeric (29)}$$

we have, by (23)—

$$\frac{b}{2} + F_n\,(a,b)_{1/\frac{b}{2}} = \sqrt{\frac{b}{a}}\left\{\frac{\sqrt{ab}}{2} + \frac{\sinh n_{,}u}{\cosh\,(n_{,}+1)\,u}\right\} \text{numeric} \quad (30)$$

$$= \sqrt{\frac{b}{a}}\left\{\sinh u + \frac{\sinh n_{,}u}{\cosh\,(n_{,}+1)\,u}\right\} \quad \text{,,} \quad (31)$$

$$= \sqrt{\frac{b}{a}}\{\cosh u \cdot \tanh\,(n_{,}+1)\,u\} \quad \text{,,} \quad (32)$$

*Application of Foregoing Formulas to Artificial Lines.*—There are two types of artificial line, the single-conductor and the double-conductor line, as shown in Figs. 1 and 2 respectively.

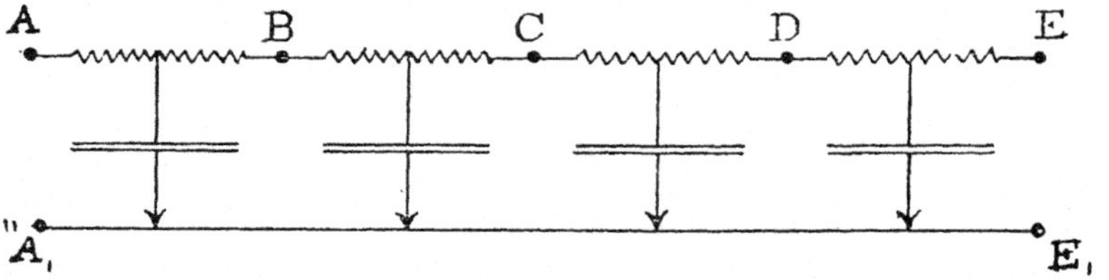


FIG. F1.—Single-Conductor Type of Artificial Line.

These are electrically equivalent, ignoring questions of lumpiness, circuit balancing, and circuit symmetry, if in each section AB of

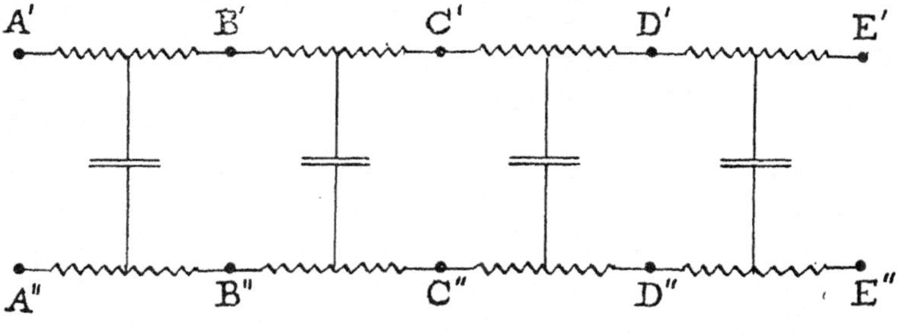


FIG. F2.—Double-Conductor Type of Artificial Line.

Fig. 1, there is the same conductor-resistance as in $\frac{A'B' + A''B''}{2}$ of Fig. 2, and that the capacity in each section of Fig. 1 should be twice the capacity in each section of Fig. 2; so that the CR product, *i.e.* the total resistance R in the line circuit, and the total capacity C across the circuit shall be the same. It is thus evident from what has been considered in earlier chapters that it is sufficient to discuss single-conductor lines only, with the understanding that the discussion applies immediately to corresponding double-conductor lines.

It is also evident, from what has preceded, that we need only consider single-conductor lines of the continuous-current type, with sections of conductor-resistance and leaks in derivation as shown in Fig. 3. The formulas which we shall

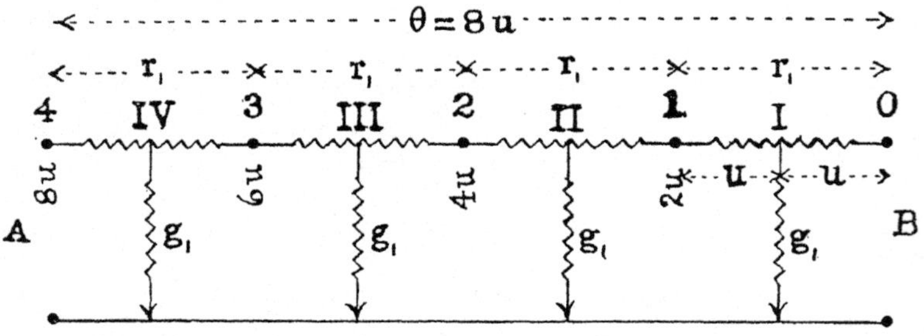


FIG. F3.—Artificial Line of Four Sections.

use will then apply to alternating-current artificial lines with condensers, by extension from real to complex numbers.

Fig. 3 shows a four-section artificial line AB. The four sections are alike. Each consists of a conductor-resistance $r_1$ ohms, with a leak of $g_1$ mhos at the centre. In other words, the artificial line is a simple succession of uniform *T*'s. The junctions are numbered from (0) at B to (4) at A, and the leaks are marked with corresponding Roman numerals.

*Sending-end Resistance of Artificial Line when freed at Distant End.*—Let it be required to find the sending-end resistance of the artificial line at A, when freed at B.

Commencing at leak I nearest to B, the free end, the conductance of this leak is $g_1$ mhos. Consequently, the resistance at and beyond point I on the line is $\frac{1}{g_1}$ ohms. The resistance in the line between leaks is $r_1$ ohms. Therefore the resistance beyond leak II is—

$$R_{fII} = r_1 + \frac{1}{g_1} \quad . \quad . \quad . \quad . \quad \text{ohms} \angle \quad (33)$$

Expressing this as a conductance beyond point II—

$$G_{fII} = \frac{1}{r_1 + \frac{1}{g_1}} = F_2(r_1, g_1) \quad . \quad . \quad \text{mhos} \angle \quad (34)$$

Adding the conductance of leak II, to obtain the total conductance at and beyond II—

$$G'_{fII} = g_1 + \frac{1}{r_1 + \frac{1}{g_1}} = g_1 + F_2(r_1, g_1) \cdot \quad . \quad \text{mhos} \angle \quad (35)$$

Expressing this as a resistance at and beyond II—

$$R'_{fII} = \frac{1}{g_1 + \frac{1}{r_1 + \frac{1}{g_1}}} = F_3(g_1, r_1) \quad . \quad \text{ohms} \angle \quad (36)$$

Now transferring attention to leak III, the total resistance beyond this leak is—

$$R_{fIII} = r_1 + \frac{1}{g_1 + \frac{1}{r_1 + \frac{1}{g_1}}} = r_1 + F_3(g_1, r_1) \quad . \quad \text{ohms} \angle \quad (37)$$

Expressing this as a conductance—

$$G_{fIII} = \frac{1}{r_1 + \frac{1}{g_1 + \frac{1}{r_1 + \frac{1}{g_1}}}} = F_4(r_1, g_1) \quad . \quad \text{mhos} \angle \quad (38)$$

Adding the conductance of leak III to obtain the total conductance at and beyond this leak—

$$G'_{fIII} = g_1 + \frac{1}{r_1 + \frac{1}{g_1 + \frac{1}{r_1 + \frac{1}{g_1}}}} = g_1 + F_4(r_1, g_1) \quad \text{mhos} \angle \quad (39)$$

Expressing this as a resistance—

$$R'_{f\mathrm{III}} = \cfrac{1}{g_1 + \cfrac{1}{r_1 + \cfrac{1}{g_1 + \cfrac{1}{r_1 + \cfrac{1}{g_1}}}}} = F_5(g_1, r_1) \quad . \quad \text{ohms} \angle \quad (40)$$

Transferring attention to leak IV, the resistance beyond this leak is—

$$R_{f\mathrm{IV}} = r_1 + \cfrac{1}{g_1 + \cfrac{1}{r_1 + \cfrac{1}{g_1 + \cfrac{1}{r_1 + \cfrac{1}{g_1}}}}} = r_1 + F_5(g_1, r_1) \quad \text{ohms} \angle \quad (41)$$

Expressing this as a conductance—

$$G_{f\mathrm{IV}} = \cfrac{1}{r_1 + \cfrac{1}{g_1 + \cfrac{1}{r_1 + \cfrac{1}{g_1 + \cfrac{1}{r_1 + \cfrac{1}{g_1}}}}}} = F_6(r_1, g_1) \quad \text{mhos} \angle \quad (42)$$

Adding in the conductance of leak IV, the total conductance at and beyond IV is—

$$G'_{f\mathrm{IV}} = g_1 + \cfrac{1}{r_1 + \cfrac{1}{g_1 + \cfrac{1}{r_1 + \cfrac{1}{g_1 + \cfrac{1}{r_1 + \cfrac{1}{g_1}}}}}} = g_1 + F_6(r_1, g_1) \quad \text{mhos} \angle \quad (43)$$

Expressing this as a resistance and adding the last half-section of resistance 4—IV, we have as the total sending-end resistance of the line in the steady state, with the distant end free—

$$R_{fA} = R_{f(4)} = \frac{r_1}{2} + \cfrac{1}{g_1 + \cfrac{1}{r_1 + \cfrac{1}{g_1 + \cfrac{1}{r_1 + \cfrac{1}{g_1 + \cfrac{1}{r_1 + \cfrac{1}{g_1}}}}}}} = \frac{r_1}{2} + F_7(g_1, r_1) \quad \text{ohms} \angle \quad (44)$$

But by (28) we have—

$$R_{fA} = \sqrt{\frac{r_1}{g_1}}(\cosh u \,.\, \coth 8u) \quad . \quad \text{ohms} \angle \quad (45)$$

If the sections of this line were indefinitely short and indefinitely numerous; that is, if the line were composed of, say, a million small sections each of conductor-resistance $r_1$, and leak $g_1$, it would be equivalent to a smooth uniform actual line of those constants and it would have, by (27), a surge-resistance $\sqrt{\frac{r_1}{g_1}}$ ohms. Consequently, in the artificial line here considered, the ratio $\sqrt{\frac{r_1}{g_1}}$ may be called the *apparent surge-resistance*, and be denoted by $z_o'$. So that—

$$R_{fA} = z_o' \cosh u \,.\, \coth 8u \quad . \quad . \quad \text{ohms} \angle \quad (46)$$

and writing $z_o$ for $z_o' \cosh u$—

$$R_{fA} = z_o \coth 8u \quad . \quad . \quad . \quad . \quad . \quad . \quad \text{ohms} \angle \quad (47)$$

where—

$$u = \sinh^{-1}\left(\frac{\sqrt{g_1 r_1}}{2}\right) \quad . \quad . \quad . \quad . \quad \text{hyps.} \angle \quad (48)$$

As an example, suppose that each section of the artificial line has a resistance $r_1 = 500$ ohms, and a leak of $g_1 = 0{\cdot}00025$ mho. Then—

$$z_o' = \sqrt{\frac{500}{0{\cdot}00025}} = \sqrt{2{,}000{,}000} = 1414{\cdot}2 \text{ ohms,}$$

and—

$$u = \sinh^{-1}\sqrt{\frac{500 \times 0{\cdot}00025}{2}} = \sinh^{-1}\left(\frac{\sqrt{0{\cdot}125}}{2}\right)$$
$$= \sinh^{-1} 0{\cdot}17678 = 0{\cdot}17586 \text{ hyp.}$$

This is the auxiliary hyperbolic angle and is also the hyperbolic angle subtended by a half-section of the artificial line. The surge-resistance $z_0$ is therefore 1414·2 cosh 0·17586 = 1414·2 × 1·01550 = 1436·1 ohms. The total angle subtended by the line is—

$$\theta_{AB} = 8u \quad . \quad . \quad . \quad . \quad . \quad . \quad . \quad \text{hyps.} \angle \quad (49)$$
$$= 1{\cdot}40688 \text{ hyps.}$$

The artificial line behaves, in fact, like a real smooth line of surge-resistance $z_0 = 1436{\cdot}1$ ohms and angle 1·40688 hyps., the sending-end resistance of which by (20) page 15, is—

$$R_{fA} = 1436{\cdot}1 \coth 1{\cdot}40688 \quad . \quad . \quad . \quad \text{ohms} \angle \quad (50)$$
$$= 1619{\cdot}38 \text{ ohms.}$$

*Sending-end Resistance of Artificial Line when grounded at B.*—If we ground the artificial line at B, the resistance beyond leak I is $\frac{r_1}{2}$ ohms. At and beyond this leak; *i. e.* including the leak, it is—

$$R'_{gI} = \cfrac{1}{g_1 + \cfrac{1}{r_1/2}} = F_1 (g_1, r_1)_{\frac{1}{r_1/2}} \quad . \quad . \quad . \quad . \quad . \quad \text{ohms} \angle \quad (51)$$

Proceeding from leak to leak as before, we find for the total sending-end resistance at A—

$$R_{gA} = \frac{r_1}{2} + \cfrac{1}{g_1 + \cfrac{1}{r_1 + \cfrac{1}{g_1 + \cfrac{1}{r_1 + \cfrac{1}{g_1 + \cfrac{1}{r_1 + \cfrac{1}{g_1 + \cfrac{1}{r_1/2}}}}}}}} = \frac{r_1}{2} + F_7 (g_1, r_1)_{\frac{1}{r_1/2}} \quad . \quad \text{ohms} \angle \quad (52)$$

By (32) this becomes—

$$R_{gA} = \sqrt{\frac{r_1}{g_1}} \cdot \cosh u \,.\, \tanh 8u \quad . \quad \text{ohms} \angle \quad (53)$$
$$= z_0' \cosh u \,.\, \tanh 8u \quad . \quad . \quad . \quad \text{,,} \quad \text{,,} \quad (54)$$
$$= z_0 \tanh 8u \quad . \quad . \quad . \quad . \quad . \quad . \quad . \quad \text{,,} \quad \text{,,} \quad (55)$$

In the case considered, with $r_1=500$ ohms, and $g_1=0{\cdot}00025$ mho, $z_0=1436{\cdot}1$ ohms, and $u=0{\cdot}17586$ hyp. the artificial line behaves like an actual smooth line of surge-resistance 1436·1 ohms, and angle 1·40688 hyps., of which by (23), page 15, the sending-end resistance at A when grounded at B is—

$$R_{gA} = z_0 \tanh 1{\cdot}40688 = 1273{\cdot}633 \text{ ohms} \quad . \quad . \quad . \quad (56)$$

Proceeding in this way, we should find that in every respect the artificial line would behave in the steady state exactly like a real smooth line of surge-resistance $z_0$ and angle $8u$ hyps. Along and within the artificial line, the distribution of potential, current, resistance, power and energy would not be the same as in the corresponding smooth line; except at section junctions, where the agreement would be complete, each section being the external counterpart of a smooth line having a surge-resistance—

$$z_0 = z_0' \cosh u = \sqrt{\frac{r_1}{g_1}} \cdot \sqrt{1 + \frac{r_1 g_1}{4}} = \frac{r_1}{2}\sqrt{\frac{4+g_1 r_1}{g_1 r_1}} \text{ ohms } \angle \quad (57)$$

and an angle—

$$\theta = 2u = 2\sinh^{-1}\left(\frac{\sqrt{r_1 g_1}}{2}\right) = 2\sinh^{-1}\left(\sqrt{\frac{r_1}{2}\cdot\frac{g_1}{2}}\right) \text{hyps. } \angle \,(58)$$

which means that each section of artificial line is the equivalent $T$ of the corresponding section of actual smooth line. See page 35, where $\rho' = r_1/2$.

If each $T$-section of the artificial line, in the case considered, represented 100 km. of actual smooth line; then the actual line of 400 km., having a surge-resistance of 1436·1 ohms, and an angle of 1·40688 hyps., would have a conductor-resistance of $1436{\cdot}1 \times 1{\cdot}40688 = 2020{\cdot}4$ ohms, and a total dielectric leakance of $1{\cdot}40688/1436{\cdot}1 = 0{\cdot}979956 \times 10^{-3}$ mho. To these values correspond linear constants of $r=5{\cdot}051$ ohms per km. and $g= 2{\cdot}44989 \times 10^{-6}$ mho per km.

If the leak in each section of artificial line is not applied at the middle of each section, the computation is only slightly altered, provided all the sections are alike. The value of $u$, the semi-section angle, remains unchanged and $z_0$ only is affected.

If the leak is applied half at each end of a section as in a symmetrical $\Pi$; then if $z_0'' = \sqrt{r_1 / g_1}$

$$z_0 = z_0'' / \cosh u \quad . \quad . \quad . \quad . \quad \text{ohms} \angle \quad (59)$$

so that the smooth line corresponding to the artificial line is readily found.

In the case of alternating-current artificial lines in the steady state, with condensers in the section leaks, the problem is the same; namely, to find the equivalent section of smooth line represented by each symmetrical $T$-section. The same formulas apply, but must be interpreted vectorially or in two dimensions. This means that the smooth actual line which is the external counterpart of a given section of artificial line comprising line and leakage impedances, varies with the frequency of operation.

Fig. 4 gives the graph of potential distribution over a five-section line of the type above considered, with the distant end free, and 100 volts applied steadily at the home end, as represented in Fig. 18, page 39. The heavy broken line represents the fall of potential in the artificial line, and the dotted curve that which would be found in the corresponding uniform smooth line. The heavy broken line is therefore a funicular polygon, or inelastic string loaded with equal weights at the equidistant points I, II, III, IV, and V; while the dotted curve is a simple catenary between the same terminal points. The catenary coincides with the polygon at the section-junctions 1, 2, 3 and 4; so that each section of the polygon is externally equivalent to the corresponding section of the catenary.

It follows from the foregoing, that the distribution of resistance potentials, currents, power and energy at the section-junction of a $T$-section artificial line, operated either with a continuous current, or with a single-frequency alternating-current, is identical with that of the corresponding points on the equivalent smooth line. Inside each $T$-section, the electric distribution is evidently different from that within the corresponding sections of the equivalent smooth line. The actual distribution at any point inside each $T$-section may, however, be readily found by Ohm's law deduction, from the distribution at the adjacent

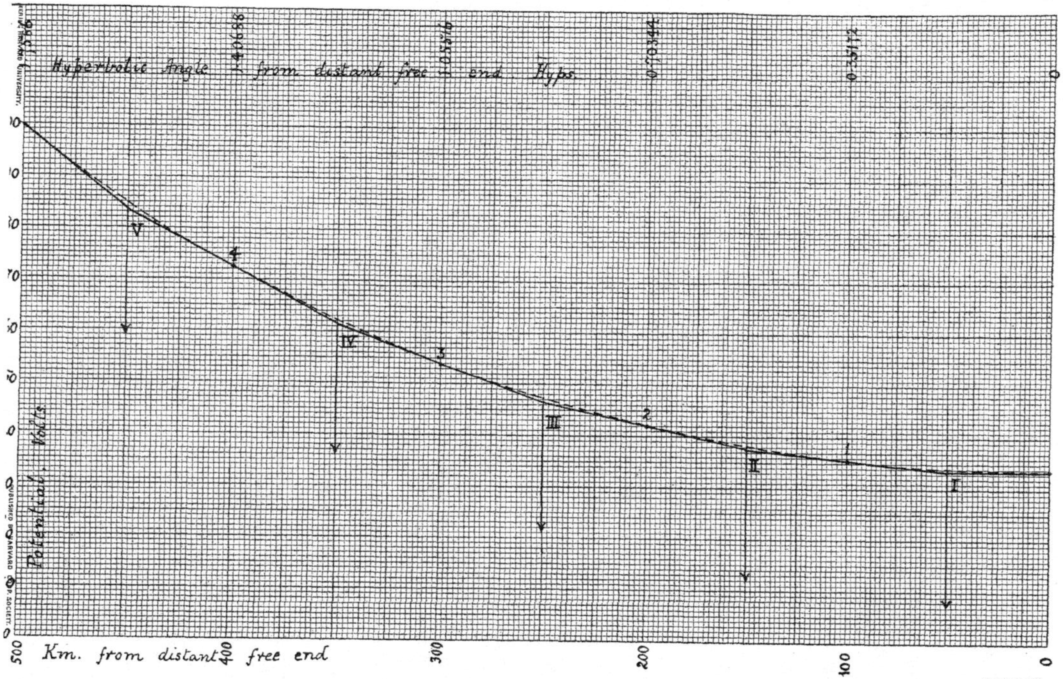


FIG. F4.—Curves of Potential along five-section artificial line, and along the corresponding smooth uniform actual line—funicular polygon and catenary.

[To face p. 238.

section-junction. The conditions at any given leak may also be readily determined in hyperbolic functions. Thus, with the distant end free, the ratio of the resistance beyond the (N+1)th leak, to that at and beyond, *i. e.* including the Nth leak, is—

$$\frac{R_{f,\,N+1}}{R'_{f,\,N}} = \frac{\cosh\,(2\,N+1)\,u}{\cosh\,(2\,N-1)\,u} \quad . \quad . \quad . \quad \text{numeric} \angle \quad (60)$$

With the distant end of the line grounded, the ratio of the resistance beyond the (N+1)th leak, to that at and beyond, *i. e.* including the Nth leak, is—

$$\frac{R_{g,\,N+1}}{R'_{g,\,N}} = \frac{\sinh\,(2\,N+1)\,u}{\sinh\,(2\,N-1)\,u} \quad . \quad . \quad . \quad \text{numeric} \angle \quad (61)$$

The line current at one side of a leak is obviously the same as the line current at the next adjacent junction on that side, whether the distant end is free, grounded, or in any intermediate state.

Moreover, whether the distant end is free or grounded, if $E_N$ is the potential on the equivalent smooth line at the position corresponding to leak N (Figs. 3 and 4); then the potential at this leak on the artificial line is—

$$E'_N = E_N \operatorname{sech} u \quad . \quad . \quad . \quad . \quad \text{volts} \angle \quad (62)$$

## APPENDIX G

*A Brief Method of Deriving Campbell's Formula:*

$$\text{Cosh } L'\alpha' = \text{Cosh } L'\alpha + \frac{\sigma}{z_0} \text{ sinh } L'\alpha \quad . \quad . \quad \textit{numeric.}$$

IN Fig. G1, let AB be a line loaded at regular distances L′ km. with series loads of $\Sigma = 2\,\sigma$ ohms. In the continuous-current case $\Sigma$ is a real numeric. In the single-frequency alternating-current case, $\Sigma$ is a complex numeric. A′B′ represents a second line which is without loads, but which has such linear constants as to be electrically equivalent to AB from any one mid-load point $a$ to any other such as $b$.

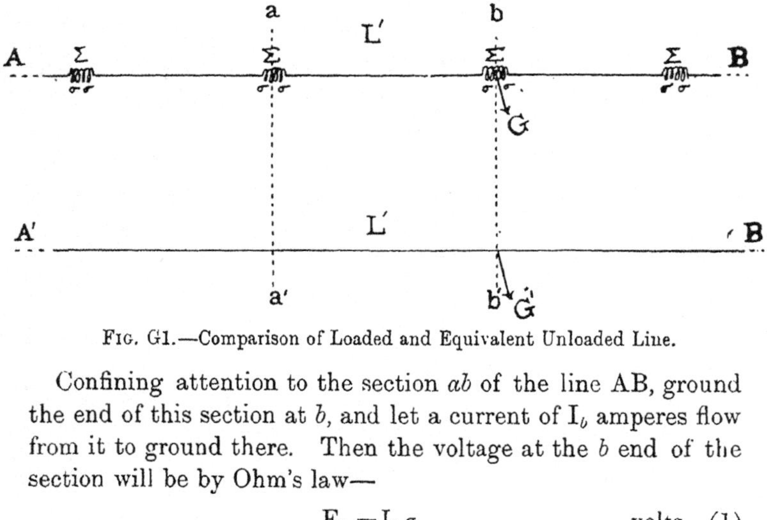


FIG. G1.—Comparison of Loaded and Equivalent Unloaded Line.

Confining attention to the section $ab$ of the line AB, ground the end of this section at $b$, and let a current of $I_b$ amperes flow from it to ground there. Then the voltage at the $b$ end of the section will be by Ohm's law—

$$E_b = I_b\,\sigma \quad . \quad . \quad . \quad . \quad . \quad . \quad \text{volts} \quad (1)$$

The current entering the section at $a$ will then be, by (15)—

$$I_a = I_b \cosh L'\alpha + I_b \frac{\sigma}{z_0} \sinh L'\alpha \quad . \quad \text{amperes} \angle \quad (2)$$

where $\alpha$ is the attenuation-constant of the line AB without loads.

Now referring to the equal length L′ km. between the

corresponding points $a'b'$ on the companion line, A′B′, grounded at $b'$ but without loads, we have, since $E'_{b'} = 0$—

$$I'_{a'} = I'_{b'} \cosh L'\alpha' \quad . \quad . \quad \text{amperes} \angle \quad (3)$$

where $\alpha'$ is the attenuation-constant of the line A′B′.

*Coils per smooth wave-length.*

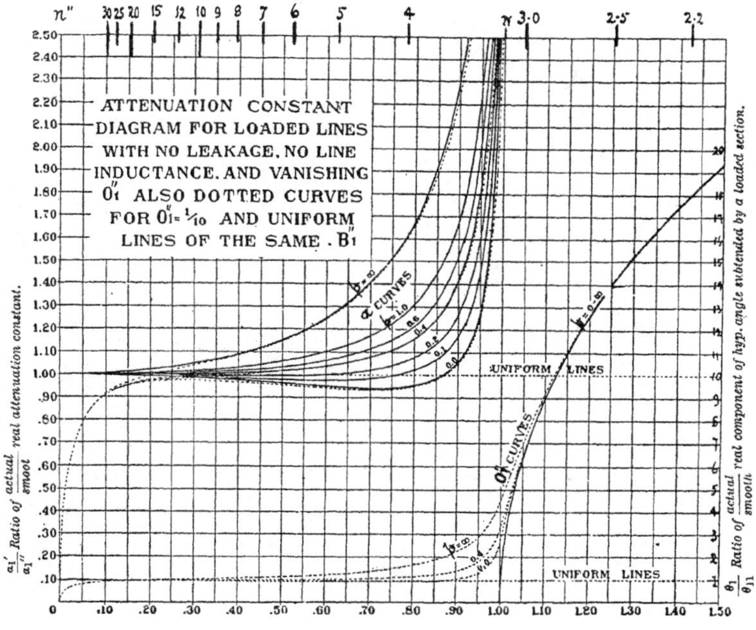


$\pi' n'' = \frac{\theta_2''}{2}$.

FIG. G2.—Curve Sheet taken from Campbell's Paper showing the increase in $\frac{a_1'}{a_1''}$, for different numbers of coils ($n''$) per smooth wave-length in a loaded cable, for various values of $b = r''/r$.

Assume that $I'_{b'} = I_b$, and $I'_{a'} = I_a$, so that the two lines are identical in regard to the entering and leaving currents; then equating (2) and (3), we have—

$$\cosh L'\alpha' = \cosh L'\alpha + \frac{\sigma}{z_0} \sinh L'\alpha \quad \text{numeric} \angle \quad (4)$$

Figs. G2 and G3 have been taken from Dr. Campbell's paper on "Loaded Lines"*, with the notation altered to conform to that here used, and with an additional scale of abscissas added. These figures relate to the properties of formula (4) as applied to lines of neglible inductance in the unloaded state. Referring to Fig. G2, and to the upper scale of abscissas, it will be seen that at $\pi$ coils per smooth wave-length $\lambda''$, as obtained by

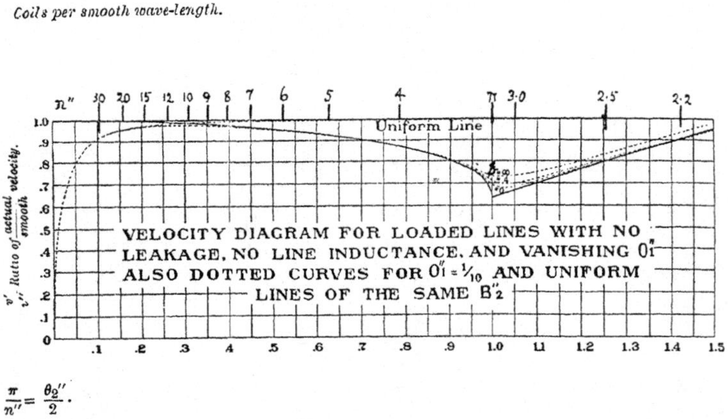


FIG. G3.—Curve Sheet taken from Campbell's Paper showing the decrease in velocity of wave propagation and the ratio $\frac{v'}{v''}$ for a loaded cable with different numbers of coils ($n''$) smooth wave-length, and various values of $b = \frac{r''}{r}$.

formula (226), the real attenuation-constant $\alpha'_1$ as found from (4) is more than 2·5 times the smooth value $\alpha''_1$ determined by (150), in which the loading is imagined to be distributed smoothly. The real attenuation-constant $\alpha'_1$ is rising very rapidly at this limit. The different heavy-line curves refer to different values of $b = \frac{r''}{r}$, the ratio of the linear smooth load resistance to the linear unloaded conductor-resistance. Thus in the case worked out on p. 150, the value of $r''$ is 1·74 ohms per wire-km. in the load coils and $r = 27{\cdot}34$ ohms per wire-km.; so that $b = 0{\cdot}064$. Interpolating in Fig. G2 between the

* *See* Bibliography, 27.

heavy lines of $b = 0$ and $b = 0{\cdot}1$ at $n'' = 8{\cdot}73$ coils per smooth wave-length, and we find $\frac{\alpha'_1}{\alpha_1''} = 0{\cdot}986$ approximately, or the real attenuation-constant is actually 1·4 per cent. less than if the loads had been distributed smoothly, the result already found on p. 151.

The dotted lines closely conforming to the heavy lines indicate the changes in the curves when the real components of the loaded section angles reach 0·1 hyp. In the case considered they only reached 0·06 hyp.

In the lower part of the figure are curves which to the right-hand scale show the ratio $\theta_1'/\theta_1''$ of the real components of loaded section angles at different coil spacings. Thus at $n'' = 2{\cdot}5$ coils per smooth wave-length $\lambda''$, the real component of the loaded section angle would be over fourteen times greater than that computed by (226), on the basis of smoothly distributed extra inductance.

Fig. G3 shows the corresponding ratio of actual velocity $v'$ to smooth velocity $v''$ (227), with different spacings of the coils in cabled wires. Thus for the case discussed on p. 152 with 8·73 coils per smooth wave-length this ratio on the curve sheet G3 is about 0·975, which agrees with that found near (229); namely, 17654/18111. It will be seen that at the critical value of $\pi$ coils per smooth wave-length, the actual velocity becomes from 0·638 to 0·73 of the smooth velocity, according to the value of $b$. That is, for $b=0$, or negligible resistance in the load coils, the velocity $v'$ is reduced to $0{\cdot}638\ v''$, and the wave-length $\lambda'$ is therefore reduced in the same ratio to $\lambda'' \times 0{\cdot}638$. Consequently, when there are apparently $\pi$ coils per smooth wave-length there are only two coils per actual wave-length.

# APPENDIX H

*Analysis of the Influence of Additional Distributed Leakance on a Loaded as compared with an Unloaded Line.*

WITH reference to Fig. 64, for any given line loaded or unloaded—

Let $\beta_1$ be the argument of the vector conductor impedance $z$.

Let $\beta_2$ be the argument of the vector dielectric admittance $y$.

Then the real component of the attenuation-constant of the line is by (150)—

$$\alpha_1 = \sqrt{zy} \cos\left(\frac{\beta_1 + \beta_2}{2}\right) \quad . \quad \text{hyp. per km.} \quad (1)$$

Differentiating with respect to $\beta_2$, we have—

$$d\alpha_1 = -\frac{\sqrt{zy}}{2} \sin\left(\frac{\beta_1 + \beta_2}{2}\right) d\beta_2 \quad . \quad \text{hyp. per km.} \quad (2)$$

$$\therefore \quad \frac{d\alpha_1}{\alpha_1} = -\tan\left(\frac{\beta_1 + \beta_2}{2}\right)\frac{d\beta_2}{2} \quad . \quad . \quad . \quad . \quad \text{numeric} \quad (3)$$

When a cabled line is unloaded, $\frac{\beta_1 + \beta_2}{2}$, at $\omega = 5000$, is near $45°$ and $\frac{d\alpha_1}{\alpha_1} = -\frac{d\beta_2}{2}$ approximately.. When such a line is loaded, $\frac{\beta_1 + \beta_2}{2}$ is usually near to $85°$ and $\tan \frac{\beta_1 + \beta_2}{2} = 11{\cdot}4$ approximately; so that $\frac{d\alpha_1}{\alpha_1} = -11{\cdot}4 \frac{d\beta_2}{2}$; that is, the percentage change in $\alpha_1$ for a given small reduction in $\beta_2$ due to extra distributed leakance is more than ten times as great for the loaded as the unloaded line.

When an aerial line is unloaded, $\frac{\beta_1 + \beta_2}{2}$, at $\omega = 5000$, is usually near to $75°$, and $\frac{d\alpha_1}{\alpha_1} = -3{\cdot}7 \frac{d\beta_2}{2}$. When such a line is loaded, $\frac{\beta_1 + \beta_2}{2}$ approximates $85°$, and $\tan \frac{\beta_1 + \beta_2}{2} \cong 11{\cdot}4$. Consequently, the percentage change in $\alpha_1$ for a given small reduction in $\beta_2$, due to extra distributed leakance, is some three times as great for the loaded as the unloaded line.

## APPENDIX J

*To Find the Best Resistance of an Electromagnetic Receiving Instrument Employed on a Long Alternating-Current Circuit.*

THE receiving instrument is assumed to be of the electromagnetic type, employing a coil or coils of fine insulated copper wire in the receiving circuit to ground. Let $\sigma$ be the resistance of this winding in ohms. Let the inductance of the receiving apparatus be considered as independent of the resistance, *i. e.* the inductance of the winding is either small relatively to the resistance; or the inductance accompanying the resistance of the winding can be modified, compensated for, or overshadowed by independent adjustments of other inductive elements in the receiving circuit.

Then following the reasoning explained in connection with (276), either the magneto-mechanical force in dynes or the magneto-mechanical torque in dyne perp. cms., exerted by the apparatus under the excitation of a certain maximum cyclic received current $I_{mB}$, is either proportional to the ampere-turns $n_1 I_{mB}$, or to some power of the ampere-turns $(n_1 I_{mB})^p$ where $p$ is a positive real exponent not greater than 2.

If the receiver winding has fixed dimensions and volume, a winding of very fine wire will make a coil of numerous turns and, therefore, great sensibility, but possessing very high resistance. On the contrary, if the coil is formed of coarse wire, the resistance will be small, but the sensibility will be low. Evidently it is advantageous to reduce the size of the insulated wire until the increase of resistance more than offsets the increase of sensibility. If throughout the range of working sizes of wire, the ratio of bare to covered wire diameter is the same, so that the same total weights of copper and of insulating material will enter the winding with each size; then halving the diameter of the wire will quadruple the number of turns, if the winding is carefully executed, but will increase the resistance sixteen-fold. Following this reasoning, the number of turns in a winding of fixed volume and dimensions varies as the square root of its resistance to continuous currents and also to low-frequency alternating-currents. Consequently formula (276) becomes—

$$F = a' (I_{mB}\sqrt{\sigma})^p \quad . \quad \text{dynes or dyne} \perp \text{cm.} \quad (1)$$

where $a'$ is a constant of the apparatus, depending upon its construction. The value of the received current by (286) is—

$$I_{mB} = \frac{{}_{mA}}{(z_0 + z_r) \sinh \theta} \cdot \quad \text{max. cy. amperes} \angle \quad (2)$$

a vector equation.

Let— $\quad z_0 = r_0 + jx_0 \quad . \quad . \quad . \quad . \quad . \quad$ ohms $\angle$ (3)

and— $\quad z_r = \sigma + R'_r + jx_r \quad . \quad . \quad . \quad$ " $\angle$ (4)

where $R'_r$ is the extra resistance and $z_r$ the total reactance in circuit with $\sigma$. The current $I_{mB}$ will always be a maximum for any given values of $z_0$, $\sigma$, and $R'_r$ when—

$$jx_r + jx_0 = 0 \quad . \quad . \quad . \quad . \quad . \quad . \quad \text{ohms} \quad (5)$$

or— $\quad x_r = -x_0 \quad . \quad . \quad . \quad . \quad . \quad . \quad . \quad$ " $\quad$ (6)

Since $x_0$ is always a negative reactance, $x_r$ must be positive or magnetic reactance. Substituting in (2) we have—

$$I_{mB} = \frac{E_{mA}}{(r_0 + \sigma + R'_r)[\sinh \theta]} \quad . \quad \text{max. cy. amperes} \quad (7)$$

where $[\sinh \theta]$ means the modulus of $\sinh \theta$ or the numerical value of $\sinh \theta$ with its argument suppressed. Formula (7) is now reduced to real numerical form and may be substituted in (1)—

$$F = a' \left\{ \frac{E_{mA}\sqrt{\sigma}}{(r_0 + \sigma + R'_r)[\sinh \theta]} \right\}^p \text{max. cy. dynes or dyne} \perp \text{cm.} \quad (8)$$

In order to find the value of $\sigma$, which will make F a maximum, we differentiate F with respect to $\sigma$ and equate to zero in the usual way—

$$\frac{dF}{d\sigma} = a'p \left\{ \frac{E_{mA}\sqrt{\sigma}}{(r_0 + \sigma + R'_r)[\sinh \theta]} \right\}^{p-1}$$

$$\times \frac{\frac{E_{mA}}{2\sqrt{\sigma}}\{(r_0 + \sigma + R'_r)[\sinh \theta] - E_{mA}\sqrt{\sigma}\,[\sinh \theta]}{(r_0 + \sigma + R'_r)^2\,[\sinh \theta]^2}$$

$$\frac{\text{max. cy. dynes}}{\text{ohm}} \text{ or } \frac{\text{dyne} \perp \text{cm.}}{\text{ohm}} \quad (9)$$

In order that this expression shall reduce to zero, it is necessary that—

$$r_o + \sigma + R'_r - 2\sigma = 0 \quad . \quad . \quad . \quad . \quad . \quad \text{ohms} \quad (10)$$

or—

$$\sigma = r_o + R'_r \quad . \quad . \quad \text{,,} \quad (11)$$

If the extra resistance-component of impedance in circuit with $\sigma$ is negligible; then—

$$\sigma = r_o \quad . \quad . \quad . \quad . \quad . \quad . \quad . \quad . \quad \text{ohms} \quad (12)$$

Consequently, the maximum cyclic magneto-mechanical force, or torque, developed in the receiving instrument in the steady alternating-current state, will be a maximum if the reactance $x_r$ in the receiving circuit is equal and opposite to the reactance-component in the surge-impedance, and if the resistance $\sigma$ of the winding is equal to the resistance-component of the surge-resistance increased by any extra resistance $R'_r$ in the receiving circuit.

# APPENDIX K

*On the Identity of the Instrument Receiving-end Impedance of a Duplex Submarine Cable, whether the Apex of the Duplex Bridge is Freed or Grounded.*

IN Fig. K, let B$a$DG be the receiving connections at either end of the cable, with the apex $a$ freed, and B′D′G′ the corre-

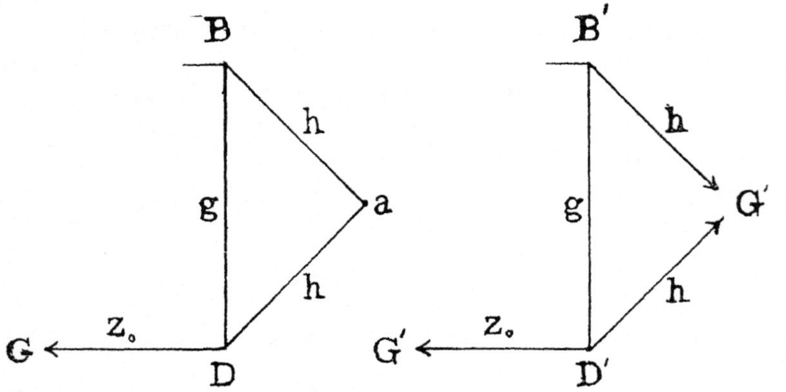


FIG. K.—Diagram of Impedances at either receiving-end in duplex connections, with the apex of the bridge disconnected in one case and connected to ground in the other.

sponding connections with the apex $a$ grounded. It is required to show that in either case the expression $(z_0 + z_r)\left(\frac{z_1 + z_2}{z_2}\right)$ has the same value, where $z_0$ is the surge-impedance, at the frequency selected, of both the real and artificial cables, $z_r$ is the impedance of the receiving apparatus to ground as a whole, $z_1$ is the impedance of the instrument branch, $z_2$ the impedance shunting the instrument.

*Apex Freed.*—With the apex free as at B$a$DG, we have—

$$z_r = \frac{2hg}{2h + g} + z_0 \quad . \quad . \quad . \quad . \quad \text{ohms} \angle \quad (1)$$

$$(z_0 + z_r) = \frac{2hg}{2h + g} + 2z_0 \quad . \quad . \quad . \quad \text{,,} \quad \text{,,} \quad (2)$$

where $h$ is the impedance of each arm in the bridge. Also $z_1 = g$ and $z_2 = 2h$. Consequently—

$$(z_o + z_r)\frac{(z_1 + z_2)}{z_2} = \left(\frac{2hg}{2h + g} + 2z_o\right) \cdot \frac{2h + g}{2h} \quad \text{ohms} \angle \quad (3)$$

$$= g + z_o\left(2 + \frac{g}{h}\right) \quad . \quad . \quad . \quad . \quad \text{"} \quad \text{"} \quad (4)$$

*Apex Grounded.*—With the apex grounded, as at B′G′D′G′, we have—

$$z_r = \frac{h\left(g + \frac{hz_o}{h + z_o}\right)}{h + g + \frac{hz_o}{h + z_o}} = \frac{hg(h + z_o) + h^2 z_o}{(h + g)(h + z_o) + hz_o} \quad \text{ohms} \angle \quad (5)$$

Also— $$z_1 = g + \frac{hz}{h + z_o} \quad . \quad . \quad . \quad \text{"} \quad \text{"} \quad (6)$$

and— $$z_2 = h \quad . \quad . \quad . \quad . \quad . \quad . \quad . \quad . \quad . \quad \text{"} \quad (7)$$

$$\therefore \quad \frac{z_1 + z_2}{z_2} = \frac{g(h + z_o) + hz_o + h(h + z_o)}{h(h + z_o)} \quad . \quad . \quad . \quad \text{"} \quad \text{"} \quad (8)$$

$$= \frac{(h + g)(h + z_o) + hz_o}{h(h + z_o)} \quad . \quad . \quad . \quad . \quad . \quad . \quad . \quad \text{"} \quad \text{"} \quad (9)$$

Hence—

$$(z_o + z_r)\left(\frac{z_1 + z_2}{z_2}\right) = \frac{z_o(h + g)(h + z_o) + hz_o^2 + hg(h + z_o) + h^2 z_o}{h(h + z_o)}$$

$$\text{ohms} \angle \quad (10)$$

$$= \frac{z_o(h + g) + hg + hz_o}{h} \quad . \quad . \quad . \quad \text{"} \quad \text{"} \quad (11)$$

$$= g + z_o\left(2 + \frac{g}{h}\right) \quad . \quad . \quad . \quad . \quad . \quad . \quad \text{"} \quad \text{"} \quad (12)$$

# APPENDIX L

*To Demonstrate the Proposition of Formula (7), page 4,*

$$\theta = \int_{s_1}^{s_2} \frac{ds}{\rho} \quad . \quad . \quad \textit{hyperbolic radians.}$$

We assume the well-known proposition that the magnitude of a hyperbolic angle is twice the sectorial area swept out by

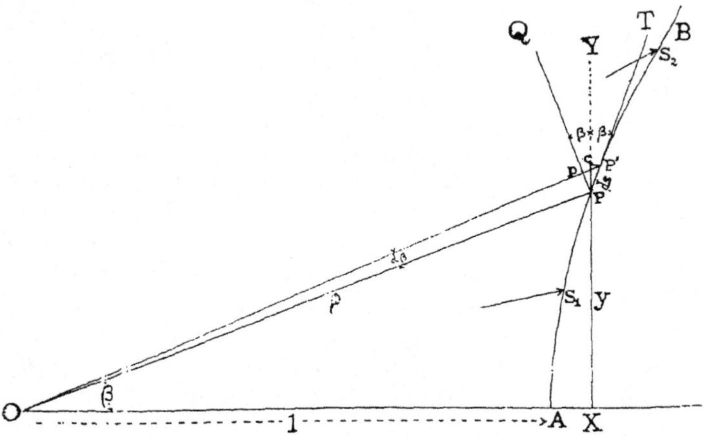


FIG. L.—Diagram for the demonstration of the Theorem $\theta = \int_{s_1}^{s_2} \frac{ds}{\rho}$.

the describing radius-vector over a rectangular hyperbola of unit radius.

In Fig. L, let the curve APB be a segment of a rectangular hyperbola whose radius OA is taken as unity. To a point P on the curve, whose cartesian co-ordinates are $x$ and $y$, draw the radius-vector from the origin OP $= \rho$, making a circular angle

$\beta$ with the initial line and radius OA. Draw PT the tangent to the curve at P. Then the equation to the curve is—

$$y^2 = x^2 - 1 \quad . \quad . \quad . \quad . \quad . \quad . \quad . \quad . \quad . \quad (1)$$

Differentiating both sides—

$$2y\,dy = 2x\,dx \quad . \quad . \quad . \quad . \quad . \quad . \quad . \quad . \quad (2)$$

$$\frac{dy}{dx} = \frac{x}{y} = \cot\beta \quad . \quad . \quad . \quad . \quad . \quad . \quad . \quad . \quad (3)$$

If then we move the point P through a differential distance $ds$ over the curve to P′, and draw P′$c$ and P$c$ parallel to the X and Y axes respectively, $c$P′ $= dx$ and P$c = dy$, so that by (3) the angle $c$PP′ $=$ YPT $= \beta$.

Join OP′, and draw PQ perpendicular thereto, intersecting the same at $p$. Then the angle QPY $= \beta$, because QP is perpendicular to OP, and YP is perpendicular to OX. Therefore the angle QPP′ $= 2\beta$.

Dividing (1) by $\rho^2$ we have—

$$\frac{y^2}{\rho^2} = \frac{x^2}{\rho^2} - \frac{1}{\rho^2} \quad . \quad . \quad . \quad . \quad . \quad . \quad . \quad . \quad (4)$$

or—

$$\sin^2\beta = \cos^2\beta - \frac{1}{\rho^2} \quad . \quad . \quad . \quad . \quad . \quad . \quad . \quad (5)$$

$$\frac{1}{\rho^2} = \cos^2\beta - \sin^2\beta \quad . \quad . \quad . \quad . \quad . \quad . \quad . \quad (6)$$

$$= \cos 2\beta \quad . \quad . \quad . \quad . \quad . \quad . \quad . \quad . \quad . \quad . \quad (7)$$

$$\therefore \quad \rho^2 = \frac{1}{\cos 2\beta} \quad . \quad . \quad . \quad . \quad . \quad . \quad . \quad . \quad . \quad . \quad (8)$$

Representing the differential angle POP′ by $d\beta$—

$$\mathrm{P}p = \rho\,.\,d\beta \quad . \quad . \quad . \quad . \quad . \quad . \quad . \quad . \quad . \quad . \quad (9)$$

$$\therefore \quad \frac{\mathrm{P}p}{\mathrm{PP'}} = \frac{\rho\,d\beta}{ds} = \cos 2\beta \quad . \quad . \quad . \quad . \quad . \quad . \quad (10)$$

or by (8)—

$$\frac{ds}{\rho\,d\beta} = \rho^2 \quad . \quad . \quad . \quad . \quad . \quad . \quad . \quad . \quad . \quad . \quad . \quad (11)$$

and—

$$\frac{ds}{\rho} = \rho^2\,.\,d\beta \quad . \quad . \quad . \quad . \quad . \quad . \quad . \quad . \quad . \quad . \quad (12)$$

But the differential area $dA$ enclosed between OP, OP′ and the element $ds$ is—

$$dA = \frac{\rho^2 . d\beta}{2} \quad . \quad . \quad . \quad . \quad . \quad . \quad . \quad . \quad . \quad (13)$$

$$\therefore \quad \frac{ds}{\rho} = 2\, dA \quad . \quad . \quad . \quad . \quad . \quad . \quad . \quad . \quad . \quad (14)$$

but if we denote the hyperbolic angle of PP′ by $d\theta$—

$$d\theta = 2\, dA \quad . \quad . \quad . \quad . \quad . \quad . \quad . \quad . \quad . \quad (15)$$

$$\therefore \quad d\theta = \frac{ds}{\rho} \quad . \quad . \quad . \quad . \quad . \quad . \quad . \quad . \quad . \quad . \quad (16)$$

Proceeding in this manner from an initial point $s_1$ whose distance along the curve from A is $s_1$ units up to a final point $s_2$ whose distance along the curve from A is $s_2$ units, we have—

$$\theta = \int_{s_1}^{s_2} \frac{ds}{\rho} \quad . \quad . \quad . \quad . \quad . \quad . \quad . \quad . \quad . \quad (17)$$

and this hyperbolic angle must be equal to the arc length $s = s_2 - s_1$ divided by a certain integrated mean radius-vector $\rho'$ or—

$$\theta = \frac{s}{\rho'} \quad . \quad . \quad . \quad . \quad . \quad . \quad . \quad . \quad . \quad . \quad (18)$$

## APPENDIX M

*Comparative Relations between T-Artificial Lines, Π-Artificial Lines, and their respective corresponding Smooth Lines.*

As has been pointed out in Chapter III, as well as in Appendix F, any symmetrical $T$ or $\Pi$ of impedances can be replaced by its equivalent smooth line, of angle $\theta$ and surge impedance $z_0$, so far as relates to external behaviour in the steady state. In other words, any symmetrical $T$ or $\Pi$, in a system of conductors, may be removed and replaced by its equivalent smooth line of distributed series-impedance and shunt-admittance, without disturbing the steady state of the system at, and outside of, the terminals of the $T$ or $\Pi$. In the continuous-current case, the equivalent smooth line is unique. There is only one "real" smooth line-angle $\theta$, and only one corresponding "real" surge-impedance $z_0$, which will satisfy the external relations of equivalence. In the alternating-current case, however, there is one "complex" or vector smooth line-angle $\theta = \theta_1 + j\theta_2$, and one "complex" or vector corresponding surge-impedance $z_0$ which will satisfy these relations at any given frequency. Consequently, for the alternating-current case, a $T$ or a $\Pi$ calls for an infinite number of equivalent pairs of "complex" line-angles and surge impedances, as the frequency of steady-state operation is varied from 0 to $\infty$. When, therefore, we construct an artificial line of a series of similar $T$s, or a series of similar $\Pi$s, and operate it by alternating currents of a single frequency, it becomes necessary to determine what is the corresponding smooth line which the artificial line simulates at that frequency.

Fig. M1 shows a single $T$-section, and also a pair of such $T$-sections connected through an ammeter A. Each $T$-section has a series impedance of $r_1$ ohms, and a shunt admittance of $g_1$ mhos applied at the middle of $r_1$.

Fig. M2 similarly shows a single $\Pi$-section, and also a pair of such $\Pi$-sections connected through an ammeter A. Each $\Pi$-section has a series or architrave impedance of $r_1$ ohms and a shunt admittance of $g_1$ mhos, divided into two equal pillar leaks of $g_1/2$ mhos.

An artificial line of four $T$-sections; *i.e.* a $T$-line of four

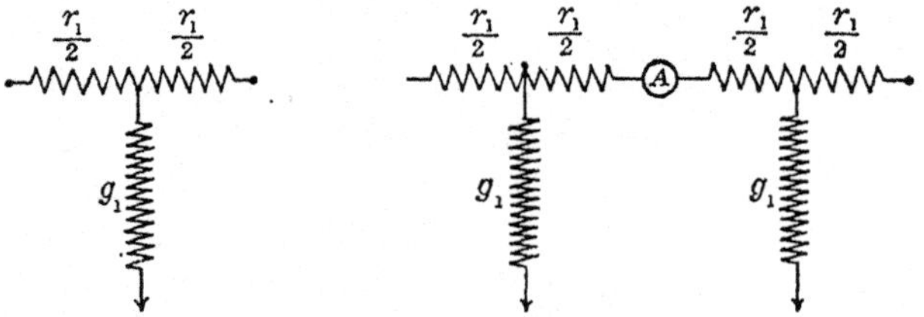


FIG. M1.—Single T-Section and pair of T-Sections connected through an Ammeter A.

sections, is indicated in Fig. M3; while a $\Pi$-line of four sections is correspondingly indicated in Fig. M4. It is evident that the only difference between the lines of Figs. M3 and 4 lies in the conditions at the ends of the line. In each case the total series impedance is the same; viz. $4r_1$ ohms, and the total shunt admittance is the same; *i.e.* $4g_1$ mhos; but in Fig. M3, the line

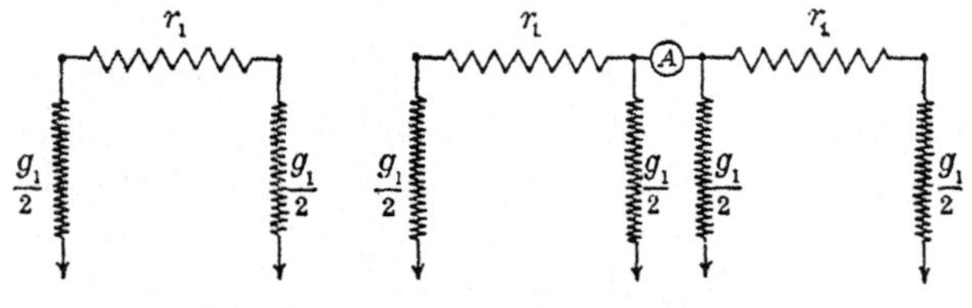


FIG. M2.—Single $\Pi$-Section and pair of $\Pi$-Sections connected through an Ammeter.

terminates in semi-impedances $r_1/2$; while in Fig. M4, the line terminates in semi-admittances $g_1/2$. In other words, any $T$-line can be converted into a $\Pi$-line of the same number of sections, and reciprocally, by changing the end-sections only.

If we compute the smooth lines corresponding to the artificial lines of Figs. M3 and 4, by the methods of Chapter III, or Appendix F, we shall find that the line-angle $\theta$ of each is the same; but that the surge-impedance $z_0$ differs. The surge impedance of the $T$-line differs from the surge impedance of the $\Pi$-line, even in the continuous-current case.

We proceed to develop the formulas which apply to such $T$- and $\Pi$-lines. These formulas may be readily deduced from the

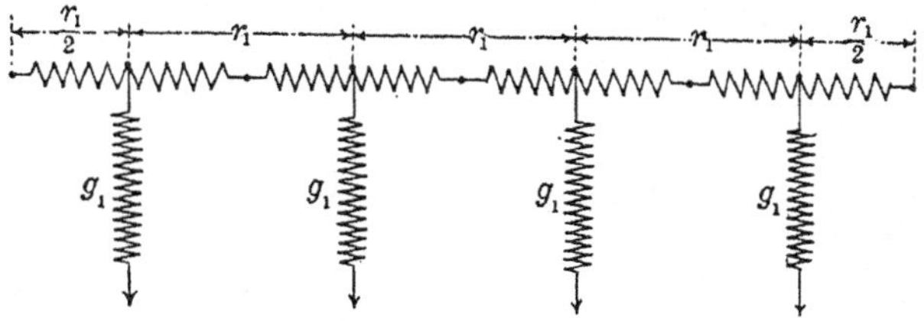


FIG. M3.—Artificial line of four T-Sections.

fundamental relations already pointed out in Chapter III and Appendix F.

*To find the Line-angle $\theta$ of any artificial Line of a given number of similar $T$- or $\Pi$-sections.*—The first step is to find the hyperbolic angle $u$ subtended by a half-section of the line.

The angle of a section of a smooth line having $r_1$ ohms of

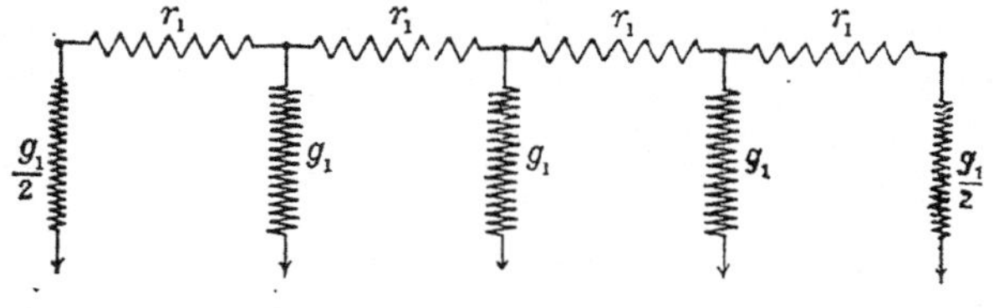


FIG. M4.—Artificial line of four Π-Sections.

distributed series impedance and $g_1$ mhos of distributed shunt admittance would be by (19), page 14—

$$2u = \sqrt{r_1 g_1} \quad . \quad . \quad . \quad . \quad \text{hyps. } \angle \quad (1)$$

Consequently the angle subtended by a half-section of such smooth line would be—

$$u = \sqrt{\frac{r_1}{2} \cdot \frac{g_1}{2}} = \frac{\sqrt{r_1 g_1}}{2} \quad . \quad . \quad \text{hyps. } \angle \quad (2)$$

But since an artificial line has alternate lumps of impedance and admittance, instead of uniformly distributed quantities, the above formulas cannot correctly apply to either a section or half-section of artificial line. We can only use (2) in relation to a half-section of a *T*- or *Π*-line by acknowledging the error due to lumpiness. For such a *T*- or *Π*-line respectively, we may therefore rewrite (2) as follows—

$$u' = u'' = \frac{\sqrt{r_1 g_1}}{2} \text{ apparent hyps.} \angle \quad (3)$$

where the term "apparent" hyps. indicates the existence of the lumpiness error. In order to correct (3) for this error, we must apply to the left-hand side the correction-factor $\frac{\sinh u'}{u'}$ or $\frac{\sinh u''}{u''}$ as the case may be, so that we have—

$$\sinh u' = \sinh u'' = \sinh u = \frac{\sqrt{r_1 g_1}}{2} \text{ hyps.} \angle \quad (4)$$

as the fundamental formula of a *T*- or *Π*-line. Expressing this proposition in language, we may say that either for a *T*-line or a *Π*-line, the sine of the hyperbolic angle $u$ subtended by a half-section is equal to half the geometric mean of the sectional series impedance and shunt admittance, each of these being a complex quantity in the alternating-current case, and depending on the frequency. Knowing the hyperbolic sine of the semi-section angle $u$, we can readily find the angle $u$ by reference to Tables of hyperbolic functions.* For continuous-

* Tables of sinh $\rho\underline{/\delta}$, cosh $\rho\underline{/\delta}$, tanh $\rho\underline{/\delta}$, $\frac{\sinh \rho\underline{/\delta}}{\rho\underline{/\delta}}$ and $\frac{\tanh \rho\underline{/\delta}}{\rho\underline{/\delta}}$ to five decimal places, or five significant digits, by steps of 0·1 in $\rho$, between $\rho = 0$ and $\rho = 0·5$, and by steps of 1° in $\delta$ between $\delta = 60°$ and $\delta = 90°$, have been published by the author in the *Proceedings of the American Institute of Electrical Engineers*, 1911, pp. 2481–2492, "Tables of Hyperbolic Functions in Reference to Long Alternating-Current Transmission Lines." These Tables have been extended to $\rho = 1$ and $\delta = 45°$ to 90°, and published, in pamphlet form, by the author in the *Harvard Engineering Journal*, Harvard University, Cambridge, Mass. At the date of this writing (October 1913), these Tables have been carried to $\rho = 3·0$; while other tables of sinh, cosh, and tanh $(x + jy)$ have been computed to either five significant figures, or five decimal places, by steps of 0·05 in $x$ and $y$, from $x = 0$ to $x = 4·0$, and from $y = 0$ to $y = \infty$. The complete set of Tables is being printed by the Harvard University Press, and also an Atlas of 23 charts 45 cm. × 45 cm. for rapid graphic interpolation.

current lines, these would be Tables of real functions, and for alternating-current lines, they would be Tables of complex functions.

The angle subtended by a single section of artificial line would then be $2u$, and if the line contained $n$ sections its total angle would be—

$$\theta = 2nu = 2n \sinh^{-1}\left(\frac{\sqrt{r_1 \cdot g_1}}{2}\right) \quad . \text{ hyps. } \angle \quad (5)$$

The next step is to find the apparent surge-impedance of the artificial line. By (27) page 16, we have for either a $T$-line or a $\Pi$-line—

$$z_0' = z_0'' = \sqrt{\frac{r_1}{g_1}} \quad \text{apparent ohms } \angle \quad (6)$$

Correcting for lumpiness, however, on a $T$-line, as in (88) page 36—

$$z_0 = z_0' \cosh u = \cosh u \sqrt{\frac{r_1}{g_1}} \quad . \quad \text{ohms } \angle \quad (7)$$

In the case of a $\Pi$-line, the lumpiness correction, as in (98) page 37, is—

$$z_0 = z_0'' / \cosh u = z_0'' \operatorname{sech} u = \operatorname{sech} u \sqrt{\frac{r_1}{g_1}} \quad \text{ohms } \angle \quad (8)$$

The surge-impedance of the smooth line corresponding to an artificial line is thus obtained by multiplying the apparent surge-impedance by $\cosh u$ for a $T$-line, and dividing by $\cosh u$ for a $\Pi$-line.

Figs. M5, 6, and 7 represent the same 5-section $T$- line as is shown in Figs. 17, 18, and 19 on page 39; but they also show the corresponding smooth-line. Here $r_1 = 500$ ohms, $g_1 = 0{\cdot}25 \times 10^{-3}$ mho. Consequently, by (4), $\sinh u = \frac{\sqrt{0{\cdot}125}}{2} = 0{\cdot}176777$; whence, by Tables, $u = 0{\cdot}175868$ hyp. The angle subtended by a section is thus $0{\cdot}351736$ hyp. and the angle subtended by the whole line in Fig. 5 is $1{\cdot}75868$ hyps. Along the line, the position-angle is indicated in Fig. M5 at each junction-point. The potentials, currents, resistances and

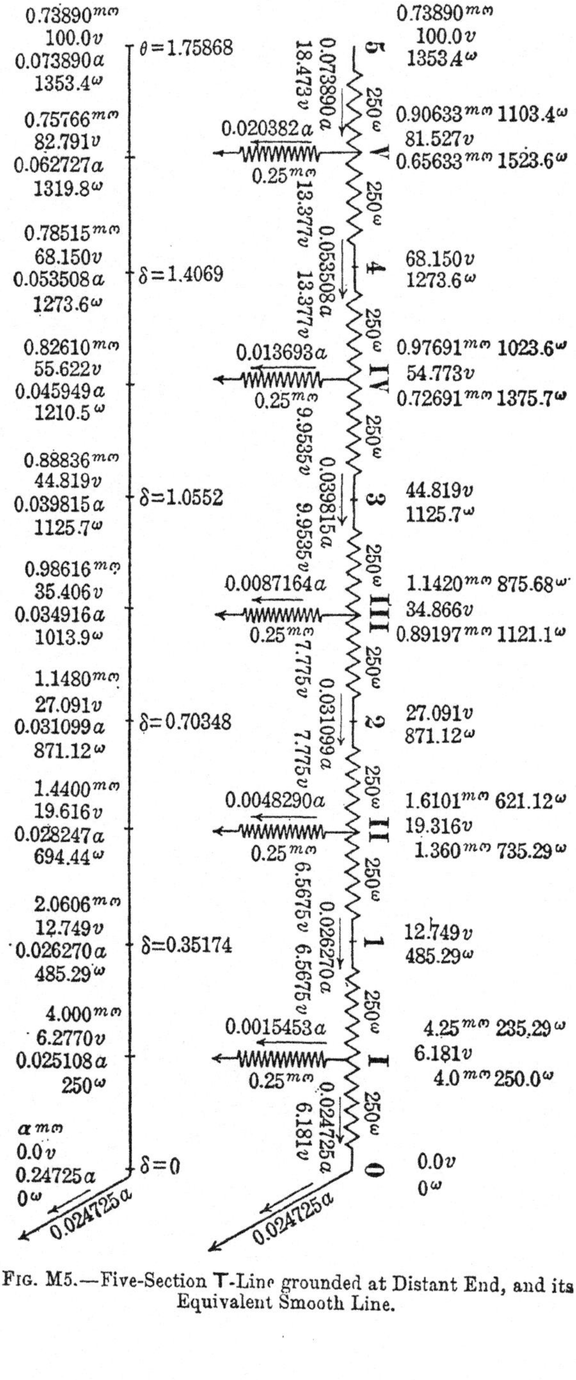


FIG. M5.—Five-Section T-Line grounded at Distant End, and its Equivalent Smooth Line.

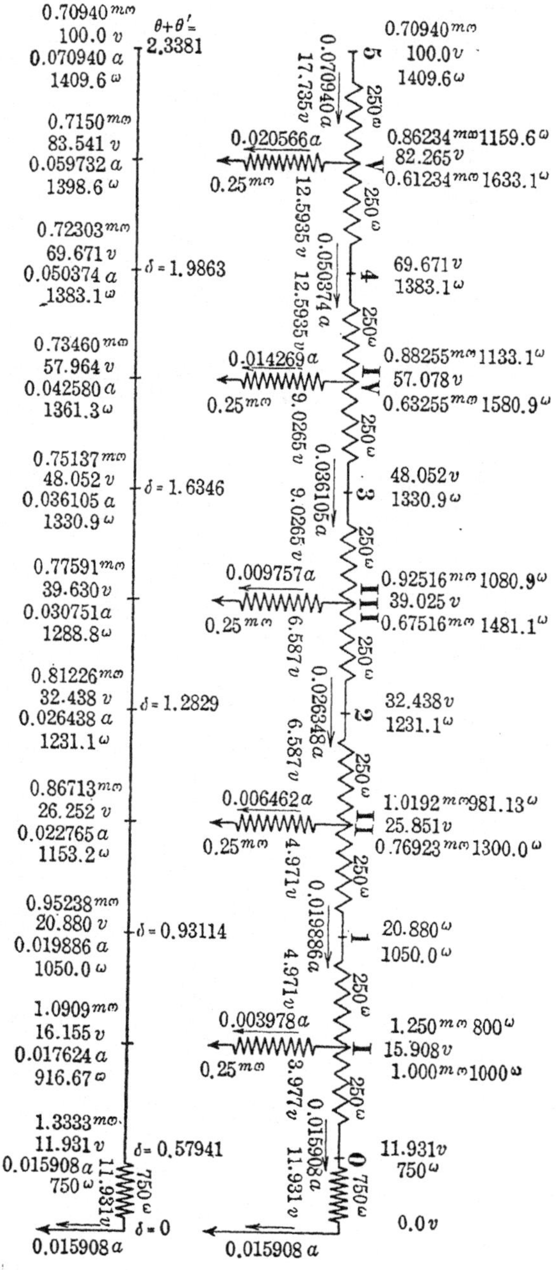

FIG. M6.—Five-Section T-Line grounded at Distant End through 750 ohms, and its Equivalent Smooth Line

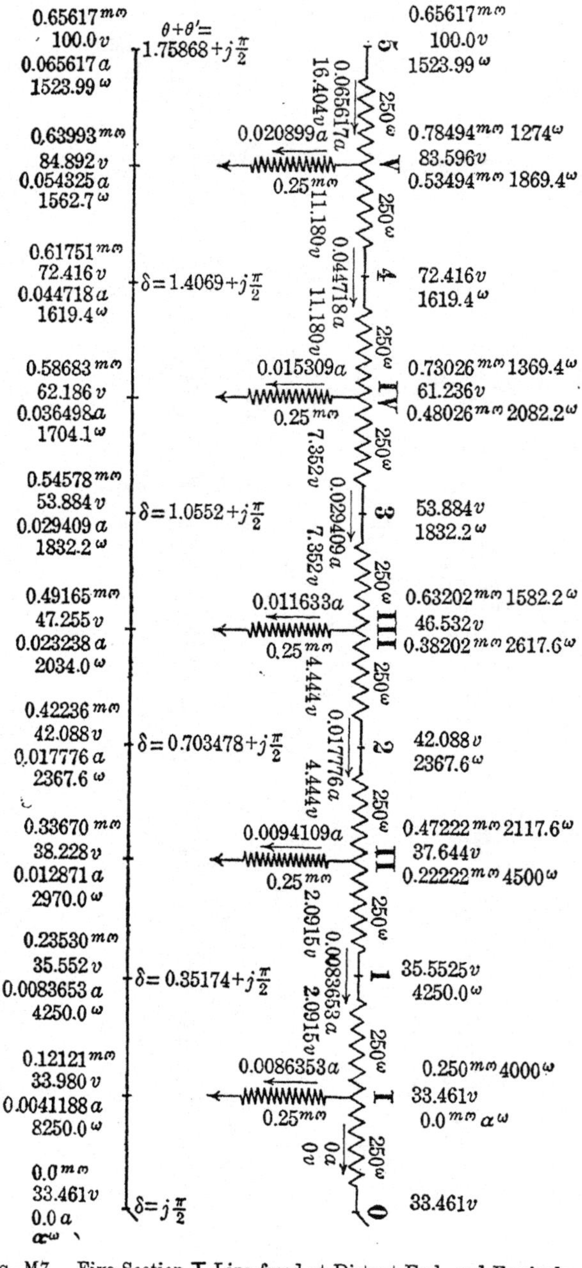


FIG. M7. Five-Section T-Line freed at Distant End, and Equivalent Smooth Line.

conductances, along the smooth line, conform at all points with formulas (40), (41) and (42) page 21.

The apparent surge-impedance of the artificial line is, by (6), $z_0' = \sqrt{500/0{\cdot}0025} = 1414{\cdot}2$ apparent ohms. Correcting this for lumpiness, we have, by (7), $z_0 = 1414{\cdot}2 \times \cosh 0{\cdot}175868 = 1414{\cdot}2 \times 1{\cdot}0155 = 1436{\cdot}1$ ohms.

The total conductor-resistance of the equivalent smooth line is by (72), page 31, $R = \theta z_0 = 1436{\cdot}1 \times 1{\cdot}75868 = 2525{\cdot}7$ ohms, as against 2500 ohms in the $T$-line. The total smooth-line leakance is also, by (73), page 31, $G = \theta/z_0 = 1{\cdot}75868/1436{\cdot}1 = 0{\cdot}0012246$ mho, as against 0·00125 mho in the $T$-line.

In Fig. M6, the same $T$-line is grounded at the distant end through 750 ohms, a terminal load subtending an angle of 0·57941 hyp., by (55) page 23. This brings the total line-angle at the sending end to 2·3381 hyps. Along the smooth line, the potential, current, resistance and conductance conform to (40), (41) and (42). Along the $T$-line, the values are correspondingly identical at junction-points, as well as at the ends of the line.

Applying (55), page 23, to the case of Fig. M7, we have the angle subtended by the open end—

$$\theta' = \tanh^{-1}\left(\frac{\infty}{z_0}\right) = j\frac{\pi}{2} \quad . \quad . \quad . \quad . \quad \text{hyps. (9)}$$

This requires $\frac{j\pi}{2}$ to be added to each of the position angles along the line of Fig. 5. But applying formulas (246) page 168, to the angles thus presented in Fig. 7, we find that formulas (40), (41) and (42) produce respectively (37), (38) and (39), page 21. Consequently, we may regard formulas (40), (41) and (42) as fundamental, and (37) (38) (39) as derived therefrom, in the case where the terminal load angle is $j\frac{\pi}{2}$ hyps., due to an open end. That is, along any line, the current is always proportional to the cosine, the potential to the sine, and the resistance to the tangent of the position-angle whether the line be freed, or grounded, or in any intermediate state, provided that the proper terminal angle be applied to the distant end, for the impedance to ground thereat, in accordance with (55).

Figs. M8, 9, and 10 represent the distributions of current, potential, resistance and conductance over a five-section $\Pi$-line of the same sectional series-resistance and shunt-leakance as before, as well as on the equivalent smooth line. The line-angles in Figs. M8 and 10 correspond precisely to those of Figs. M5 and 7. Owing, however, to the different surge-impedances of the two lines, the line-angles in Fig. M9 are not identical with those of Fig. M6. The surge-impedance of the line is $z_0 = \frac{1414{\cdot}2}{1{\cdot}0155} = 1392{\cdot}6$ ohms. The total line resistance in the equivalent smooth line is thus $\theta z_0 = 1392{\cdot}6 \times 1{\cdot}75868 = 2449$ ohms, and the total line leakance $\theta / z_0 = 1{\cdot}75868/1392{\cdot}6 = 1{\cdot}263 \times 10^{-3}$ mho, as against 2525·7 ohms and $1{\cdot}2246 \times 10^{-3}$ mho, in the equivalent smooth line with $T$-sections.

The fall of potential on the $\Pi$-line, $T$-line, and equivalent smooth lines is indicated graphically in Fig. M11. Referring to the curves of "Line Free," the broken curve is a simple catenary, or curve of hyperbolic cosines, corresponding to the fall of potential along either equivalent smooth line. The internally contacting series of chords represents the fall of potential along the $\Pi$-line. The externally contacting series of chords represents the fall of potential along the $T$-line. These are respectively the internal and external string polygons contacting with the catenary in the theory of non-extensible weighted strings hanging between terminal supports, in the science of statics.

The general relation between the electric conditions at a junction, or a mid-section, on a $T$-line, and at the corresponding points on the equivalent smooth line, are indicated in Fig. M12. The lines are supposed to be voltaged at the sending end only, and to be either freed, grounded or in any intermediate condition, at the receiving end. It will be seen that, at junctions, the potential, current, line-resistance, and line-conductance are

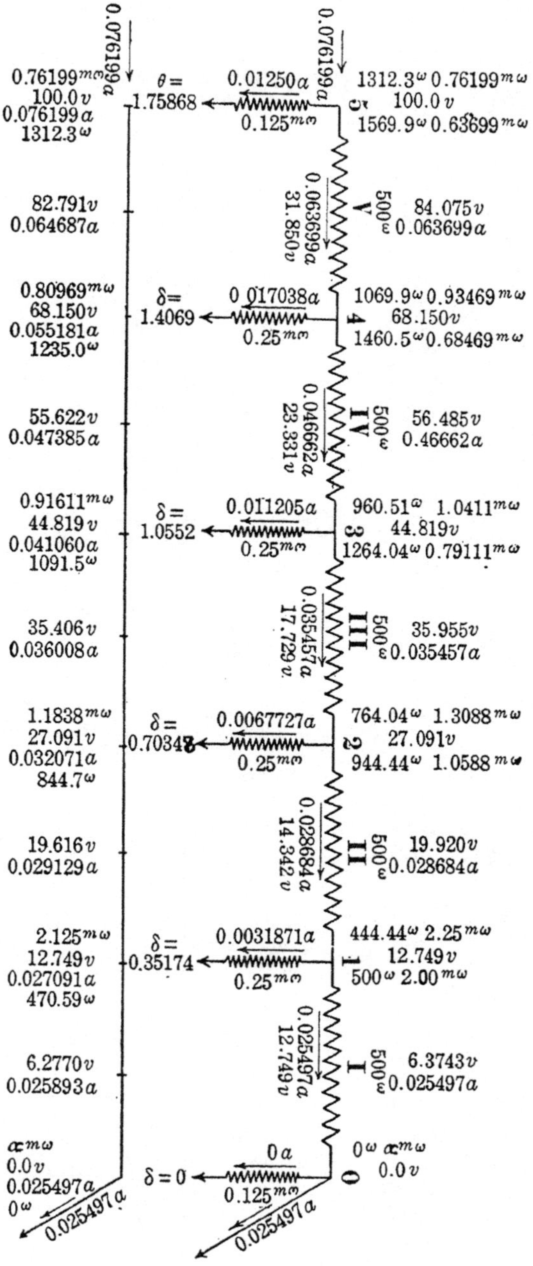

Fig. M8.—Five-Section Π-Line grounded at Distant End, and Equivalent Smooth Line.

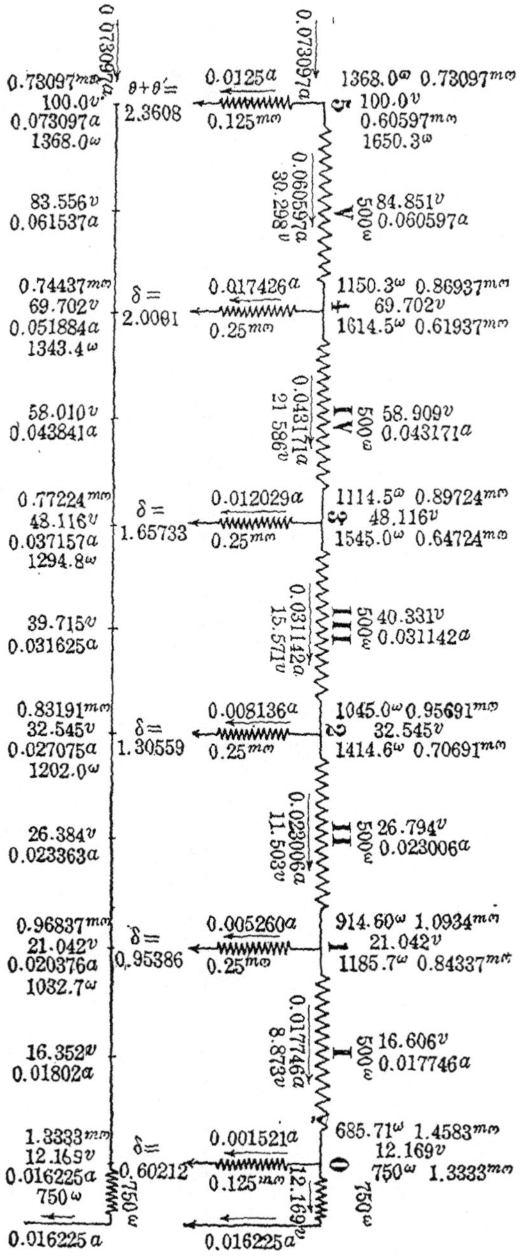


FIG. M9.—Five-Section Π-Line grounded at Distant End through 750 ohms, and its Equivalent Smooth Line.

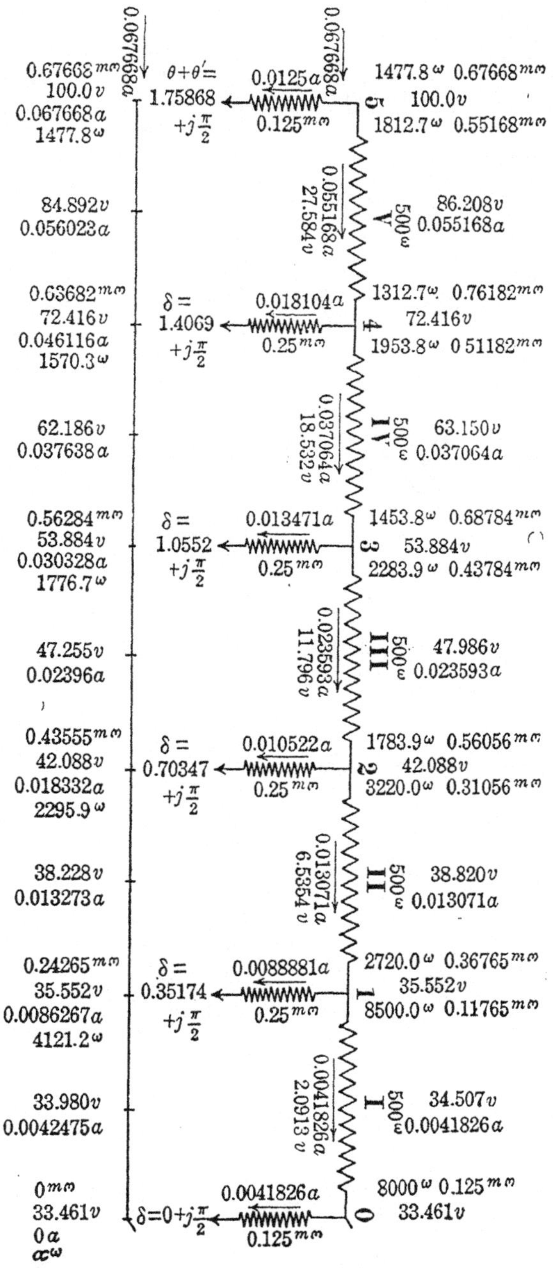

FIG. M10.—Five-Section Π-Line freed at Distant End, and its Equivalent Smooth Line.

all the same on the $T$-line and on the smooth line. At a mid-section, where the leak appears, the $T$-line potential is equal to the corresponding smooth line potential multiplied by

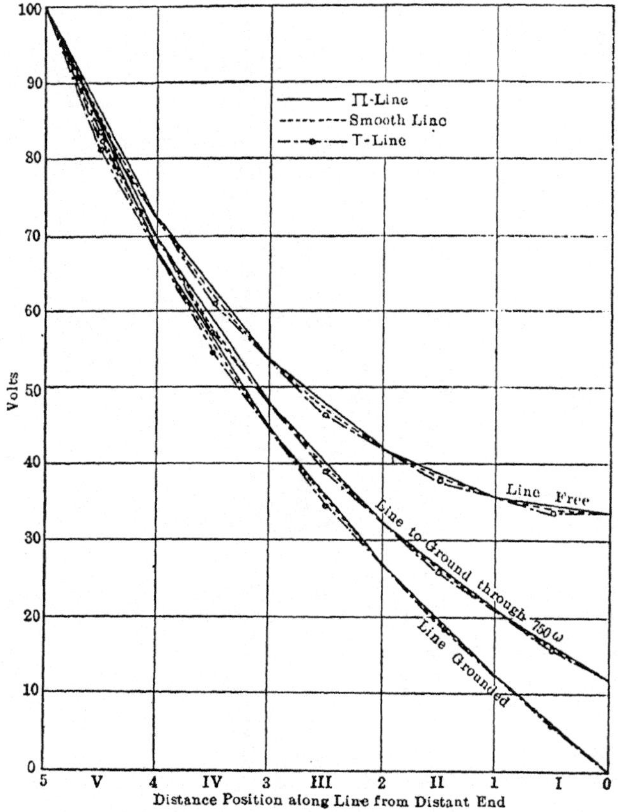


FIG. M11.—Fall of Potential along Π-Line, T-Line, and Equivalent Smooth Lines, with Distant End freed, grounded directly, and grounded through 750 ohms.

sech $u$, while the mean of the currents in the line on each side of the leak is equal to the current on the mid-section point of the smooth line multiplied by cosh $u$. In the alternating-current case, this mean value would be a vector mean value, instead of an arithmetical mean value.

The corresponding general relations between a $\Pi$-line and its equivalent smooth line, at section-junctions and mid-sections,

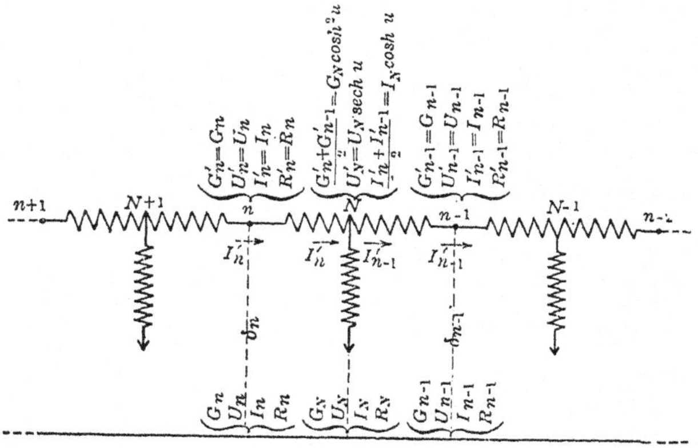


FIG. M12.—Comparative relations of Potential, Current, Conductance and Resistance along a T-Artificial Line and along its Equivalent Smooth Line.

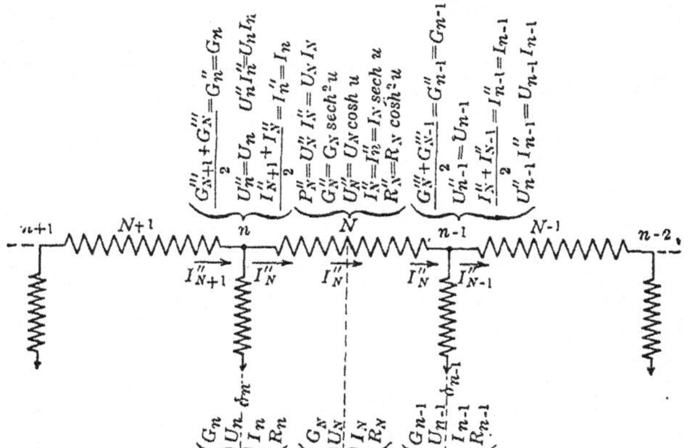


FIG. M13.—Comparative relations of Potential, Current, Conductance and Resistance along a $\Pi$-Artificial Line and along its Equivalent Smooth Line.

are indicated in Fig. M13. At junction-points, where the leaks occur, the potentials agree on both, and the currents, which should be measured theoretically between half-leaks as in Fig. M2 agree also, if we take the mean of the two currents on each side of any leak. At any mid-section, the $\Pi$-line potential is equal to the corresponding smooth-line potential multiplied by $\cosh u$; while the $\Pi$-line current is equal to the corresponding smooth-line current multiplied by $\operatorname{sech} u$. The last relation is important in the practical use of $\Pi$-lines; because the currents along the line, at junction-points, have usually to be measured on one side, or the other, of any leak, which is equivalent to measuring them at the adjacent mid-section.

The facts may be summed in regard to Figs. M12 and 13 for both $T$- and $\Pi$-lines, by saying that at either a junction-point, or at a mid-section, the power UI, in the line, is the same on the artificial line as on the equivalent smooth line, if the line-current I at a leak is interpreted as the mean of the line-currents on each side thereof.

As an example of the preceding propositions applied to an alternating-current artificial line, let there be a line of 10 sections, each consisting of a series resistance of 92·78 ohms, with an included inductance of 0·2241 henry, and with shunt leaks of 1-microfarad pure condensers, having negligible losses of energy, the frequency of operation being $f = 125{\cdot}84$ cycles per second, or $\omega = 790{\cdot}67$ radians per second. Required the equivalent smooth line.

$$\text{Here } r_1 = 92{\cdot}783 + j0{\cdot}2241 \times 790{\cdot}67 = 92{\cdot}783 + j177{\cdot}18$$
$$= 200 \; \underline{/62^\circ{\cdot}36} \text{ ohms.}$$

$$\text{and } \quad g_1 = j10^{-6} \times 790{\cdot}67 = 7{\cdot}9067 \times 10^{-4} \; \underline{|90^\circ} \text{ mho.}$$

$$r_1 g_1 = 0{\cdot}15813 \underline{\backslash 152^\circ{\cdot}36}, \text{ and by (4), } \sinh u = \frac{\sqrt{0{\cdot}15813 \underline{\backslash 152^\circ{\cdot}36}}}{2}$$
$$= \frac{0{\cdot}39766 \; \underline{/76^\circ{\cdot}18}}{2} = 0{\cdot}19883 \; \underline{/76^\circ{\cdot}18}$$

By reference to Tables of $\sinh u$ already published, we find $u = 0{\cdot}2 \; \underline{/76^\circ{\cdot}0}$ hyp., and this is the angle of a half section. The angle subtended by the whole line at this frequency is therefore $4{\cdot}0 \; \underline{/76^\circ{\cdot}0}$ hyps.

The apparent surge-impedance is by (6)

$$\sqrt{200\ \underline{/62^\circ{\cdot}36}\ /\ 7{\cdot}9067 \times 10^{-4}\ \underline{|90^\circ}} = \sqrt{25{\cdot}29 \times 10^{4}\ \overline{\backslash 27^\circ{\cdot}64}}$$
$$= 502{\cdot}94\ \overline{\backslash 13^\circ{\cdot}82} \text{ apparent ohms}$$

By (8), however, it becomes necessary to divide by cosh $u$, in order to correct for lumpiness, and by Tables already published, $\cosh u = \cosh 0{\cdot}2\ \underline{/76^\circ{\cdot}0} = 0{\cdot}98242\ \underline{/0^\circ{\cdot}545}$; so that the surge-impedance of the equivalent smooth line at the frequency considered is $502{\cdot}94\ \overline{\backslash 13^\circ{\cdot}82}\ /\ 0{\cdot}98242\ \underline{/0^\circ{\cdot}545} = 511{\cdot}94\ \overline{\backslash 14^\circ{\cdot}365}$ ohms. The total series-impedance in the equivalent smooth line is thus $4{\cdot}0\ \underline{/76^\circ{\cdot}0} \times 511{\cdot}94\ \overline{\backslash 14^\circ{\cdot}365} = 2047{\cdot}8\ \underline{/61^\circ{\cdot}635}$ ohms, as against $2000\ \underline{/62^\circ{\cdot}36}$ in the actual artificial line; while the total shunt admittance in the equivalent smooth line is $4{\cdot}0\ \underline{/76^\circ{\cdot}0}\ /\ 511{\cdot}9\ \overline{\backslash 14^\circ{\cdot}365} = 7{\cdot}8138 \times 10^{-3}\ \underline{|90^\circ{\cdot}365}$ mho, as against $7{\cdot}9067 \times 10^{-3}\ \underline{|90^\circ}$ actually existing in the artificial line at the frequency considered.

By means of an experimental artificial line comprising resistances only, in five sections, with either *Π*- or *T*-connection, and $r_1 = 500$ ohms; $g_1 = 0{\cdot}00025$ mho, operated by a single storage cell, all of the distributions in Figs. M5 to 10 have been verified in the laboratory with a potentiometer.

If the impressed emf. on the artificial line exceeds 2 volts, a potential multiplier leak has ordinarily to be employed with the potentiometer. This introduces a slight error, which, however, can be corrected for. See Bibliography 83.

## APPENDIX N

*Solutions of the Fundamental Steady State Differential Equations for any Uniform Line in terms of a single Hyperbolic Function.*

It is shown in Appendix C that the fundamental differential equations for potential and current—page 217 (5) and (6)—have solutions which were given originally by Heaviside for the continuous-current case, but which are also immediately applicable to alternating-current cases through the use of complex numbers. These solutions are presented on page 12 (15) and (16) in the following form—

$$E_A \cosh L_1 a - I_A z_0 \sinh L_1 a = e = E_B \cosh L_2 a + I_B z_0 \sinh L_1 a \quad \text{volts} \angle \quad (1)$$

$$I_A \cosh L_1 a - \frac{E_A}{z_0} \sinh L_1 a = i = I_B \cosh L_2 a + \frac{E_B}{z_0} \sinh L_2 a \quad \text{amperes} \angle \quad (2)$$

In each of these four equations, the potential or current is expressed in terms of two hyperbolic functions: namely, the cosine and the sine of one hyperbolic angle $L_1 a$ or $L_2 a$, according as reference is made to the sending or receiving terminal conditions. But, just as in circular functions, we have the well-known relation—

$$A \cos \beta \pm B \sin \beta = \sqrt{A^2 + B^2} \cos \left\{\beta \mp \tan^{-1}(B/A)\right\} \quad \text{numeric} \angle \quad (3)$$

so it can be demonstrated that—

$$A \cosh \theta \pm B \sinh \theta = \sqrt{A^2 - B^2} \cosh \left\{\theta \pm \tanh^{-1}(B/A)\right\} \quad \text{numeric} \angle \quad (4)$$

whatever may be the relative magnitudes of the constants A and B. It thus becomes possible to express the solutions (1) and (2) in terms of a single hyperbolic function, viz. a sine or a cosine, as may be the more convenient, of a single hyperbolic angle. This form of the solution has especial interest;

not only in presenting clearly the fundamental conditions of the problem, but also in assisting the computer to find arithmetical solutions.

In (1) and (2), let $L_1 a = \theta_1$ . . . . . . . hyps. ∠

$L_2 a = \theta_2$ . . . . . . . " "

and, in conformity with the notation of Figs. 13, 14, and M5 to M10, let $\theta_1 + \theta_2 = \theta$ . . . . . . . hyps. ∠

Also, the angle subtended by the terminal load of $\sigma$ ohms ∠ at B is by (55), $\theta' = \tanh^{-1}\left(\frac{\sigma}{z_0}\right)$ . . . hyps. ∠

Moreover, the position-angle $\delta$ of the point P, Figs. 10 and 13, is—

$$\delta = \theta_2 + \theta' \quad . \ . \ . \ . \ \text{hyps. } \angle$$
$$= (\theta + \theta') - \theta_1 \quad . \ . \quad " \ "$$

Then, applying (4) to the left-hand equation of (1), and using (272) for expressing $I_A$, we have—

$$e = E_A \cosh \theta_1 - \frac{E_A}{z_0 \tanh(\theta + \theta')} \cdot z_0 \sinh \theta_1 \quad . \quad \text{volts } \angle$$
$$= E_A\{\cosh \theta_1 - \coth(\theta + \theta') . \sinh \theta_1\} \quad . \ . \quad " \ "$$
$$= E_A\sqrt{1 - \coth^2(\theta + \theta')} . \cosh [\theta_1 - \tanh^{-1} \{\coth(\theta + \theta')\}] \quad " \ "$$
$$= E_A . \frac{j}{\sinh(\theta + \theta')} . \cosh\left[\theta_1 - \tanh^{-1}\left\{\tanh\left(\theta + \theta' \pm j\frac{\pi}{2}\right)\right\}\right] \quad . \quad " \ "$$
$$= E_A . \frac{j}{\sinh(\theta + \theta')} . \cosh\left[\theta_1 - \left(\theta + \theta' \pm j\frac{\pi}{2}\right)\right] \quad " \ "$$
$$= E_A . \frac{j}{\sinh(\theta + \theta')} . -j \sinh(\theta + \theta' - \theta_1) \quad . \quad " \ "$$
$$= E_A . \frac{\sinh \delta}{\sinh(\theta + \theta')} = E_B \frac{\sinh \delta}{\sinh \theta'} \quad . \ . \ . \ . \ . \quad " \ " \quad (5)$$

which agrees with (40), (60), and the results given in Appendix M. This means that along any section of uniform line, in the steady state, whether complete in itself, or forming a sectional part of a composite line, the potential at any position is directly proportional to the sine of the position-angle, and is equal to the potential impressed at the generating end of the section multiplied by the ratio of the sine of the position-angle to the sine of the total angle subtended by the section at the sending end. This rule applies whatever may be the load applied at the receiving end of the section.

Similarly, considering the right-hand equation in (1),

$$e = E_B \cosh \theta_2 + \frac{E_B}{\sigma} z_0 \sinh \theta_2 \quad . \quad . \quad . \quad . \quad \text{volts } \angle$$

$$= E_B (\cosh \theta_2 + \coth \theta' \sinh \theta_2) \quad . \quad . \quad . \quad \text{"} \quad \text{"}$$

$$= \frac{E_A \sinh \theta'}{\sinh (\theta + \theta')} . \sqrt{1 - \coth^2 \theta'} . \cosh \{\theta_2 + \tanh^{-1} (\coth \theta')\} \quad \text{"} \quad \text{"}$$

$$= \frac{jE_A}{\sinh (\theta + \theta')} . \cosh \left\{\theta_2 + \theta' \pm j\frac{\pi}{2}\right\} \quad \text{"} \quad \text{"}$$

$$= \frac{jE_A}{\sinh (\theta + \theta')} . -j \sinh (\theta_2 + \theta') \quad . \quad \text{"} \quad \text{"}$$

$$= E_A . \frac{\sinh \delta}{\sinh (\theta + \theta')} = E_B \frac{\sinh \delta}{\sinh \theta'} \quad . \quad . \quad \text{"} \quad \text{"} \quad (6)$$

which is the same formula as (5). That is, the result is the same, whichever end of the section is started from.

Considering the left-hand side of the equation (2) for current strength at the Point P—

$$i = I_A \cosh \theta_1 - \frac{E_A}{z_0} \sinh \theta_1 \quad . \quad \text{amperes } \angle$$

and by (56) $I_A = \dfrac{E_A}{z_0 \tanh (\theta + \theta')}$

so that—

$$i = I_A \cosh \theta_1 - I_A \tanh (\theta + \theta') \sinh \theta_1 \quad . \quad \text{"} \quad \text{"}$$

$$= I_A \{\cosh \theta_1 - \tanh (\theta + \theta') \sinh \theta_1\} \quad . \quad . \quad \text{"} \quad \text{"}$$

$$= I_A \sqrt{1 - \tanh^2 (\theta + \theta')} . \cosh \{\theta_1 - (\theta + \theta')\} \quad \text{"} \quad \text{"}$$

$$= I_A . \frac{1}{\cosh (\theta + \theta')} . \cosh \{(\theta + \theta') - \theta_1\} \quad . \quad \text{"} \quad \text{"}$$

$$= I_A . \frac{\cosh \delta}{\cosh (\theta + \theta')} = I_B \frac{\cosh \delta}{\cosh \theta'} \quad . \quad . \quad . \quad . \quad \text{"} \quad \text{"} \quad (7)$$

which agrees with (41). That is, the current at any position P of a line, or composite-line-section, with any load at the receiving end defined by the angle $\theta'$, is proportional to the cosine of the position-angle, and is equal to the current entering at the sending end multiplied by the ratio of the cosine of the position-angle to the cosine of the angle subtended by the entire line up to the sending end.

Similarly, considering the right-hand equation (2)—

$$i = I_B \cosh \theta_2 + \frac{E_B}{z_0} \sinh \theta_2 \quad . \quad . \quad . \quad \text{amperes } \angle$$

But by (54) $E_B = I_B \sigma$, and $\frac{E_B}{z_0} = I_B \tanh \theta'$.

$$\therefore i = I_B \cosh \theta_2 + I_B \tanh \theta' . \sinh \theta_2 \quad . \quad . \quad \text{amperes } \angle$$

$$= I_B(\cosh \theta_2 + \tanh \theta' . \sinh \theta_2) \quad . \quad . \quad \text{,,} \quad \text{,,}$$

$$= I_A \frac{\cosh \theta'}{\cosh (\theta + \theta')}\left(\cosh \theta_2 + \tanh \theta' . \sinh \theta_2\right) \quad \text{,,} \quad \text{,,}$$

$$= I_A . \frac{\cosh \theta'}{\cosh (\theta + \theta')} . \sqrt{1 - \tanh^2 \theta'} . \cosh (\theta_2 + \theta') \text{,,} \quad \text{,,}$$

$$= I_A . \frac{\cosh \theta'}{\cosh (\theta + \theta')} . \frac{1}{\cosh \theta'} . \cosh (\theta_2 + \theta') \quad \text{,,} \quad \text{,,}$$

$$= I_A . \frac{\cosh \delta}{\cosh (\theta + \theta')} = I_B . \frac{\cosh \delta}{\cosh \theta'} \quad . \quad . \quad . \quad \text{,,} \quad \text{,,} \quad (8)$$

This is the same result as in (7).

We may thus conclude that formulas (40), (41) and (42), page 21, are fundamental, while (37), (38) and (39) are subsidiary thereto. In other words, the receiving end of a line may always be conceived of as connected to ground, or to zero potential, through some load $\sigma$, possessing a terminal angle $\theta'$. If the line is directly grounded, $\sigma = 0$, and $\theta' = 0$. If the line is freed, $\sigma = \infty$, and $\theta' = j\frac{\pi}{2}$. Whatever intermediate condition may exist at B, the potential and current, there delivered, jointly determine an apparent load impedance $\sigma$ ohms $\angle$, and a corresponding apparent terminal angle $\theta' = \tanh^{-1} (\sigma/z_0)$ hyps. $\angle$, which must be added to the line angles $\theta_2$ and $\theta$, in order to assign the corrected position angle $\delta$, and total angle $(\theta + \theta')$ subtended at A, as indicated by the results of Appendix M.

# LIST OF SYMBOLS EMPLOYED AND THEIR BRIEF DEFINITIONS

| | |
|---|---|
| A . . . | Area of a hyperbolic sector (cm.$^2$). |
| A, $a$ . . . | Fourier-series constants (volts). |
| A, $A_1$, $A_2$ . . | Moduli of complex numbers (cm. or length units). |
| a . . . | In diagrams, an abbreviation for current-strength in amperes $\angle$. |
| $a$, $a'$ . . . | Torque constants of receiving instruments (dyne $\perp$ cm. per ampere). |
| $a$ . . . | Silence-interval in Kelvin theory of submarine cables (seconds). |
| $a$.-$c$. . . . | An abbreviation for alternating-current. |
| $a$ . . . | A constant in the hyp. theory of continued fractions (numeric $\angle$). |
| $\alpha$ . . . | Attenuation-constant of a line (hyps. per wire-km.). |
| $\alpha_{,}$ . . . | Attenuation-constant of one line of a loop circuit (hyps. per wire-km. $\angle$). |
| $\alpha_{,,}$ . . . | Attenuation-constant of a loop line (hyps. per loop-km. $\angle$). |
| $\alpha_1$, $\alpha_2$ . . | Real and imaginary components respectively of an attenuation-constant (hyps. per wire-km.). |
| $\alpha_1$, $\alpha_2$, $\alpha_3$ . . | In hyp. theory of composite lines, the attenuation-constants of successive sections (hyps. per wire-km. $\angle$). |
| $\alpha'$, $\alpha'_1$, $\alpha'_2$ . . | Attenuation-constant and its components for a loaded line including lumpiness effects (hyps. per wire-km. $\angle$). |

$\alpha''$, $\alpha''_1$, $\alpha''_2$ . Attenuation-constant and its components for a smoothed loaded line (hyps. per wire-km. ∠).

B, $b$ . . . Fourier-series constants (volts).

$b$ . . . A constant in the hyp. theory of continued fractions (numeric ∠).

Also in the theory of loaded lines, the ratio $r''/r$ (numeric).

$\beta$, $\beta_1$, $\beta_2$ . . Circular angles (radians or degrees).

C, $c$ . . . Fourier-series constants (volts).

C . . . Total capacitance of a line (farads).

$c.c.$ . . . An abbreviation for continuous current.

$c$ . . . Linear capacitance of a line (farads per wire-km.).

$c_{\prime}$ . . . Linear capacitance of one line of a loop (farads per wire-km.).

$c_{\prime\prime}$ . . . Linear capacitance of a loop-line (farads per loop-km.).

$c'$ . . . Linear capacitance of one line of a loop (farads per wire-mile).

$c''$ . . . Linear capacitance of a loop-line (farads per loop-mile).

$\Gamma$ . . . Admittance of a leak load (mhos ∠).

$\gamma = \Gamma/2$ . . Admittance of a semi-leak load (mhos ∠).

$\gamma$ . . . Also the power-factor circular angle at the sending end of a transmission-line (degrees or radians.

D, $d$ . . . Fourier-series constants (volts).

$D$ . . . Distance between the axes of two parallel eccentric cylinders (cm.).

$d$ . . . Distance between a plane and the axis of a parallel cylinder (cm.). Also a constant in the hyp. theory of continued fractions (numeric ∠). Also sign of differentiation.

$d_1, d_2$ . . Distances between an inferred plane and the axes of eccentric cylinders (cm.).

$\triangle = \sigma_1 - \sigma_2$ Difference of radii of two conducting cylinders (cm.).

$\delta_A, \delta_B, \delta$ . . Hyperbolic angles of junctions of a series-load, and of a selected point on a line (hyps.).

E, $e$ . . . Fourier-series constants (volts).

E . . . E.M.F. (volts ∠).

$E_A$, $E_B$, $E_P$ . E.M.F. at the sending end, receiving end, and intermediate point of a line (r.m.s. volts ∠).

$E'_N$, $E_N$ . . E.M.F. at a leak of an artificial line, and at the corresponding point of the equivalent uniform line (r.m.s. volts ∠).

$E_{,,}$, $E_{,}$ . . E.M.F. impressed on a pair of looped lines, and on either component single line (r.m.s. volts ∠).

$E_m$, $E_{mA}$ . . Maximum cyclic e.m.f. at alternator terminals, and at sending terminal of a submarine cable (max. cy. volts ∠).

$\overline{E}_o$ . . . In oscillating-current theory, the initial value of a vector e.m.f. of self-induction (volts ∠).

$e$ . . . Instantaneous value of an e.m.f. (volts ∠).

Also the value of e.m.f. at an intermediate point on a line (volts ∠).

$\varepsilon = 2{\cdot}71828$ . . (numeric).

e.m.f. . . An abbreviation for electromotive force.

F . . . Magneto-mechanical force, or torque developed in a receiving instrument (dynes or dyne ⊥ cm.).

$F_n$ ( ) . . Expression for an alternate or constant continued fraction of $n$ stages.

$f$ . . . Frequency of an alternating-current or e.m.f. (cycles / sec.).

$f' = 2f''$ . . Frequency of dot-signalling on a submarine cable (dot-cycles / sec.).

$f''$ . . . Frequency of reversal-signalling on a submarine cable (reversal-cycles / sec.).

$f_0$ . . . Limiting maximum frequency of alternation on a submarine cable (cycles per sec.).

G . . . Total dielectric leakance of a line (mhos ∠).

$G_{gA}$ . . . Admittance of a line at sending end when grounded at receiving end (mhos ∠).

$G_{fp}$ . . . Admittance at and beyond a point on an artificial line, free at far end (mhos ∠).

$G'_{fp}$ . . . Admittance at and including a leak on an artificial line, free at far end (mhos ∠).

$g$ . . . Linear leakance of a line (mhos per wire-km. ∠). Also, impedance of a receiving instrument in submarine cable duplex theory (ohms ∠).

$g_1$ . . . Admittance of the leak in an artificial-line section (mhos ∠).

$g_{\prime}$ . . . Linear leakance of one of a pair of looped lines (mhos per wire-km. ∠).

$g'$ . . . Linear leakance of one of a pair of looped lines (mhos per wire-mile ∠).

$g_{\prime\prime}$ . . . Linear leakance of a loop-line (mhos per loop-km. ∠).

Also the admittance of a $\Pi$ pillar after adding a leak load (mhos ∠).

$g''$ . . . Linear leakance of a loop-line (mhos per loop-mile ∠).

$g'$ . . . Admittance of the leak in the staff of a $T$ (mhos ∠).

$g''$ . . . Admittance of the leak in the pillar of a $\Pi$ (mhos ∠). Also in the theory of loaded lines, the total linear leakance of a smoothed loaded line (mhos per wire-km.).

$g^1$ . . . Apparent linear leakance of a line, uncorrected for line resistance (mhos per km. ∠).

$h$ . . . Impedance of a duplex telegraph bridge arm (ohms ∠).

hyp. . . An abbreviation for hyperbolic radian (numeric ∠).

$\theta$ . . . Hyperbolic angle of a line, or section (hyps. ∠).

$\theta'$ . . . Auxiliary hyperbolic angle of an appended impedance or line section (hyps. ∠).

$\theta_{,}$ . . . Hyperbolic angle of a line-section after being loaded (hyps. ∠).

$\theta_{,,}$ . . . Hyperbolic angle of a looped line (hyps. ∠).

$\theta''$ . . . Hyperbolic angle of a loaded and smoothed line-section (hyps. ∠).

$\theta_1, \theta_2$ . . Real and imaginary components of a hyperbolic angle (hyps.).

$\theta_1, \theta_2, \theta_3$ . . In hyp. theory of composite lines, hyp. angles of successive sections (hyps. ∠).

$\theta_0$ . . . Limiting hyperbolic angle subtended by a submarine cable at maximum working speed (hyps.).

$\theta'_0, \theta''_0$ . . Values of $\theta_0$ for dot-signalling and reversal-signalling respectively (hyps.).

$I_{mA}, I_{mB}$ . . Maximum cyclic current-strengths at sending and receiving ends of line (amperes ∠).

$I_A, I_B, I_P$ . . Current-strengths at sending end, and receiving end, and assigned point of a line (r.m.s. amperes ∠).

$I'_A, I'_B$ . . Condenser currents at sending and receiving ends of a line $\Pi$ (r.m.s. amperes ∠).

$I_f$ . . . Current-strength at sending end of line with far end free (r.m.s. amperes ∠).

$I_g$ . . . Current-strength at sending end of line with far end grounded (r.m.s. amperes ∠).

$i$ . . Instantaneous value of an alternating current (amperes ∠). Also the current-strength at an intermediate point on a line (amperes ∠).

$\overline{I}_0$ . . . In oscillating-current theory, the initial value of the vector discharging current (amperes $\angle$).

$j = \sqrt{-1}$ . The quadrantal operator (quadrantal versor).

$k_{g,}$ . . . Correcting factor for reducing a nominal $T$ staff to an equivalent $T$ staff (numeric $\angle$).

$k_{g,,}$ . . . Correcting factor for reducing a nominal $\Pi$ pillar to an equivalent $\Pi$ pillar (numeric $\angle$).

$k_{\theta,}$ . . . Correcting factor for correcting the line angle pertaining to a $T$ (numeric $\angle$).

$k_{\theta,,}$ . . . Correcting factor for correcting the line angle pertaining to a $\Pi$ (numeric $\angle$).

$k_{r_0'}$ . . . Correcting factor for correcting the surge-impedance to a $T$ (numeric $\angle$).

$k_{r_0''}$ . . . Correcting factor for correcting the surge-impedance to a $\Pi$ (numeric $\angle$).

$k_{\rho,}$ . . . Correcting factor for reducing a nominal $T$ arm to an equivalent $T$ arm (numeric $\angle$).

$k_{\rho}$ . . . Correcting factor for reducing a nominal $\Pi$ architrave to an equivalent $\Pi$ architrave (numeric $\angle$).

$k_L$ . . . Normal attenuation-factor of a line of L km. (numeric $\angle$).

km. . . . An abbreviation for kilometer.

L . . . Length of a line or section (km.).

L′ . . . Length of a section of loaded line, or distance between loads (km.).

$L_1$, $L_2$ . . Distances along a line as measured from sending and receiving ends respectively (km.).

$L_1$, $L_2$, $L_3$ . . In the theory of composite lines, lengths of successive sections (km.).

$L_\epsilon$ . . . Length of line in which the normal attenuation factor is $1/\varepsilon$ (km.).

$L_{\frac{1}{2}}$ . . . Length of line in which the normal attenuation factor is $1/2$ (km.).

$l$ . . . Linear inductance of a line (henrys per wire-km.).

$l_,$ . . . Linear inductance of one line of a loop (henrys per wire-km.).

$l'$ . . . Linear inductance of one line of a loop (henrys per wire-mile).

$l_{,,}$ . . . Linear inductance of a loop line (henrys per loop-km.).

$l''$ . . . Extra linear inductance in the smoothed loads of a line (henrys per wire-km.).

$\lambda$ . . . Wave-length on a line (km.).

$\lambda''$ . . . Wave-length on a loaded smoothed line (km.).

$\lambda'$ . . . Wave-length on a loaded line including lumpiness effects (km.).

$m$ . . . Transmission coefficient of a current wave at a point of discontinuity (numeric $\angle$). Also a numerical constant in the hyp. theory of continued fractions (numeric $\angle$).

$\mu f$ . . . An abbreviation for microfarad.

N″ . . . Number of load coils encountered by an advancing wave per second on a single loaded smoothed line (numeric / sec.).

N′ . . . Number of load coils encountered by an advancing wave per second on a single loaded line, including lumpiness effects (numeric / sec.).

N . . . In the theory of artificial lines, the number of a leak counting from the far end of an artificial line (numeric).

$n$ . . . Number of letters per second in maximum working speed of a submarine cable (numeric / sec.).

Also in the hyp. theory of continued fractions, the number of stages of such a fraction (numeric).

Also in the theory of complex numbers and of the Ferranti-effect, any integer (numeric).

$n_1$ . . . Number of turns in a receiving-instrument winding (numeric).

$n''$ . . . Number of load coils per smoothed wave-length of a loaded line (numeric / km.).

$n'$ . . . Number of load coils per wave-length of a loaded line, including lumpiness effects (numeric / km.).

$\nu$ . . . Ratio of transformation of a transformer (numeric).

$\xi$ . . . Hyperbolic angle of a point on a line measured from sending end (hyps. $\angle$).

$P_A$, $P_B$ . . Alternating-current power, to sending, and at receiving, end of line (watts $\angle$).

$P_{fA}$, $P_{fB}$ . . Effective components of $P_A$ and $P_B$ respectively (watts).

$P_{kA}$, $P_{kB}$ . . Reactive components of $P_A$ and $P_B$ respectively ($j$ watts).

$p$ . . . A real exponent between 0 and 2 (numeric).

$\Pi$ . . . A delta connection of three impedances externally equivalent to a line (ohms $\angle$).

$\pi = 3{\cdot}14159$ ..... (numeric).

R . . . Total conductor resistance of a line (ohms $\angle$).

$R_A$, $R_P$ . . Resistance of a line at its sending end, and at a selected point (ohms $\angle$).

$R_{fA}$, $R_{fB}$ . . Resistance of a line at A or B when free at far end (ohms $\angle$).

$R_f$ . . . Resistance of a line when free at far end (ohms $\angle$).

$R_g$ . . . Resistance of a line when grounded at far end (ohms $\angle$).

$R_{gA}$, $R_{gB}$ . . Resistance of a line at A or B when grounded at far end (ohms $\angle$).

$R_r$ . . . Resistance between receiving end of a line and ground (ohms $\angle$).

$R'_r$ . . . Additional resistance inserted in receiving-end circuit of a submarine cable (ohms $\angle$).

$R'_g$ . . . Resistance offered by a line when grounded through apparatus at far end (ohms ∠).

$R_{fP}$ . . . Resistance at and beyond a point on an artificial line free at far end (ohms ∠).

$R'_{fP}$ . . . Resistance at and including a point on an artificial line free at far end (ohms ∠).

$R_l$ . . . Receiving-end resistance of a line (ohms ∠).

$R'$ . . . Resistance in the staff of a $T$ (ohms ∠).

$R''$ . . . Resistance in the pillar of a $\Pi$ (ohms ∠).

$r$ . . . Linear resistance of a line (ohms per wire-km.).

$r^1$ . . . Apparent linear resistance of a line, uncorrected for leakance (ohms per wire-km. ∠).

$r_{,}$ . . . Linear resistance of single line in a loop (ohms per wire-km. ∠).

$r_1$ . . . In hyp. artificial-line theory, the resistance of a conductor section of artificial line (ohms ∠).

$r''$ . . . In the theory of loaded lines, the extra linear conductor-resistance due to the load coils smoothed (ohms per wire-km.).

$r_{,,}$ . . . Linear resistance of loop line (ohms per loop-km. ∠).

$r_0$ . . . Surge-resistance of a line (ohms ∠).

$r_{0,}$ . . . Surge-resistance of one wire of a loop (ohms ∠).

$r_{0,,}$ . . . Surge-resistance of a loop-line (ohms ∠).

$r_0'$ . . . Apparent surge-resistance of a $T$ (ohms ∠).

$r_0''$ . . . Apparent surge-resistance of a $\Pi$ (ohms ∠).

*r.m.s.* . . . Abbreviation of root-mean-square.

$\rho$ . . . Radius-vector to a point on a curve (cm.).

Also Resistance in each arm of a nominal $T$, or in architrave of nominal $\Pi$ (ohms ∠).

$\rho'$ . . . Resistance in each arm of a $T$ (ohms ∠).

Also, in the theory of hyperbolic angles, the integrated mean value of a radius-vector (cm.).

$\rho''$ . . . Resistance in architrave of a $\Pi$ (ohms $\angle$).

$\rho$ . . . In oscillating-current theory, half the resistance of the circuit (ohms).

Also in plane-cylinder theory, the resistivity of the environing medium (absohm-cm.).

S . . . A Fourier-series, equivalent to an impressed rectangular e.m.f. (volts $\angle$).

$s, s_1, s_2$ . . Distances measured along a curve (cm.).

$\Sigma$ . . . Resistance, in one line, of a regular series line-load (ohms $\angle$).

$\Sigma$ . . . In plane-cylinder theory, the sum of the radii of two parallel eccentric cylinders (cm.).

$\Sigma_y$ . . . Sum of the three admittances of a triple star (mhos $\angle$).

$\Sigma_z$ . . . Sum of the three impedances of a delta (ohms $\angle$).

$\sigma$ . . . Resistance inserted between line and ground at receiving end of line (ohms $\angle$).

$\sigma = \Sigma/2$ . . In theory of loaded lines, half of the resistance of a regular load inserted in each line (ohms $\angle$).

$\sigma'$ . . . Resistance inserted between line and ground in each wire of a loop (ohms $\angle$).

$\sigma_{''}$ . . . Resistance inserted between the receiving ends of a loop-line (ohms $\angle$).

$\sigma$ . . . In plane-cylinder theory, the radius of a cylinder (cm.).

$T$ . . . A triple star, or Y, connection of three impedances, externally equivalent to a line (ohms $\angle$).

$t$ . . . Time elapsed since a certain epoch (seconds).

$\tau = \text{CR}$ . . Time-constant of a submarine cable (seconds or farad-ohms).

$\overline{\text{U}}_\circ$ . . . In oscillating-current theory, the initial value of a vector discharging e.m.f. (volts $\angle$).

| | |
|---|---|
| $U_A$, $U_P$ . . | Potential at sending end and at a selected point on a line (volts ∠). |
| $u$, $u'$, $u''$ . . | Auxiliary hyperbolic angles in theory of artificial lines and continued fractions hyps. ∠). |
| $u$, $v$ . . . | Real and imaginary components of a complex number (numeric). |
| $v$, $w$ . . . | Cartesian co-ordinates of a point in a plane (cm.). |
| $v$ . . . | In diagrams, an abbreviation for potential in volts ∠. |
| $v$ . . . | Velocity of propagation of waves on a line (km/sec.). |
| $v''$ . . . | Velocity of propagation of waves on a smoothed loaded line (km/sec.). |
| $v'$ . . . | Velocity of propagation of loaded line, including lumpiness effects (km/sec.). |
| X . . . | Total reactance of a line-wire ($j$-ohms). |
| $x$ . . | Distance from the sending end of a line (km.). |
| | Also the linear reactance of a line ($j$-ohms/km.). |
| | Also the inductive reactance in an oscillating-current circuit (ohms). |
| X, Y, $x$, $y$ . | Cartesian co-ordinates of a point in a plane (cm.). |
| Y . . . | Total admittance of a line (mhos ∠). |
| $Y_A$, $Y_B$ . . | Admittance of semi-line condensers at ends of a $\Pi$ (mhos ∠). |
| $y_A$, $y_B$, $y_C$ . . . | Admittances of the branches of a star (mhos ∠). |
| $y_1$, $y_2$, $y_3$ . . | In composite-line theory, the surge-admittances of successive sections (mhos ∠). |
| | Also admittances of the sides of a delta (mhos ∠). |
| $y_0 = 1/z_0$ . . | Surge-admittance of a line per wire (mhos ∠). |
| $y'' = 1/\rho''$ . | Admittance of the architrave of a $\Pi$ (mhos ∠). |
| $y = g + jc\omega$ . | Linear admittance of a conductor (mhos per km. ∠). |

| | |
|---|---|
| $Z$ . . . | Total impedance of a line-wire (ohms $\angle$). |
| $Z_c$ . . . | Impedance of a half-line condenser in a $\Pi$ (ohms $\angle$). |
| $Z_l$ . . . | Receiving-end impedance of a line (ohms $\angle$). |
| $Z'_l$ . . . | Receiving-end impedance of a line corrected for the shunting of the receiving instrument (ohms $\angle$). |
| $z$ . . . | In oscillating-current theory, the surge-impedance of the circuit, $\sqrt{l/c}$, in the absence of resistance (ohms). |
| $z = r + jl\omega$ | Linear impedance of a conductor (ohms per km. $\angle$). |
| $z_o$ . . . | Surge-impedance of a line-wire (ohms $\angle$). |
| $z_{o''}$ . . . | Surge-impedance of a looped line (ohms $\angle$). |
| $z_{oo} = \sqrt{l/c}$ | Surge-impedance of a line-wire with no dissipation (ohms). |
| $z_o'$ . . . | Apparent surge-resistance of a $T$-section artificial line (ohms $\angle$). |
| $z_o''$ . . . | Apparent surge-resistance of a $\Pi$-section artificial line (ohms $\angle$). |
| $z_1$ . . . | Impedance of a receiving instrument when shunted (ohms $\angle$). |
| $z_2$ . . . | Impedance of the shunt to a receiving instrument (ohms $\angle$). |
| $z_A$, $z_B$, $z_C$ . | Impedances of the branches of a star (ohms $\angle$). |
| $z_1$, $z_2$, $z_3$ . | In composite-line theory, the surge-impedances of successive line-sections (ohms $\angle$). |
| | Also impedances in the three branches of a delta (ohms $\angle$). |
| $z$ . . . | Impedance of receiving apparatus to ground (ohms $\angle$). |
| $z_s$ . . . | Impedance in the sending-circuit of a line (ohms $\angle$). |

| | |
|---|---|
| $\Omega$ . . . | Hyperbolic angular velocity in an ultra-periodic circuit (hyps. / sec.). |
| $\omega = 2\pi f$ . . | Angular velocity of an alternating e.m.f. or current (radians / sec.). |
| | Also, in diagrams, a sign for ohms $\angle$. |
| $\omega', \omega''$ . . | Angular velocities of dot- and reversal-frequencies impressed on a long submarine cable (radians / sec.). |
| $\omega_\circ$ . . . | Limiting maximum angular velocity impressed on a submarine cable (radians / sec.). |
| ℧ . . . | In diagrams, a sign for mhos $\angle$. |
| $\cong$ . . . | "Approximately equals." |
| $\angle$ . . . | Sign of an angle, to indicate the existence of a vector or complex quantity, either actually or potentially. |
| $\not<$ . . . | "Is not less than." |
| 0·5″ . . . | 0·5 inch. |
| ~ . . . | Cycles per second. |

# BIBLIOGRAPHY

## PUBLICATIONS DEALING WITH HYPERBOLIC FUNCTIONS OR WITH COGNATE SUBJECTS


1. LORD KELVIN—"On the Theory of the Electric Telegraph." *Proc. Roy. Soc.* London, May 1855. "Mathematical and Physical Papers," Vol. II, p. 71.

2. A. B. STREHLKE—"Über Periodische Kettenbrüche." Grunert's *Archiv der Mathematik und Physik*, Vol. XLII, p. 341. 1864.

3. H. R. KEMPE—"On Testing by Received Currents." *Journ. Soc. of Tel. Engrs.*, Vol. IX, pp. 222–231. 1880.

3A. W. LIGOWSKI—"Tafeln der Hyperbelfunctionen." Berlin: Ernst & Korn, 1890.

3B. T. H. BLAKESLEY. "Alternating Currents of Electricity." London, 1889. Chapter VIII and Table A.

4. J. J. THOMSON—"On the Heat produced by Eddy Currents in an Iron Plate exposed to an Alternating Magnetic Field." *The Electrician*, Vol. XXVIII, p. 599. 1891.

5. O. HEAVISIDE—Reprinted "Electrical Papers." London. Vol. II, p. 247. 1892.

6. A. E. KENNELLY—"Impedance." *Trans. Am. Inst. El. Engrs.*, Vol. X, p. 175. April 1893.

7. O. HEAVISIDE—"Electromagnetic Theory." London. Vol. I, p. 450. 1893.

8. F. BEDELL and A. C. CREHORE—"Alternating Currents." New York. Chap. xiii. 1893.

9. C. P. STEINMETZ—"Complex Quantities and their Use in Electrical Engineering." *Proceedings of the Int. El. Congress, Chicago*, pp. 33–76. August 1893.

9A. A. E. KENNELLY—"Impedance of Mutually Inductive Circuits." *The Electrician*, Vol. XXXI, pp. 699–700. Oct. 27, 1893.

10. A. E. KENNELLY—"On the Fall of Pressure in Long-Distance Alternating-Current Conductors." *Elec. World*, Vol. XXIII, No. 1, p. 17. Jan. 1894. Abstr. *The Electrician, Jan.* 5, 1894.

11. A. E. KENNELLY—"A Contribution to the Theory of Telephony." *Elec. World*, Vol. XXIII, No. 7, p. 208. Feb. 1894; also *The Electrician*, June 29, 1894.

12. W. E. AYRTON and C. S. WHITEHEAD—"The Best Resistance for the Receiving Instrument on a Leaky Telegraph Line." *Journ. Inst. of El. Engrs.* London. Vol. XXIII, Part 3, p. 327. March 1894.

13. A. BLONDEL—"Inductance des Lignes Aériennes pour Courants Alternatifs." *L'Eclairage Electrique*, Oct.-Nov. 1894.

14. E. J. HOUSTON and A. E. KENNELLY—"Resonance in Alternating-Current Lines." *Trans. Am. Inst. El. Engrs.*, Vol. XII, pp. 133–169. April 1895.

15. J. McMAHON—"Hyperbolic Functions." Chap. IV. of "Higher Mathematics," by Merriman and Woodward, pp. 107–168. Tables of sinh and cosh $(x + j\,y)$ to $x = 1{\cdot}5, y = 1{\cdot}5$. Wiley & Sons, New York. 1896.

16. E. J. HOUSTON and A. E. KENNELLY—"Alternating-Current Machinery." *El. World and Engineer.* Oct. 30, 1897.

"Electrical Engineering Leaflets." Advanced Grade. *The Electrical Engineer*, N.Y. 1897.

17. C. P. STEINMETZ—"Theory and Calculation of Alternating-Current Phenomena." W. J. Johnston Co., New York. 1897.

18. A. MACFARLANE—"Application of Hyperbolic Analysis to the Discharge of a Condenser." *Trans. Am. Inst. El. Engrs.*, Vol. XIV, p. 163. 1897.

19. C. GODFREY—"Wave Propagation along a Periodically Loaded String." *Phil. Mag.*, Vol. XVI, p. 356. 1898.

20. M. I. PUPIN—"Propagation of Long Electrical Waves." *Trans. Am. Inst. El. Engrs.*, Vol. XVI, pp. 93–142. March 1899.

21. A. E. KENNELLY—"On the Predetermination of the Regulation of Alternating-Current Transformers. *Elec. World and Engineer*, N.Y. Vol. XXXIV, p. 343. Sept. 2, 1899.

22. A. E. KENNELLY—"The Equivalence of Triangles and Three-Pointed Stars in Conducting Networks." *El. World and Engineer*, N.Y. Vol. XXXIV, No. 12, pp. 413–414. Sept. 16, 1899.

23. A. C. CREHORE and G. O. SQUIER—"A Practical Transmitter using the Sine Wave for Cable Telegraphy and Measurements with Alternating-Currents upon an Atlantic Cable." *Trans. Am. Inst. El. Engrs.*, Vol. XVII, p. 385. May 18, 1900.

24. M. I. PUPIN—"Wave-Transmission over Non-Uniform Cables, and Long-Distance Air-Lines." *Trans. Am. Inst. El. Engrs.*, Vol. XVII, pp. 445–513. May 1900.

25. M. I. PUPIN—"Wave Propagation over Non-Uniform Conductors." *Trans. Am. Math. Soc.*, Vol. I, No. 3. pp. 259–286. July 1900.

25A. A. E. KENNELLY—"The Reactance Drop and Reactance-Factor of Transformers." *El. World and Engineer*, Vol. XXXVIII, No. 3, pp. 92–94. July 20, 1901.

25B. M. I. PUPIN—"A Note on Loaded Conductors." *El. World and Engineer*, Vol. XXXVIII, No. 15, pp. 587–588. Oct. 12, 1901.

25C. A. E. KENNELLY—"Surges in Transmission Circuits." *El. World and Engineer*, Vol. XXXVIII, No. 21, pp. 847–849. Nov. 23, 1901.

26. M. LEBLANC—"Formula for Calculating the Electromotive Force at any Point of a Transmission Line for Alternating-Current." *Trans. Am. Inst. El. Engrs.*, Vol. XIX, pp. 759–768. June 1902.

27. G. A. CAMPBELL—"On Loaded Lines in Telephonic Transmissions." *Phil. Mag.*, Series VI, Vol. V, p. 313. March 1903.

28. J. HERZOG and C. FELDMAN—"Die Berechnung Elektrischer Leitungsnetze." Julius Springer, Berlin. Vol. I, chap. v. April 1903.

29. A. E. KENNELLY—"On Electric Conducting Lines of Uniform Conductor and Insulation Resistance in the Steady State." *Harvard Engineering Journal*, pp. 135–168. May 1903.

30. A. E. KENNELLY—"On the Mechanism of Electric Power Transmission." *Elec. World*, N.Y. Vol. XLII, p. 673. Oct. 24, 1903.

31. A. E. KENNELLY—"Two Elementary Constructions in Complex Trigonometry." *Am. Annals of Mathematics, Salem Press*, 2nd Series. Vol. V, No. 4, pp. 181–184. July 1904.

32. H. V. HAYES—"Loaded Telephone Lines in Practice." *Proc. Int. Elec. Congress, St. Louis*, Sec. G., p. 638. Vol. III, 1904.

33. A. E. KENNELLY—"The Alternating-Current Theory of Transmission Speed over Submarine Telegraph Cables." *Proc. Int. El. Congress, St. Louis*, Sec. A. Vol. I, pp. 68–105, with Table of sinh, cosh, tanh, coth, sech, cosech $\rho \angle 45^\circ$, to $\rho = 20{\cdot}50$. 1904.

34. J. A. FLEMING—"A Model illustrating the Propagation of a Periodic Current in a Telephone Cable, and the Simple Theory of its Operation." *Phil. Mag.*, Aug. 1904, and *Proc. Phys. Soc.* London. Vol. XIX. 1904.

35. A. E. KENNELLY—"High-Frequency Telephone Circuit Tests." *Proc. Int. El. Congress, St. Louis*, Sec. G. Vol. III, pp. 414–437. 1904.

36. A. RUSSELL—"A Treatise on the Theory of Alternating Currents." *Cambridge University Press.* 1904.

37. J. HERZOG and C. FELDMANN—"Die Berechnung Elektrischer Leitungsnetze." Julius Springer, Berlin. Vol. II, chap. vii. Sept. 1904.

38. G. ROESSLER—"Die Fernleitung von Wechselströmen." J. Springer, Berlin. 1905.

39. A. E. KENNELLY—"The Distribution of Pressure and Current over Alternating-Current Circuits." *Harvard Engineering Journal.* Vol. IV, No. 3, pp. 149–165, No. 4, Oct. 1905, pp. 206–225, Jan. 1906; Vol. V, No. 1, pp. 30–56, April 1906.

40. C. V. DRYSDALE—"The Measurement of Phase Differences." *The Electrician*, Vol. LVII, pp. 726–783. 1906.

41. BÉLA, GÁTI—"On the Measurement of the Constants of Telephone Lines." *The Electrician.* Nov. 2, 1906.

42. C. V. DRYSDALE—"Some Measurements on Phase Displacements in Resistances and Transformers." *The Electrician*, Vol. LVII. Nov. 16 and 23, 1906.

43. B. S. COHEN and G. M. SHEPHERD—"Telephonic Transmission Measurements." *Journ. of Proc. Inst. Elec. Engrs.*, London. Vol. XXXIX, p. 503. 1907.

44. A. E. KENNELLY—"The Process of Building-up the Voltage and Current in a Long Alternating-Current Circuit." *Proc. Am. Ac. Arts & Sc.*, Vol. XLII, No. 27, pp. 701–715. May 1907.

45. C. V. DRYSDALE—"The Theory of Alternate-Current Transmission in Cables." *The Electrician.* Dec. 6, 13, 20, 27, 1907, and Jan. 10, 1908.

46. O. LODGE and B. DAVIES—"On the Measurement of Large Inductances containing Iron." *Journ. Inst. El. Engrs.*, London. Vol. XLI, pp. 515–526. March 1908.

47. A. E. KENNELLY—"The Expression of Constant and of Alternating Continued Fractions in Hyperbolic Functions." *Am. Annals of Mathematics*, Salem Press. Vol. IX, No. 2, pp. 85–96. Jan. 1908.

48. A. E. KENNELLY—"Artificial Lines for Continuous Currents in the Steady State." *Proc. Am. Ac. of Arts & Sc.*, Vol. XLIV, No. 4, pp. 97–130. Aug. 26, 1908.

49. B. S. COHEN—"On the Production of Small Variable-Frequency Alternating Currents." *Phil. Mag.*, Sept. 1908; or, *Proc. Phys. Soc.*, London. Vol. XXI, p. 283. 1909.

50. BÉLA, GÁTI—"Description et Utilisation de la Méthode pour la Mesure des Constants de Ligne au Moyen du Barretter." Oct. 1908.

51. C. V. DRYSDALE—"The Use of a Phase-shifting Transformer for Wattmeter and Supply Meter Testing." *The Electrician*, Vol. LXII, p. 341. Dec. 11, 1908.

52. G. DI PIRRO—"Sui Circuiti non uniformi." *Atti dell' Assoc. Elettrotecn, Ital.*, Vol. XII, No. 6, 1909; also *La Lumière Electrique*, Series 2, Vol. VII, p. 227. 1909.

53. A. E. KENNELLY—"The Influence of Frequency on the Equivalent Circuits of Alternating-Current Lines." *The Elec. World*, Vol. LIII, p. 211. Jan. 21, 1909.

54. C. V. DRYSDALE—"The Use of the Potentiometer on Alternating-Current Circuits." *Phil. Mag.*, Vol. XVII, p. 402, March 1909; also *Proc. Phys. Soc.*, London, Vol. XXI, p. 561, 1909; also *The Electrician*, Vol. LXIII, p. 8, April 16, 1909.

55. A. E. KENNELLY—"The Linear Resistance between Parallel Conducting Cylinders in a Medium of Uniform Conductivity." *Proc. Am. Phil. Soc.*, Vol. XLVIII, pp. 142–165. April 1909.

55A. G. F. BECKER and C. E. VAN ORSTRAND—"Hyperbolic Functions. Tables of Real Values, Smithsonian Mathematical Tables." *Smithsonian Institution*, Washington, D.C. - 1909.

56. A. RUSSELL—"The Coefficients of Capacity and the Mutual Attractions or Repulsions of Two Electrified Spherical Conductors when Close Together." *Proc. Roy. Soc.* A., Vol. LXXXII. June 1909.

57. BÉLA, GÁTI—"Das C. R. Gesetz und die Kabelschnelltelegraphie." *Elektrotechnik und Maschinenbau.* Heft 37. 1909

58. Béla, Gáti—"Wechselstrom als Träger von Telephonströmen." *Elektrotechnische Zeitschrift.* Heft 39. 1909.

59. B. S. Cohen—"The Impedance of Telephone Apparatus." *National Telephone Journal*, p. 113. Sept. 1909.

60. A Blondel and C. Le Roy—"Calcul des Lignes de Transport d'Énergie à Courants Alternatifs et tenant compte de la Capacité et de la Perditance Réparties." *Revue d'Electricité.* Sept. 18 and 25; Oct. 23 and 30, 1909.

61. A E. Kennelly—"The Equivalent Circuits of Composite Lines in the Steady State." *Proc. Am. Ac. Arts & Sc.*, Vol. XLV, No. 3, pp. 31–75. Sept. 1909.

62. F. Rusch—"Über die Wirbelstromverluste im Leitungskupfer der Wechselstromarmaturen." *Elektrotechnik und Maschinenbau.* Jan. 23 and 30, 1910.

63. J. Perry—"Telephone Circuits." *Proc. of the Phys. Soc.*, London. Vol. XXII, pp. 674–684. Feb. 25, 1910. *Phil. Mag.* May 1910. With Table of sinh and cosh, $\rho \underline{/45°}$, up to $\rho = 1$, in small steps.

64. W. E. Miller—"Formulae, Constants and Hyperbolic Functions for Transmission-Line Problems." Table of sinh and cosh $(x+jy)$ up to $x=1$, $y=1$. *General Electric Review, Supplement*, Schenectady, N.Y. May 1910.

65. Béla, Gáti—"Über die Anwendung hyperbolischer Functionen auf weite Entfernungen wirkenden Telegraphen- und Telephonströmen." *Elektrotechnik und Maschinenbau.* Heft 33. 1910.

66. A. E. Kennelly—"Vector Power in Alternating-Current Circuits." *Proc. Am. Inst. El. Engrs.*, pp. 1023–1057. June 27, 1910.

67. H. Pender and H. S. Osborne—"The Electrostatic Capacity between Equal Parallel Wires." Vol. LVI, No. 12, pp. 667–670. Sept. 12, 1910.

68. A. E. Kennelly—"Graphic Representations of the Linear Electrostatic Capacity between Equal Parallel Wires." *Elec. World.* Oct. 27, 1910.

69. K. W. Wagner—"Über die Frequenz der Fernsprechströme." *Phys. Zeit.*, 1122–1127. 1910.

70. W. A. J. O'Meara—"Submarine Cables for Long-Distance Telephone Circuits," and Discussion on the same. *Journ. Proc. Inst. El. Engrs.*, Part 206, Vol. XLVI. Dec. 1910.

71. A. E. Kennelly—"Vector Diagrams of Oscillating-Current Circuits." *Proc. Am. Ac. Arts & Sc.*, Vol. XLVI. No. 17, pp. 373–421. Jan. 1911.

72. C. V. Drysdale—"Propagation of Magnetic Waves in an Iron Bar." *The Electrician*, April 28, 1911.

73. E. Hausmann—"Electric Wave Propagation and Distortion along Conductors." Doctorate Thesis, New York University. April 1911.

74. J. A. Fleming—"The Propagation of Electrical Currents in Telephone and Telegraph Conductors." Constable & Co., London. May 1911.

75. F. Breisig—"Über die Energieverstellung in Fernsprechkreisen." *Elektrotechnische Zeitschrift*, Heft 23, pp. 558–561; Heft 24, pp. 590–593. June 1911.

76. A. E. Kennelly—"Tables of Hyperbolic Functions in Reference to Long Alternating-Current Transmission Lines." *Proc. Am. Inst. El. Engrs.*, pp. 2481–2492. Dec. 1911. sinh, cosh, and tanh of $\rho\underline{/\delta}$ up to $\rho = 0{\cdot}5$, by steps of 0·1 and between $\delta = 60°$ and $\delta = 90°$ by steps of 1°.

77. A. E. Kennelly and H. Tabossi—"Artificial Power-Transmission Line." *Electrical World*, Vol. LIX, pp. 359–361. Feb. 17, 1912.

78. A. E. Kennelly—"Tables of Sines, Cosines, Tangents Cosecants, Secants, and Cotangents of Real and Complex Hyperbolic Angles." *Harvard Engineering Journal*. Jan. 1912. Vol. X. Functions of $\rho\underline{/\delta}$ up to $\rho = 1{\cdot}0$ and $\delta$ between 45° and 90°.

79. K. W. Wagner—"Eine neue künstliche Leitung zur Untersuchung von Telegraphienströmen und Schaltvorgängen." *Elektrotechnische Zeitschrift*, Vols. L and LI. 1912.

80. H. W. Malcolm—"The Theory of the Submarine Cable." *The Electrician*, March to September 1912.

81. A. E. Kennelly and F. W. Lieberknecht—"Measurements of Voltage and Current over a long Artificial Power-Transmission Line at 25 and 60 cycles per second." *Proc. Am. Inst. El. Engrs.*, Vol. XXXI, pp. 805–837. June 1912.

82. A. E. Kennelly—"The Distribution of Voltage and Current over *Π*-Artificial Lines in the Steady State." *Electrical World*, pp. 306–311. Aug. 10, 1912.

83. A. E. Kennelly—"Disturbances of Potential and Current Produced in an Active Network by the Application of a Leak Load." *Electrical World*. Dec. 28, 1912.

84. J. H. Morecroft—"Hyperbolic Functions and their Application to Problems in Electrical Engineering." *The School of Mines Quarterly*, Vol. XXXIV, No. 3. April 1913.

85. A. J. Aldridge—"Practical Application of Telephone Transmission Calculations." *Journ. Inst. El. Engrs.*, Vol. LI, pp. 390–442. May 1913.

86. A. E. Salazar—"Las Funziones Iperbolicas i su Applicazion á los Problemas de Ingenieria Elektrika." *Universidad de Chile*. 1913.

87. A. E. Kennelly—"A Convenient Form of Continuous-Current Artificial Line." *Electrical World*, June 14, 1913.

88. A. E. Kennelly and F. W. Lieberknecht—"Test of an Artificial Aerial Telephone Line at a Frequency of 750 cycles per second."—*Proc. Am. Inst. El. Engrs.*, pp. 1281–1301. June 24, 1913.

89. D. C. and J. P. JACKSON—"Alternating Currents and Alternating-Current Machinery." Macmillan & Co. New Edition. Chapter XIII. 1913.

90. G. M. B. SHEPHERD—"A Note on High-Frequency Wave Filters." *The Electrician,* June 13, 1913. pp. 399–401.

91. C. V. DRYSDALE—"The Use of the Alternate-Current Potentiometer for Measurements on Telegraph and Telephone Circuits." *The Electrician,* Aug. 1, 1913.

92. A. E. KENNELLY—"Tables of Complex Hyperbolic and Circular Functions." *Harvard University Press.* In press November 1913. Functions of $\rho \underline{/\delta}$ to $\rho = 3{\cdot}0$ by steps of $0{\cdot}1$ and to $\delta = 45°$ to $90°$; also functions of $x + iq$ to $x = 10$ by steps of $0{\cdot}05$, and of $q$ virtually to infinity by steps of $0{\cdot}05$.

93. A. E. KENNELLY—"Chart Atlas of Complex Hyperbolic and Circular Functions." *Harvard University Press.* In press November 1913. Large charts, 48 cm. × 48 cm., presenting curves for all of the Tables published in 92, for rapid graphic interpolation.

94. J. A. FLEMING—"Predetermination of the Current and Voltage at the Receiving End of a Telephone or other Alternating-Current Line." *Journ. Inst. Elec. Engrs.* Vol. LII, pp. 717–723, June 15, 1914; also *The Electrician,* Vol. LXXIII, pp. 691–693, July 31, 1914.

95. A. E. KENNELLY and H. PENDER—"Resonance Tests of a long Transmission Line." *Electrical World,* Aug. 8, 1914. pp. 278–282.

96. A. E. KENNELLY—"The Computation of Composite Alternating-Current Lines." *Journ. Franklin Inst.,* Sept. 1914. pp. 287–297.

97. H. B. DWIGHT—"Simplified Hyperbolic Calculations for Transmission Lines." *Electrical World*, Sept. 5, 1914. Vol. LXIV, p. 474.

98. C. E. MAGNUSSON, J. GOODERHAM and R. RADER—"A 200-Mile Artificial Transmission Line." *Electrical World*, June 22, 1915.

99. M. TONEGAWA and T. ARAKAWA. "Transmission Losses in Telephone Lines." Report No. 17 of Electrotechnical Laboratory, Tokyo, Japan, June 1915.

100. H. E. CLIFFORD and C. L. DAWES—Sec. 11, *Standard Handbook for Electrical Engineers*, 1915, p. 900.

101. A. E. KENNELLY—"The Receiving End Impedance of a Conducting Line loaded at both Ends." *Electrical World*, July 24, 1915.

102. C. W. RICKER—"Analysis of Electric Waves for Harmonics." *Electrical World*, Sept. 15, 1915.

103. D. ROBERTSON—"A Mode of Studying Damped Oscillations by the Aid of Shrinking Vectors." *Journ. Inst. Elect. Engrs.*, Dec., 1915. Vol. LIV, pp. 24–34.

104. A. E. SALAZAR—"El Kalkulo Esakto de las Líneas de Trasmision kon Admitanzia dielektrika rrepartîda i el método iperboliko komplejo." *Proc. of the Pan Am. Scientific Congress at Washington D.C.*, Jan. 1916. 26 pp.

105. G. W. O. HOWE—"The Application of Telephone Transmission Formulae to Skin-Effect Problems." *Journ. Inst. Elect. Engrs.*, April 1, 1916. Vol. LIV, pp. 473–480.

# INDEX

Printed in Great Britain for the UNIVERSITY OF LONDON PRESS, Ltd.,
by RICHARD CLAY & SONS, Ltd., London and Bungay.

Printed in Dunstable, United Kingdom

91219490R00181